PLANT SCIENCE RESEARCH AND PRACTICES

LEAF EPIDERMAL ANATOMICAL CHARACTERS AND ANATOMICAL TOOLS FOR SYSTEMATICAL STUDIES OF SOME MEDICINALLY IMPORTANT ANGIOSPERMIC FAMILIES

PLANT SCIENCE RESEARCH AND PRACTICES

Additional books in this series can be found on Nova's website under the Series tab.

Additional e-books in this series can be found on Nova's website under the e-book tab.

PLANT SCIENCE RESEARCH AND PRACTICES

LEAF EPIDERMAL ANATOMICAL CHARACTERS AND ANATOMICAL TOOLS FOR SYSTEMATICAL STUDIES OF SOME MEDICINALLY IMPORTANT ANGIOSPERMIC FAMILIES

ZUBAIDA YOUSAF
AFIFA YOUNAS
AND
ARUSA AFTAB
EDITORS

nova publishers
New York

Library of Congress Cataloging-in-Publication Data

Leaf epidermal anatomical characters and anatomical tools for systematical studies of some medicinally important angiospermic families / editors: Zubaida Yousaf, Afifa Younas, and Arusa Aftab (Department of Botany, Lahore College for Women University, Lahore, Pakistan).
pages cm. -- (Plant science research and practices)
Includes index.
ISBN 978-1-63483-190-1 (hardcover)
1. Leaves--Anatomy. 2. Angiosperms--Classification. I. Yousaf, Zubaida, editor. II. Younas, Afifa, editor. III. Aftab, Arusa, editor. IV. Series: Plant science research and practices.
QK649.L44 2015
581.4'8--dc23

2015021911

Published by Nova Science Publishers, Inc. † New York

CONTENTS

PREFACE

Plants are vital for life on earth as it exists nowadays. They produce our world with oxygen and food, donate an essential part of water and nutrient cycling in ecosystems, and add beauty to our environment. Botany at the moment plays a special role in the welfare of both major and non-major students. For example, in this text, topics such as systematic studies, epidermal characteristics and many other topics are discussed. To intelligently pursue such topics, one needs to understand how plants are constructed and how they function.

The chapters on anatomy have been altered to incorporate the modifications of older views which, are based on recent scientific examination. More space has been committed to plant anatomical markers, in which different anatomical characters are momentarily discussed. It may be well to state that the experiments recommended in these chapters are mostly written simply enough that every student can readily understand them. Anatomical facts have been freely introduced in the discussion of characters to familiarize the students with the relationships betweenstructures and functions of plants and habitats. The text is intended to be clear and accurate, and an attempt has been made to present facts in a logical sequence. Like a good instructor, a book should constantly remind students of certain facts previously dealt with, and thus connect the subject-matter under discussion with content presented previously.

Since this book is intended primarily for students who have had relatively limited experience in the study of plants, a few words about the significance of plant anatomy among the plant sciences should be helpful. In these interrelationships of sciences, anatomy plays a major role. A realistic interpretation of plant function by the physiologist must rest on thorough knowledge of cell and tissue structures associated with that function.

Plant anatomy is interesting for its own sake. It is a gratifying experience to follow the ontogenetic and evolutionary development of structural features and gain the realization of the high degree of complexity and remarkable orderliness in the organization of the plants. This book contains a glossary which was prepared with special regard to the latest developments of concepts and terminology. These chapters are used for introducing the concept of plant anatomical structures and description of the families of plants.

Zubaida Yousaf, PhD

ACKNOWLEDGMENT

First of all we would like to thanks Allah Almighty for not only making the lovely area, in which we could do study, but also for helping us, and enabling us to carry out this study. Secondly we would like to thanks our Vice Chancellor Prof. Dr. Sabiha Mansoor, Prof. Dr. Bushra Khan Dean of Sciences, Prof. Dr. Farah Khan, Chairperson, Department of Botany Lahore College For Women University Lahore, for their kindness and patience encouragement and continual help and advice. The Information in this handbook has been compiled from many sources. We sincerely appreciate the time and effort that everyone spent preparing and reviewing this material. Many of the past authors and contributors have been identified in the 2013 "Leaf epidermal Anatomical tools for systematical studies of some medicinal Angiospermic Families." Our attempt in this section is to identify all individuals who worked on the 2013 version of the handbook whether their effort was great or small. Those acknowledge as authors or contributors have written completely new sections or conducted major rewrites of existing section. Special Thanks go to our Friends for their moral support without their encouragement we can't fulfill our piece of work.

Zubaida Yousaf, PhD
Afifa Younas, MS
Arusa Aftab, MS

LIST OF ABBREVIATIONS

Elemental dispersive spectrophotometers	(EDS)
Formalin propionic acid	(FPA)
Formic-acetic acid	(FAA)
Forestry herbarium Ibadan	(FHI)
Grain research unit	(GRU)
Hexa methyl disilazane	(HMDS)
Infra-Red gas analyzer	(IRGA)
Light microscope	(LM)
Post Harcourt herbarium	(UPH)
Scanning electron microscopy	(SEM)
Transmission electron microscope	(TEM)
Tertiary butyl alocohal	(TBA)
Unweight pair group method with arithmetic average	(UPGMA)
Water-use efficiency	(WUE)

In: Leaf Epidermal Anatomical Characters...
Editors: Z. Yousaf, A. Younas and A. Aftab
ISBN: 978-1-63483-190-1

Chapter 1

Introduction to Leaf Anatomical Characters for Systematical Studies

Z. Yousaf* and N. Saleh
Department of Botany,
Lahore College for Women University,
Lahore, Pakistan

Abstract

Plant anatomy or phytotomy is the general term for the study of the internal structure of plants. Plant anatomy refers to the description of the structure and parts of plants. Plant anatomy is required in order for individuals to fully participate in the diversity of modern interdisciplinary studies. The subject of plant anatomy is basically the same way, discussing cells, tissues, tissue systems, and organs in sequence. In other words, anatomists employ critical and extensive observation, resulting in the compilation, codification, and analysis of descriptive data, but they also use the analytical methods of the experimental sciences. The role of anatomical data in traditional taxonomy has been long recognized. Anatomical features are of particular value to scientists who need to identify small scraps of plant material. The functional characters which are exposed to the strong selection during evolution are more or less important taxonomically than all non-functional (or apparently non- functional) characters should be examined and utilized in classifying plants. These conserves are the basis for further study of the subject, or provide students of botany with an introduction to the field.

Introduction

Plant anatomy or phytotomy is the general term for the study of the internal structure of plants [1]. While originally it include plant morphology, which is the description of the physical form and external structure of plants, since the mid-20th century the investigations of plant anatomy are considered a separate, distinct field, and plant anatomy refers to just the

* z4zubaida@yahoo.com.

internal plant structures [2]. Plant anatomy refers to the description of the structure and parts of a plant. There are some standard terms that are used to describe leaves, stems, flowers, roots, etc. These terms are often used in plant, weed, and wildflower field guides in order to help people identify the plants they find.

Plant anatomy, or the developmental and comparative study of plant cells, tissues, and organs, is a botanical discipline with a long tradition. Many individuals have emphasized the fact that anatomy is both a descriptive and an experimental science. In other words, anatomists employ critical and extensive observation, resulting in the compilation, codification, and analysis of descriptive data, but they also use the analytical methods of the experimental sciences.

The Foliar epidermis is one of the most striking taxonomic characters from the biosystematic point of view and the taxonomic studies of a number of families are made on the basis of leaf epidermis [3, 4, 5, 6]. The role of anatomical data in traditional taxonomy has been long recognized since the variations within the species, genera or a family is usually reflected in anatomical features as well. Leaf epidermal anatomical features such as stomata, trichomes and other characters are useful anatomical tools. For example; Measurements of leaf epidermal cell length and width are regarded useful aids in distinguishing varieties with similar flowering date in perennial ryegrass [7].

Anatomical features are of particular value to scientists who need to identify small scraps of plant material. Epidermal cells are quite variable in their configuration. The characters, which have been proven to be of systematic value, are cuticular characters, epidermis, stomata, subsidiary cells and trichomes [8]. There has been a remarkable evolution in the past 30 years or so in the investigation of vascular plant anatomy and its uses in classification [9]. The taxonomic significance of epidermal morphology is well documented in botanical literature [10].Some particular groups of plants or taxa seem to be characterized by specific type of epidermal features, which are the epidermis, stomata, gland and trichomes [11, 12].

NATURE OF LEAF EPIDERMAL CHARACTERS

The nature of characters are divided into categories.

1. Functional Characters

The functional characters which are exposed to the strong selection during evolution are more or less important taxonomically than non-functional (or apparently non- functional) ones. In the twentieth century most plant taxonomists have avoided the use of characters which have obvious and well- defined functions, on the grounds that they have become channeled by evolutionary process. This channeling must have taken the form of convergence, parallelism and extremely rapid divergence in many cases, so that the possession of such characters in common (or the reverse) is not a very good measure of overall similarity. Succulence, the presence of vessels in pteridophytes and gymnosperms, lack of fusion of petals, and explosive fruits are obvious examples. The case for considering this sort of features to have little taxonomic value has been put very strongly by Willmott

(481), among others. It is a view which is particularly related to taxonomy which attempting to construct phylogenetic classification, and less so to ones among at a phonetic system. Nevertheless, it is still a view upheld by many plant taxonomists, despite attempts by several recent workers to show that there is no a prior basis for ascribing a greater or lesser value to functional characters. Thus, there is little evidence for the view stated by smith 386 that characters.

It is dangerous to pay too much attention to the apparent functional significance of taxonomic characters, especially as often we do not know whether or not the features are adaptive. But, equally, a character which has clearly evolved more than once in response to a particularly evolutionary pressure must be accepted as of limited value in indicating either phenetic or phyletic relationships. The elimination of parallelism and hence of at least some highly functional characters is a features of compatibility analysis.

2. Conservative Characters

Although flowers are no longer regarded as essential and therefore taxonomically particularly important, they still provide the bulk of information contained in the diagnoses of angiosperm taxa. This is because in general the flowers appear to be more conservative than do most other organs. It is also often stated, with some justification that endomorphic vegetative characters are more conservative than exomorphic ones and the same are frequently stated for many chemical characters. Conservative organs or characters are those which tend to remain relatively unchanged over a long period of evolutionary development, and hence tend to vary little between closely related taxa. Conservative characters are therefore most useful in delimiting the higher taxa, where the emphasis is on the recognition of similarity between the members of a taxon. This is true whether one is consciously constructing a phyletic or a phonetic system. At the lowest levels of species the emphasis is often on the distinction between taxon rather than the clustering of taxon into larger ones. Therefore, non-conservative characters, i.e., those which show the greatest diversification, are often more useful.

It must be emphasized that in most cases it is not possible to conclude that a particular character is always more or less conservative than another - in one group of plants it might be more so, in another less so. Also, a character or organ in which conservatism has been demonstrated in a taxa might be lacking altogether in another. In such instances conservative characters have to be sought anew. In lower plants, for example, in which flowers are lacking and the reproductive organs are often short-lived or even rare, vegetative characters have assumed a far greater prominence in classification than in angiosperms.

All characters should be examined and utilized in classifying plants. There is, in fact, no contradiction so long, as conclusions on the relative values of different organs or characters are made a posteriori, from experience gained after analysis of a very wide range of possibilities. The floral characters which are predictive, i.e., taxa delimited by floral characters are in general more natural than those delimited by vegetative characters.

In the present book an overview of plant anatomical structure description have been provided. This can serve as the basis for further study of the subject, or provide students of botany with an introduction to the field. Later chapters discuss these anatomical concepts and approaches at greater length in relation to more specialized topics about Plant families' anatomical structures.

REFERENCES

[1] Raven, P.H., Evert, R.F. and Eichhorn, S.E. (2005) *Biology of Plants* (7th edition) W. H. Freeman, New York, Page 9, ISBN 0-7167-1007-2.

[2] Hagemann, W. (1992). ''The Relationship of Anatomy to Morphology in Plants: A New Theoretical Perspective''. *International Journal of Plant Sciences* 153 (3(2)): S38-S48.

[3] Bhatia, C.R. (1984). *Foliar epidermal studies of Heliotropiym Supinum L. folia Geobotanica and Phytotaxonomica.,* 19(4): 381-385.

[4] Stace, C.A. (1984). *The taxonomic importance of leaf surface.* In: Heywood V.H., Moore B.M., (eds). Current concept of plant taxonomy. London: Academic press, 67-94.

[5] Jones, J.H. (1986). Evaluation of the Fagaceae: the implications of foliar features. *Annals of the Missouri Botanical Garden* 73: 228-275.

[6] Baranova, M. (1972). *Systematic anatomy of the leaf epidermis in the Magnoliaceae and some related families.* Taxon, 21: 447-469.

[7] Ellis, R.P. (1976). A procedure for standardizing comparative leaf anatomy in Poaceae. The leaf blade as viewed in transverse section. *Bothalia,* 12: 65–109.

[8] Wilkins, P.W. and Sabanci, C.O. (1990). Genetic variation in leaf epidermal cell size and shape in *Lolium perenne*. *Euphytica,* 47: 233–239.

[9] Behnke, H.D. (1975). The basis of Angiosperm phylogeny; ultra structure. *Annl. Mol. Bot. Gdn.,* 62: 647–663.

[10] Dehgan, B. (1980). Application of epidermal morphology to taxonomic delimitations in the genus *Jatropha* L. Euphorbiaceae. *J. Bot. J. Linnaean. Soc.,* 80: 257–278.

[11] Park, K.W. (1994). A taxonomy study of the Magnoliaceae. *J. Res. Rep. For. Res. Inst.*, 50: 173–190.

[12] Hong, S.P. and Oh, I.C.(1999). The taxonomic study of leaf epidermal micro structure in the genus *Polygonum* L. s. Str. and *Polygonella* Michx. Polygoneae-Polygonaceae J. *Korean J. Plant Taxon.,* 29: 75–90.

In: Leaf Epidermal Anatomical Characters…
Editors: Z. Yousaf, A. Younas and A. Aftab

ISBN: 978-1-63483-190-1

Chapter 2

Protocols for Leaf Epidermal Anatomical Studies

A. Younas**, *I. Afzal and H. S. Ahmad
Department of Botany,
Lahore College for Women University,
Lahore, Pakistan

Abstract

The necessary method used in plant anatomy, or the study of internal plant structure is the removal of chlorophyll pigments from leaf epidermis, which are considered microscopically. From this, the skill derives its name (in Greek, *anatomë* means "dissection"). New research techniques are in use as well as, microtome sections, ultrathin sections, luminescence, scanning electron microscopy; electron microscopy and histo-chemical methods examination. In present writing, authors explored different types of plants like herbs, shrubs, dicots, monocots and trees.

Protocol for Leaf Epidermal Anatomy of Herbs

Leaves of herbaceous species are soft; therefore fresh leaves should be preserved in 70% alcohol [1]. Transversal and superficial sections were made free hand. (T.S) & (S.S). Then they were stained and fixed in Sartur Reagent and chloral hydrate (50%) solution. As the density of indumenta of adaxial surface was higher, only superficial sections were used. Photographs were taken using Leica DM 400B type microscope attatched to the camera in JPEG format. Drawings were made using Olympus BX50 microscope attatched with U-Da 2K 17149 sketching prism.

Márcia do Rocio Duarte and Maria do CarmoDebur [2] worked on "Characters of the leaf and stem morpho-anatomy of Alternanthera brasiliana (L.) O. Kuntze, Amaranthaceae." They collected the plant material and the mature leaves from the fourth node onwards, and the stem

* botanistafifa@yahoo.com.

fragments were obtained 10cm from the apex of stem progressing towards the base. They were fixed in FAA 70 [3] and kept in 70% ethanol. Free hand longitudinal and transverse sections were stained with Fuchsine/astra blue combination [4] or with toluidine blue. Hydrochloric phlorogucin was applied to observe lignified elements [5], iodine-iodine was applied for starch [6], Sudan IV for lipophilic substances [7] and ferric chloride was applied to reveal phenolic compounds [3].

Pompelli et al., [8] carried out their work on "Leaf anatomy and plasticity of Coffea arabica L. in response to light and nitrogen." Leaves were oven-dried for 72 hours at 70° C. Leaf mass per area was calculated with leaf mass to surface area ratio. For studying anatomy of leaf, the mid-laminar areas without the central vein were taken and fixed in FAA for 48 hours in 70% ethanol (v/v). Then dehydration was done and fixed in methacrylate. (Historesin-Leica Microsystems Nussloch, Heidelberg, Germany). Cross sections 7µm and paradermic 3µm sections were made with rotary microtome. (Model RM2155, Leica Microsystems Inc., Deerfield, USA). Toluidine blue was used to stain them at pH 4.0 and fixed in synthetic resin (Permount). The data was quantified using software (Image Pro-Plus, version 4.5, Media Cybernetics, Silver Spring, USA). Images were captured using Olympus Microscope (AX70TRF, Olympus Optical, and Tokyo, Japan). Drawings were made from photo micrograph. The data which was collected was based on total leaf thickness, palisade and spongy parenchyma thickness, adaxial and abaxial epidermis thickness and air space calculated as cross-section of total mesophyll tissue, for an overall per treatment 300 measures.

To study leaf characters under electron microscopy, leaves should be embedded in 2.5% Glutaraldehyde with 0.08m sucrose in a 0.05m phosphate buffer having pH 7.2. After washing the buffer, leaves were again fixed in OsO4 which was 1% buffered. It was again washed with phosphate buffer. It was embedded in spurr through a graded acetone series. Sections were cut with Ultra cut E ultra-microtome, staining was done with 2.5% uranyl acetate and Reynold's lead citrate [9] afterwards. Examination was done under JEOL 1200 EX transmission electron microscope at 80kV. Thick sections were also observed under the light microscope (LM) and were viewed in detail with the help of Nikon Microphot-FX-Light Microscope.

Rodriguez, E.G. et al., (10), observed "Texture, composition and anatomy of spinach leaves in relation to nitrogen fertilization." For these, 6, 8, 12 and 16 number leaves were used to observe their cross-section and variation was observed by comparison. Images of 10 leaves per treatment of nitrogen were observed. Younger leaves (12 and 16) and older ones (6 and 8) were captured digitally.

PROTOCOL FOR LEAF EPIDERMAL ANATOMY OF SHRUBS

Dzomeku Beloved Mensah [11] had studied "Leaf Anatomical variation in relation to stress tolerance among some woody species on the Accra plains of Ghana." Leaf epidermal strips were fixed in Canada balsam for one night. Leaves had been collected and kept in Formal Acetic Alcohol (FAA). Sections were made using a microtome for anatomical studies. At least 3µm transverse section of leaf epidermis was made. These thin sections were screened under a microscope which was stained with safranin along the different series of ethanol with which they were washed. After this, they were stained with light green (1/2-1min) and then washed with clove oil for 5 min. They were again fixed with Canada balsam in the end.

Rakic T. et al., [12] had carried out their study on "Eco physiological and anatomical characteristics of the subtropical shrub, Zantho xylum acanthopodium (Rutaceae) in conditions of a temperate continental climate (Serbia)." Leaf sections had been cut and permanent slides were made. Leaves were embedded in Navashin's fluid for one night. Through a series of alcohol, they were washed and fixed in paraplast. Transverse sections were made of 5μm thickness by using a Reichert sliding microtome after removing the paraplast. Then they were stained with hematoxylin and safranin. They were observed under a light microscope (Leica DMLS with attached Canon digital camera). To examine leaf surface structure, a scanning electron microscope was used (JEOL JSM-35).

Nwachukwu, C. U. and Mbagwu F. N [13] had studied the "Anatomical features of the roots and leaves of Hibiscus rosa sinensis and Abelmoschus esculenta." Mature and fresh leaves of Hibiscus rosa sinensis of 1 x 0.5cm in length were taken. They were embedded in FAA and 70% ethanol for 72 hours. They were then washed in distilled water for several times and through different series of alcohol and were unfiltered with wax and passed through different series of alcohol and chloroform and later. Wax was added into the bottles in which the leaves were kept. After the wax had been removed, sections were made on a Reichert rotary microtome of 20 - 24cm. Thin sections were placed on slides along the Haupt's albumen which had been dried and few drops of water added before fixation. Some drops of alcian blue were added for 5min and then washed off. They were stained with safranin and then washed with a series of alcohol and pure xylene, afterwards, they were fixed in Canada balsam. Through the slides, the photo micrographs were taken using Leitz Wetzlarartholus microscope attached to a vivitarV-335 camera.

Ronald A. Balsamo et al., [14] observed "Leaf biomechanics, morphology and anatomy of the deciduous mesophyte Prunusserrulata (Rosaceae) and the evergreen sclerophyllous shrub Heteromelesar butifolia (Rosaceae)." Prunusserrulata and Heteromelesar butifolia leaves were collected for the experiment. Lamina length, width and thickness were observed using a digital caliper Fowler Promax; (Newton, Massachusetts, USA). For study under Scanning Electron Microscope, leaves were cut into quarter pieces and embedded in a solution of glutaraldehyde in 50mmol/L Na-cacodylate buffer with a pH of 7.0. They were then fixed in 1% OsO4 in 50mmol/L Na-cacodylate buffer (pH 7.0) for half a day i.e., 12 hours at room temperature. After getting it through 50mmol/L Na-cacodylate buffer they had been dehydrated in acetone. Pieces were kept under liquid nitrogen keeping it perpendicular to the mid-rib. Cross-sections were made clear using a razor. These pieces were then dried (Polaron Jumbo Critical Point model E3100, London, UK) and gold sputter coater (Polaron SC7640). Samples were photographed and viewed using a Hitachi S-570 SEM (Tokyo, Japan).

Pereira, D. C et al., [15] carried out their study on "In situ variation in leaf anatomy and morphology of Andiralegalis (Leguminosae) in two neighboring but contrasting light environments in a Brazilian sandy coastal plain." Five leaves were collected from each of the 50 samples for the anatomical studies. Leaves were collected from the upper node and then fixed in formaic-acetic acid alcohol (FAA). They were dehydrated in ethanol and fixed in glycol methacrylate resin (Johansen 1940). Transverse sections of 6mm were made with a rotary microtome and stained with 0.1% toluidine blue and safranin (Feder & O'Brien, 1968).

Quantification of tissue layers was done using an Olympus BH2 microscope with an ocular micrometer. Epidermal dissociation was made using Jeffrey's reagent to study under light microscope, the stomata, common epidermal cells and trichome density (number per mm^2) were measured. Histological tests were made to identify tannins and calcium oxalate. (3).

Lakusic B. et al., [16] studied "Morpho-anatomical characteristics of the raw material of the herbal drug Olivae folium and its constituents." Permanent slides were made for anatomical studies of the leaves under light microscopy. 10μm thick cross-sections of leaf were made using a Reichert sliding microtome. These sections were washed in Parazone and were embedded in safranin (1% w/v in 50% ethanol) and alcian blue (1% w/v, aqueous). For stomatal and epidermal analysis, Jeffery's solution was used and stained with alcian blue and safranin. Slides were dehydrated and fixed in Canada balsam. Leaf hairs and paradermal features of epidermis were studied by first coating them with gold and observed using scanning electron microscope JOEL JSM-6460 instrument.

Zhuang Li et al., [17]. had "Anatomical and morphological characteristics of *Populuseuphratica* in the lower reaches of Tarim River under extreme drought environment." Leaves for this study were collected and fixed in formal acetic-acid alcohol (FAA). Dehydration with alcohol was then done by using different strengths. These dehydrated leaves were cleared with ultrasonic wave for 10min. Using liquid nitrogen, the samples were frozen by means of a gold sputter coater, SCT620. Cross-sections and epidermis stood were observed under scanning electron microscope (LEO1430VP SEM). For study under Transmission electron microscope (TEM), leaves were fixed in FAA and washed with phosphate buffer and embedded with osmic acid and again washed with phosphate buffer. Dehydration was done using varying strengths of alcohol from 30% to 100%. Alcohol-acetone to pure acetone series was made and the slides were passed through them. Samples were fixed in epoxy resin and baked in oven up to 60 °C starting from 35 °C with an interval of 5 degrees. Samples were made by means of a microtome. Sections were then stained with uranium acetate and citric acid lead. Hitachi H600 X TEM (Tokyo, Japan) was used to study cellular structures.

Souza L.A. et al., [18] carried out their study on "Comparative morphology and anatomy of the leaf and stem of Peper omiadahlstedii, Ottoniamartiana and Piper diospyrifoliumkunth (Piperaceae)." Plant material was collected and fixed in FAA 50 (Johansen 1940). Rotatory microtome was used for making free hand sections. Staining of the sections was done using astra blue, safranin and hematoxylin [3 and 19]. Special tests were carried out to identify calcium oxalate (10% hydrochloric acid), silica (phenol) [3] and lipid substances (Sudan IV) [3]; Cutler [20]. A reflex camera was used to count stomatal, epidermal cell and trichome frequency with quadrates of $0.04mm^2$. Optical microscope was used to make drawings while attached to a reflex camera. To calculate stomatal index, Salisbury formula was used.

Aysegu Guvenc and Gulcen Kendir [21] studied "The leaf anatomy of some Erica taxa native to Turkey." Plant parts were kept in 70% ethanol. From the preserved material, cross-sections of the leaves were obtained by hand. Material was then preserved in chloral hydrate solution R and Sartur reagent. Using Olympus BX50 microscope attached to a camera, photographs were made.

Marques A.R et al., [22] observed "Variations in leaf characteristics of two species of Miconia in the Brazilian cerrado under different light intensities." Leaves were first fixed in 70% alcohol to assess leaf thickness. Out of the ten selected plants, five leaves per plant were collected and embedded in 70% alcohol. After this, they were transferred to Jeffrey's solution

[3]. Ten fields of unknown area of leaf were examined per leaf and stomata and trichome density was observed. Transverse sections were made and mesophyll comparison was made. 10μm thick cross- sections were made using a microtome. Staining was done with astra blue and Fuschin [3]. Quantification of leaf anatomical features and thickness was made using a microscope with a micrometer eye piece.

Scott R. Koukol and Curtis Clark studied "Comparative anatomy of the leaves and stems of Encelia (Asteraceae-Heliantheae)." Plants were collected and leaves were fixed in FPA (Formalin propionic acid). TBA (tertiary butyl alcohol) was then used for dehydration. Fixation was done in paraffin. Before sectioning, tissues were made soft using Aerosol OT, glycerin and water in 10:3:90 solutions (volume wise). Staining was done using zinc chloride, Safranin, orange G/tannic acid mixture, tannic acid and iron alum. Then they were fixed on slides and photographs were taken.

Protocol for Leaf Epidermal Anatomy of Trees

Yunus et al., [23] considered the anatomical characteristics of *Vitexagnus-castus* (*Verbenaceae*). In this study, leaf sample materials of *V. agnus-castus* investigated were collected from the localities determined from five different cities of Western Anatolia from different altitudes according to the Grid system used by Davis [24]: A1 Canakkale–Ayvacik,B1 Balikesir–Savastepe, B1 Izmir–Beydag,C2 Denizli–Honaz andB2 Mugla–Bodrum. The collected materials were identified taxonomically according to Davis [24]. Then, leaves were placed into a 70% alcohol-water solution to store for later examination. Root, stem and leaf anatomical characteristics were investigated in cross–sections which were taken in laboratory conditions. Then cross-sections were been dyed by "sartur reactive" through microscopic investigation, and then coloured pictures were been taken by a microphotography apparatus.

Nwachukwu and Mbagwu [25] premeditated the Anatomical features of the roots and leaves of *Hibiscus rosa sinensis* and *Abelmoschus esculentus.* Section of mature and fresh roots and leaves of *Hibiscus rosa sinensis* not longer than 1 × 0.5 cm each of the two taxa were collected from the cultural garden of IMSU and Songhi farms, Nekede Owerri-West Local Government in Imo State were set into labeled vials and then fixed in FAA (1. 1. 18) 40% formaldehyde, 70% ethanol (v/v) for at least 72 hours. These were then cleaned in several variations of distilled water and passed through various alcohol series (30%, 50%, 95% and 100%). The dehydrated materials were infiltrated by wax passing through different proportions of alcohol and chloroform. Gradually, the alcohol, pure chloroform and wax were added into the bottles. The idea was to slowly infiltrate the tissues with wax that would be hard enough to microtomy. Later, the metal molds were removed and the specimens within the wax cube trimmed section on Reichert rotary microtome at 20 – 24 cm. The ribbons were then sited on clean slides smeared with a film - Haupt's albumen and allowed to dry. Prior to mounting, drops of water were added and then drops of "alcian blue" were dropped on the specimen for five minutes, washed off by water and counter stained with safranin for two minutes, then dehydrated in a series of alcohol - 50%, 70%, 80%, 90%, xylene/absolute alcohol solution (i.e., 1: 3 and 1: 1 v/v) and pure xylene at intervals of a few seconds and mounted in Canada Balsam. Photomicrographs were taken from the slides using a Leitz Wetzlarartholus microscope fitted with a vivitarv-335 camera.

Hussein, [26] studied the "Anatomical Structure of Jatropha Leaves Affected by Proline and Salinity Conditions." Experiments were carried out on *Jatropha curcas* seeds grown in pots, 30 cm in diameter, that were filled with sandy soil under greenhouse conditions of National Research Centre, Dokki, Giza, Egypt. Seeds of *Jatropha curcas* were sown during the two summer seasons of 2010 and 2011. The seeds were irrigated regularly with tap water for three weeks until seedling emergence to avoid the effect of salinity on seedlings. Irrigation with diluted Mediterranean Sea water was carried out after twenty-one days from sowing to attain salinity levels of 2500, 5000 and 10,000 ppm (Three irrigation by salt water, and the next was by tap water alternatively) and tap water was used as control. Amino acid proline at the rate of 0 and 100 ppm were sprayed twice - 156 days from sowing and the second, two weeks later and the control sprayed with distilled water. The experiment included 8 treatments which were the combination of four levels of salinity 0, 2500, 5000 and 10000 ppm and two concentrations (0 and 100 ppm) of proline. After 186 days of sowing representative samples, 3 replicas of treatments were used for anatomical measurements. The samples were then taken from the upper forth leaves of the plant and cleaned with tap water and cut into suitable parts and fixed in FAA (10 ml formalin, 5 ml acetic acid and 85 ml. of ethyl alcohol 70%) then dehydrated in different concentrations of ethyl alcohol, cleared in different concentrations of ethyl alcohol + xylene, infiltrated and embedded in pure paraffin wax (M.P.58-60 co) [3]. Sectioning at thickness of 20 μm was performed by using a rotary microtone. Paraffin ribbons were mounted on slides and sections stained with "safranine and fast green" [27]. Sections was mounted in Canada Balsam, then examined microscopically and micro photographed.

Souza et al., [28] studied the "Comparative Morphology and Anatomy of the Leaf and Stem of *Peperomia dahlstedtii* c.Dc. *Ottonia martiana* Miq. And *Piper diospyrifolium* Kunth (Piperaceae)." The collection of the botanical material and the field observations of *P. dahlstedtii*, *O. martiana* and *P. diospyrifolium* were carried out in "HortoFlorestalde Maringá," Paraná State (Brazil), a 37 ha forest remnant. The morphological description of the leaves of the species was based on Rizzini's study [29]. The analysis of the stem and leaf structure was done using freshly-collected material or material fixed in FAA 50 [3]. This material was then cut freehand or in a rotative microtome to make semipermanent and permanent slides for the microscopic studies. Sections were stained in "astra blue, safranin and hematoxylin" [3]. Specific micro-chemical tests were done in stems and leaves for Calcium oxalate (10% hydrochloric acid) (Chamberlain 1932), Silica (phenol),and lipid substances (sudam IV).

Abdual Khalil [30] studied the chemical composition, anatomy, lignin composition and cell wall structure of Malasyian plant waste fibre. Plant material including oil palm frond, coconut, pineapple leaf and banana stem were used for that purpose. For determining chemical composition of plant material, agro fibers were grounded and then 40 mesh fractions selected. The samples were first submitted to Soxhlet extraction for 6 hours consuming ethanol/ benzene in 1:2(v/v). To determine alpha – cellulose, lignin and ash content standard methods were used. For anatomical study of cell wall structure and lignin distribution, samples were chosen randomly and cut into 2×3 mm blocks. Samples were hydrated in ethanol series and embedded in Epoxy resin, polymerized for 24 hours at 60 °C. Transverse section sections was made and stained with " touledine blue" and observed under polarized microscope (Olympus BX50) for lignin distribution. Ultra-thin sections were also prepared (0.1μm) from the embedded samples and stained with " 2% uranyl acetate and lead citrate" and then observed under transmission electrone microscope (Phillips CM12).

Comparative Impacts of Water Stress on the Leaf Anatomy of a Drought-Resistant and a Drought-Sensitive Olive Cultivar

For leaf anatomy, small pieces (1 cm^2) were taken from the centre of ten leaves of both WW- and UW-treated olive plants of each cultivar and fixed in FAA solution [50% (v/v) ethanol, 5% (v/v) acetic acid, 10% (v/v) formaldehyde] for $\geq$ 6 h at 4°C. The samples were dehydrated in an ascending series of ethanol solutions followed by infiltration with paraffin (CP-EMB44026; Melun, France) in which they were finally embedded. Semi thin (1 – 2 μm-thick) sections were cut with an ultra-microtome (RM2135; Leica, Nussloch, Germany). All sections were stained with 1% (v/v) "toluidine blue O," and observed under a light microscope.

Hoyam and Kordofani [31] considered the morphological and anatomical studies of *Cordia abyssinica*R.Br. and *Cordia sinensis* Lam. (Boraginaceae) in Sudan. For that purpose, fresh specimens of the two species were collected from Khartoum River Nile bank and Shambat; Faculty of Forestry, University of Khartoum. For anatomical study, Specimens were prepared by cutting the stem in 2 cm length and leaf in 1cm length, and fixed in formalin: glacial acetic acid: 70% (5:5:90%) for 6 days. Then the fixed parts of the plants were washed with distilled water and dehydrated by serial of ethyl alcohol (50%, 70%, 90% and 100% respectively). The plant organs were then transferred to mixture of "1:1 Cedar wood oil absolute ethyl alcohol and now into pure cedar wood oil, followed by a mixture of cedar wood oil and xylene and finally left over night in pure xylene." Wax embedded was carried out in an oven adjusted at 60ºC where the plant organs were transferred from a mixture of wax and xylene into pure wax. Each change took place for about 20 minutes and was finally put into another container of pure wax. The melted wax containing the plant organs which were poured into a mold cooled in water and trimmed, and thin sections were cut using rotary microtome. Sections were collected on glass slides and covered with egg albumin to keep attached to the slides and then left for about 15 minutes on a hot plate at temperature below 60ºC to give maximum expansion of the tissue on to slides. Then the slides were left overnight to ensure complete dryness before staining, and staining required successive process of dewaxing, rehydration, then staining and dehydration. The process was carried out by passing the slides through a series of coupling jars containing xylene, absolute ethanol, 95%, 90%, 70%, 50% ethanol, safranine stain - 50%, 70%, 50%,95%, fast green stain, xylene at them, a drop of D.P.X (mountant) was added to the slide before placing the cover slips (JE Sass,1958).

Aguiar and Preisinger [32] studied the traits of leaf Anatomy of *Croton lanjouwensis* Jablonski (Euphorbiaceae) in different Strata of the Plant. The study was carried out in a 12-year secondary forest that was situated in the experimental site of Embrapa, 29 km north of Manaus, Amazonas, Brazil. In this area, 50 leaves were taken from the peripheries of the upper (17 m) and the lower stratum (9 m) of an adult tree and from a young tree (2 m) of *C.*

lanjouwensis. The leaf areas of the collected leaves were then measured and 5 leaves of each stratum were selected for anatomical studies (transverse sections and front views). Transverse cuts were made of the petiole and in the region of the central nerve. The epidermis was then surveyed in three regions of the leaf (apex, middle, base), after dissociation in Jeffrey's solution. The material died in astral blue and basal fuscin, and mounted in Canada balsam (transverse cuts) and glycerine (epidermis). The numbers of stomata and hairs per mm^2 were determined for each of the leaf regions. The microscopic photos were made with a photo microscope Axioscop Zeiss MC80.

Muhammad, S. [33] studied the Anatomical studies of stems, roots and leaves of selected citrus root stock varieties relative to their vigour. Seeds of 7 citrus root stock varieties such as V_1 Troyercitrange (*Citrus sinensis*×*Poncirus trifoliata*), V_2 Rough lemon(*Citrus jambhiri*) V_3Swinglecitrumelo (*Citrus paradisi*×*Poncirus trifoliata*),V_4 Sweet lime (*Citrus limettioides*),V_5 Carrizocitrange(*Citrus sinensis*×*Poncirus trifoliata*),V_6 Sour orange (*Citrusaurantium*) and V_7 Flying dragon (*Poncirustrifoliata*) were imported from the USA to provide plants for experimentation. For anatomical study of leaves, the fourth or fifth leaf down from the terminal leaf was used. From each leaf, single portions of about 10 × 5 mm were taken from the area mid-way between the leaf apex and base. These portions were then kept in separate and labeled vials containing excess F.A.A (formaldehyde-acetic acid-alcohol). All manipulations were carried out quickly to avoid wilting of leaves. The subsequent processes together with dehydration in ascending concentrations of ethanol followed by Histoclear, infiltrated by wax and embedded in wax and mounting were done in the standard way as described by Purvis et al., [34]. Transverse sections 12 μm thick were cut with a sledge microtome and mounted on glass slides. All slides were then labeled with variety and replication number. Slides were dipped in Histoclear for 3 min to dissolve the wax followed by rehydration in descending strengths of ethanol (70 - 100%) for 1 min each. The samples were stained with Safranin O (biological stain used in histology) (3% solution made up in 50% ethanol) for one hour followed by 5 min gentle washing in running tap water and transferred to crystal violet (0.25% solution made up in 90% ethanol) for 2 min followed by a dip in distilled water for 1 min and several dips in 90% ethanol until no more violet stain was detached. Numerous dips in fast green (1% solution made up in distilled water) followed until no more violet stain was removed when they were been transported into two separate isopropanol solutions for 3 min each respectively. Earlier covering with cover slips of the slides were been put in histo-clear for 5 min again.

Naima [35] studied the anatomical adaptations in vegetative structures of apricot tree (*Prunusarmenaica. L.*) cv. 'Amor El Euch' grown under water stress. For leaf anatomical study, the materials were disinfected using 5% sodium hypochlorite followed by three washes with distilled water. After drying in open air, samples were fixed in FAA, [3] then dehydrated in ascending concentrations of ethanol and then in toulene. Samples were embedded in paraffin wax and histological sections then made using rotary microtome. For clear observation of vegetative plant material e.g., leaves, blades had been stained in a mixture of haematoxylin and eosin. After adding a drop of Canada balsam, sections adhered to the blade. Cover slips were added for anatomical observation.

PROTOCOL FOR LEAF EPIDERMAL ANATOMY OF DICOT

Sayantan and Mondal [36] studied "Comparative (Quantitative and Qualitative) Studies of Stomata of Selected Six Medicinally Viable Species of *Cassia* L." for the isolation of leaf epidermis. The Plant Material was used both fresh and dried. For this purpose, they were mature leaves, and then fixed for 24 hours in FAA solution (acetic acid: alcohol: formalin: water = 2:5:1:12) and after washing with 70% ethanol. Three samples were taken along the midrib of each leaf in form of circular disc. Then these samples were boiled in KOH which was 5% aqueous solution for 5 -10 minutes. The epidermis were then peeled off, stripped, and stained with 1% in 50% aqueous ethanol, saffranine and temporarily mounted in glycerin. In this way, the slides were prepared for further study.

Morpho-Anatomical Diversity of Leaves in *Lagerstroemia* L. [Lythraceae] collected species from variety of areas like Dehradun (U.A.) Kolhapur, Kannad, and locally from Botanical garden. After this, all the collected materials were preserved in 70% alcohol. The plant material used in experiment which belonged to herbarium specimens were primarily boiled in water for about 5-10 minutes. Before passing the herbarium material to the next step of process, some drops of acetic acid were added to soften the tissues and help out the tissues to natural state after few minutes of boiling. Later, transaction was taken from middle of the leaves by using both sources from free hand and microtome. Sections were stained by means of fast green (1%) and safranin (1%) and mounted in Canada balsam following the normal dehydration. Some of the hand sections used in experiment were also observed in glycerin. In the experiment, Sudan IV was additionally used to distinguish the cuticle. Transaction at the middle of the leaf was taken by both free hand and microtome. Sections were stained in safranin (1%) and fast green (1%) and mounted in Canada balsam after the customary dehydration. Some hand sections were also examined in glycerine, Sudan IV was used to demarcate the cuticle. The phloroglucinol and aniline sulphate tests were also conducted for the lignified tissues. And by using camera Lucida, the cellular sketches of the figures were drawn.

Gilani et al., [37] studied anatomy of leaf epidermis of *Digitaria* species belonging to family paniceae which is poaceae of Pakistan. For anatomical study, they used both fresh and dried leaves of eight spicies of *digitaria*. For epidermal scraping, the dried leaves were boiled until they became soft and unfolded. Fresh leaves were used directly without passing through any treatment. Both fresh and dried leaves were put in a test tube containing 88% lactic acid kept hot by being placed in hot water bath for 50-60 minutes. Lactic acid made leaf peeling easy by softening its tissues, and adaxial surfaces were prepared by flooding with 88% cold lactic acid. While the adaxial epidermis were scrapped by sharp scalped blade.

Rahman, et al., [38] studied the leaf epidermal features in the family, Cucurbitaceae under the research topic "Diagnostic Significance of Leaf Epidermal features in the Family Cucurbitaceae." For the purpose of anatomical study, leaves of 14 species of Curcurbitaceae family were gathered by the botnists. Then the cuticles of these leaves were dipped in concentrated nitric acid or trioxonitrate (v) acid and washed with distilled water. After this process, these processed leaf cuticles were stained in 1% aqueous safranin solution and mounted in dilute glycerine. After all these processes, they measured the stomatal density, stomatal index and stomatal size. The percentage occurrences of complex type were expressed by the frequency of each complex type which was based on all occurrences. The product of length and breadth of a cell based on sample size determined the size of epidermal cell. Then the index of similarity of leaf was determined by Philips formula. Analysis of variance and Duncan Multiple Range Test were considered under Statistical analysis.

Ultra structure of Stomata Development in Arabidopsis (Brassicaceae) Leaves was the research topic of Liming Zhao and Fred D. Sack to study the stomatal development in Arabidopsis leaves. Firstly, seeds of Arabidopsis were planted in plastic pots filled with peat, perlite and vermiculite. The continuous light from ''Cool White'' fluorescent lamps (General Electric Co.) at a light intensity of 150 μmol· m^-2· s^-1 and a room temperature of 23 °C were the conditions implemented on the growing plants. Hand-cut sections of fresh leaves were examined by differential interference contrast microscopy. Leaves from 3 to 21 day old seedlings, that were cleared according to Geisler, Yang and Sack were used to reconstruct the possible division patterns from fixed tissue as in palia and Landre. Cleared material was also used to calculate cell size. By tracing the paradermal outline of cells of different stages from a bright-field video microscope image onto an acetate overlay on a video monitor, the cell size was estimated. Then using NIH image analysis software and by scanning outlines digitally, areas were determined for 60 cells of each stage. Leaf pieces from 5- to 15-days old seedlings were fixed in 3% (v/v) glutaraldehyde in 75 mol/L phosphate buffer at 4°C for 6–8 hither the tissue was rinsed, postfixes in 2% (w/v) osmium tetroxide for at least 3 h, dehydrated in an acetone series, and embedded in Spurr's epoxy resin. Thin sections cut with a diamond knife were stained with aqueous uranyl acetate and lead acetate. and examined at 60 kV with a CM-12 Phillips electron microscope.

Ghazalah et al., [39] studied leaf epidermal anatomy and their research topic was Taxonomic significance of leaf epidermal anatomy of selected *Persicaria* Mill. species of family *Polygonaceae*. They took dried leaves of selected species of family polygonacae and placed them in boiling water for purpose of softening and unfolding and made them ready for epidermal scrapping. Leaf samples were prepared by following Clark's technique. Then leaves were placed in a tube filled with 88% lactic acid kept hot in boiling water bath for about 30 to 40 min. Leaves became soft due to lactic acid in the tube and leaf surface scrapped by sharp scalpel. After preparing the abaxial and adaxial surface, the leaves were again mounted in 88% clean lactic acid. Both qualitative and quantitative micro morphological foliar characteristics were observed using light microscope. Micro histological photographs of both surfaces were taken by Nikon (FX-35) Camera equipped light microscope. A dendrogram was constructed by μm weighted paired group method with EUCLIDEAN option, using MVSP software version 3.13 (Kovach, 2007).

Lalita and Paliwal [40] studied leaves under the research topic "Studies on the Leaf Anatomy of Euphorbia." The epidermal peels were used in this research. They were obtained from fresh or fixed leaves whereas the dried herbarium leaves were softened by keeping in hot water overnight prior to obtaining the peel. Then the peels were stained with Delafield's haematoxylin. Young leaves fixed in Carnoy's fixative and their peels stained with propinocarmine were used for studying the ontogeny of stomata. By counting the stomata and epidermal cell in unit area, stomata frequency and stomata indices could be calculated. Three counts were made at random from the peels of leaf and it was repeated at least five leaves of a species.

Nels et al., [41] studied leaf mesophyll cell and their research topic was "Oil Bodies in Leaf Mesophyll Cells of Angiosperms: Overview and a Selected Survey." During summer, plants or plant parts of 302 species and hybrids representing 113 angiosperm families were collected at various times of day and from various places. These collected leaves were sectioned with a razor blade cleaned with ethanol and few sections were mounted immediately in a drop of water on a standard microscope slide. Sudan IV dye (Biological Stain Commission Certification #AcZ-2) dissolved in 95% ethanol was added to test for oil. This reagent turns oil drops bright orange just before they gradually dissolve in the ethanol. The presence or absence of oil bodies in intact cells was observed with bright-field optics (see end of Results section), and selected photomicrographs were taken to show the range of variation among the 113 families. The film negative from each selected taxon was digitized using a Power Look (www.umax.com) 3000 scanner, processed in Adobe (www.adobe.com) Photoshop 7.0, and made into a plate using Adobe Illustrator 10.0. Figure 9 was downloaded from the Angiosperm Phylogeny Group website and modified in Photoshop and Illustrator.

Anjum et al., [42] studied stomatal morphology of 69 dicot species distributed in 64 genera and 28 families, and the title of their research was "Stomatal Types Some Dicots within Flora Of Karachi, Pakistan." For this Purpose From A Variety Of District Of Karachi Fresh Materials Of 69 Species Were Collected. For About Half an Hour to 72 Hour, Leaves were dipped in methanol for the study of stomata depending upon the texture of the leaves and then observed under compound microscope (Nikon type -102). Occasionally, leaves were examined without passing through any procedure for stomata study. Impression technique was used in few cases which incorporated the appliance of a film of clear nail polish directly to the lower surface of leaf. After drying, epidermis left its impression on nail film which was applied on the leaf and observed stomata by using scanning electron microscope.

Sheteol and Ayodele [43] Studied Epidermal morphology of the genus *Dialium* (Fabaceae:Caesalpinoideae). For this purpose, the author examined at least five specimens which were taken by each taxon having wide geographical and ecological range. For the analysis, plant materials were attained from Department of Biological Sciences, University of Lagos, Nigeria (LUH) and herbaria of the Forestry Research Institute of Nigeria. Ibadan (FHI). For epidermal preparation near the midrib, middle segment of about one cm square of the leaflet lamina was cut out. Each plant sample was put in a beaker containing 5% sodium hypochlorite (Commercial bleach). The sample in beaker was then boiled for 10-20 minutes by placing the beaker in water bath at 100ºC and as a result, each sample were bleached and separated from the mesophyll. Afterwards, the beaker was drawn out of the water bath and leaflet sample extracted out carefully, washed throughout with distilled water. Afterwards, two drops of glacial acetic acid were added and then upper and lower epidermises were finally isolated from the mesophyll by means of forceps and dissecting needles. Later on, to solidify, the membranes were transferred to 50% ethanol for 5 minutes and then stained in aqueous safranin for 5 minutes. By passing through different grades of alcohol, (viz. 50%, 70%, 90%) each membrane was dried out. By using xylene for 2 minutes, each membrane were cleared and then mounted in Canada. Photomicrographs were made by Reichert Microstar IV microscope just before a camera is connected and at the same time as drawing were made from Wild MI2 microscope fitted through a camera lucida.

Kehinde and Ariwaodo [44] studied the epidermal under the topic, Comparative foliar epidermal studies of genus *Pericopsis* (*Papilionaceae*) in Nigeria. For this purpose, fresh specimens of the species were obtained from the Forestry Research Institute of Nigeria, Ibadan and by the side of Eruwa road, also in Ibadan, Oyo State where the species are well distributed. Fresh plant specimens were macerated in trioxonitrate acid (concentrated) for 2–4 hours. After that, specimens were transferred in Petri dish which contained water. By using forceps and dissecting needle, adaxial and abaxial epidermises were separated. The isolated mesophyll tissue of leaves were carefully cleaned with a Carmel hair brush and washed in water several times. To harden the tissues, 50% alcohol was used for 1 to 2 minutes. After this, the excess water was drained off and the specimens were stained with safranin, transferred on clear microscopic glass slides. Afterwards, the preparation was mounted in glycerin on a slide and examined under the light microscope whereas photomicrographs were taken (×400) by cx31 Olympus Biological microscope.

The leaf epidermis of *Abruse* species were studied by Ikechukwu et al., [45] and title of their research was Leaf epidermal micromorphology in the systematic of *Abrus* (*Papilionaceae*) in parts of tropical West Africa. For this purpose fresh specimens *A. precatorius*, *A.pulchilus* and new collection *Abrus sp.* were collected between 1998 and 2000, deposited in university of Port Harcourt Herbarium (UPH). While the herbarium specimens *A. precatorius*, *A. pulchellus*, and *A. canescens* were atttained from Forestry Herbarium Ibadan (FHI). The fresh samples were fixed into FAA, Formol acetic alcohol (1 part formaldehyde, 1 part glacial acetic acid and 18 parts 70% ethanol v/v) directly. While for the herbarium specimens first scraping method was applied, this began with being soaked in water for 8-12 h. Following this, the surfaces were cleared by flooded with 5% sodium hypochlorite for at least 2 minutes. Cleared epidermal layers achieved were then washed in several changes of distilled water and then stained in 1% safranin and temporarily mounted in glycerin. The preparations were studied and observed with a LETIZ microscope. Photomicrographs were obtained from LETIZ diaplan fitted with LEICA WILD MPS 52 camera at 10 40 magnification of objective lens.

Farooq et al., [46] studied leaf epidermal anatomy under research topic Leaf epidermal anatomy as an aid to the identification of genus *Setaria* weeds, tribe *paniceae (Poaceae)*, from the salt range of Pakistan. For this research, they used leaves of five *setaria* species. For the purpose of softening, unfolding and making them ready for epidermal scraping, dried leaves were placed in boiling water. According to the modified method of cotton that followed Clark technique, leaf samples were prepared. The fresh or dried leaves were placed in a tube filled with 88% lactic acid, kept hot in boiling water bath for about 50 to 60 min. Lactic acid softens the tissues of leaves due to which its peeling off is made easy. Leaf was placed keeping its adaxial surface upward and it was flooded with 88% cold lactic acid for preparing the abaxial surface. Using a sharp scalpel blade, the adaxial epidermis was cut across the leaf and scrapped away together with the mesophyll cells until only the abaxial epidermis of the leaf remained on the tile. Then the epidermis was placed outside and mounted in clean 88% acetic acid. To prepare the adaxial epidermis, same procedure was followed but the leaf was placed in the uppermost part of the abaxial surface. A camera (35 mm) mounted on the microscope was used to take photographs of these mounted materials. On available representative specimens for the taxa, anatomical observations were made. The specimens of 5 species of the genus Setaria are preserved in the herbarium of Quaid-i-Azam University, Islamabad. A high variability was found in the length and width of different characters, therefore the typical size is given. To obtain the typical size, numerous measurements were made on each of the representative specimen.

Oladele [47] studied leaf epidermal features and his research topic were leaf epidermal features in Vernonia amygdalina and Vernonia cinerea. He collected the specimen of fresh leaf of *Vernonia amygdlina* and *vernonia cinerea* from the campus of University of Ilorin, Ilorin. Voucher herbarium specimens prepared were deposited in the Ilorin University

Herbarium. The materials were fixed in acetic acid-alcohol mixture (F.A.A), dehydrated through an ethanol series and embedded in paraffin-wax. Transverse sections were cut on a Cambridge rotary microtome at 15 micro-meter thickness. The sections were stained in safranin/fast green stain combination. The mid region, mid rib and petiole of the lamina were examined by the maceration of tissue in 20% chromium trioxide for one hour at room temperature, cuticles were examined. For the study of leaf surface, paradermal sections of leaves were also cut. With the help of camera lucida drawings, observations were recorded. Terminologies of the stomatal complex followed Dilcher.

PROTOCOL FOR LEAF EPIDERMAL ANATOMY OF MONOCOTS

Inamdar and Shenoy [48] studied the leaf architecture in some convolvolaceae. They collected the plant material of 7 genera and 10 species of the following plants namely Calonyction speciosum CHOISY, Convolvulus arvensis L., Convolvulus microphyllus (Roth) SIEB., Ericibe paniculata ROXB., Evolvulus alsinoidus (L.) L., Evolvulus nummularius (L.) L., Hewittia sublobata (L. f.) O.K. Jacquemontia paniculata (Brum.f.) Hall f., Jacquemontia pententha (JacQ.) G. DoN., and Porana paniculata RoXB were collected from Gujrat and Kerala states. The mature leaves were cleared in 10% sodium hydroxide solution followed by sodium hypochlorite solution (PAYNE, 1969) and secondly in solution of trichloro-acetic acid and phenol crystals in the ratio of 2:1 (by weight) at 60°C (Mohan Ram & NAYYER, 1978). Cleared leaves were stained with fast green or Kores stamp pad purple ink (manufactured by Kores India Ltd., Bombay, 400 018).

Ramadoss et al., [49] studied the morpho anatomical studies of leaves of *Abuiliton indicum* (linn.) sweet. The required samples of different organs were cut and removed from the plant and fixed in FAA (Formalin 5ml+Acetic acid -5ml+70% ethyl alcohol -90ml). After 24 hours of fixing, the specimens were dehydrated with graded series of tertiary-butyl alcohol as per the schedule given by [5]. Infiltrations of the specimens were carried by gradual addition of paraffin wax (melting point 58-60°C) until TBA solution attained super saturation. The specimens were cast into paraffin blocks.

Pimentel and Sahuquillo [50] studied the leaf anatomy of the Iberian species of genus *Anthoxanthum (Poaceae).* In the study, 110 samples from 22 populations of *Anthoxanthum* were analyzed. For each sample, the abaxial epidermal surface and the leaf transverse section were analyzed using fragments of the lower third of the basai leaf that were taken from dried specimens. To get suitable samples for microscope observation, the method of Devesa (1992) was used. For a more detailed study, the technique applied was a double staining, using alcian blue and safranine following Ruiz et al., [51]. The terminology employed was that of Devesa [52] and Ellis [53, 54]. The obtained data were stored in a database, afterwards, a dendrogram was performed using the program SPSS 10.0. The distance measure used was the "Phi" Coefficient, and the grouping method was the UPGMA (Unweighted pair-group method, arithmetic average). Finally, to calibrate the influence of the environment in the anatomical characters, a Canonical Correspondence Analysis was performed using the CANOCO program 4.0 [55]. The environmental parameters included in the analysis were: soil pH, annual mean temperature, mean temperature of the hottest and the coldest months, annual rainfalls and existence of a drought period during the year.

Leaf anatomical studies of the annual species of *Polygonum* s.l. (*Polygonaceae*) were done in Iran. For the study, 38 populations of 15 annual taxa of *Polygonum* s.l. were gathered from different habitants in Iran and studied accordingly. From each population, three specimens were taken and three replications were used for each specimen. Samples were taken from middle part of the leaf blade. Leaf material was prepared by manual cutting. Cross sections were immersed in 10% of Hydrogen peroxide for 20 minutes for bleaching, then washed and stained with methyl green and Congo red. Photos were taken by a light microscope with an Olympus DP12 digital camera. The quantitative characteristics were measured by UTHSCSA Image tool, version 3.0 (2002). The terminology of Stearn [56] was followed for the general outline of leaf cross section.

Manik Baral [57] carried out the Anatomical and Histological Study of Stem, Root and Leaf of the Medicinal Plant, *Amaranthus spinosus* Linn. Leaf anatomy showed kranz mesophyll. Endodermal wall is covered with casperian strips. The stomata (S) occurred on both adaxial and abaxial leaf surfaces. The stomata were found to be anomocytic type. The plant *Amaranthus spinosus* Linn was collected from district Purulia, West Bengal in the month of June-July 2012. It was identified by the Department of Botany, BB College, Asansol, West Bengal, India. The leaves were dissected with the help of 7'O clock blade using potato and were kept in watch glass in water to prevent drying. Two stains were used namely, Safranine to detect cell wall tissue and where as soft tissue took light green colour. The fine thin sections were allowed to immerse in 50% alcohol for 3-5 min. Then, they were allowed to dip in safranin for 3 min. The sections were then washed with 50% alcohol. After washing, they were then placed in 70%, 80% and 90% alcohol successively for 5 min each. After that, the sections were placed in light green for 2 min and then washed with absolute alcohol. Finally, they were mounted and were observed under microscope and photographed (camera model COOLP1*L3) with (×400) magnifications.

Brian et al., [58] studied the diversity in leaf anatomy, and stomatal distribution and conductance, between salt marsh and freshwater species in the C_4 genus *Spartina* (Poaceae). Leaf anatomy, stomatal density, and leaf conductance were studied in 10 species of *Spartina* (Poaceae) from low versus high salt marsh, and freshwater habitats. *Spartina* grasses were collected from field sites. Species identities were verified against herbarium specimens and Flora of North America keys. All species were grown in potting soil in 11 × 11 cm pots, and were maintained under glasshouse conditions on the campuses of both Fort Hays State University (Hays, KS, USA) and Washington State University (Pullman, WA, USA). Plants were watered as needed with tap water Light microscopy and transmission electron microscopy (TEM):

All microscopy work on leaves was performed on the middle of the first fully-expanded mature leaf, usually the third leaf from the apex. Samples were fixed at 4°C in 2% (v : v) paraformaldehyde and 2% (v : v) glutaraldehyde in 0.1 m phosphate buffer (pH 7.2), postfixed in 2% (v : v) OsO4, and then, after a standard acetone dehydration procedure, embedded in Spurr's epoxy resin. Cross-sections were made on a Reichert Ultracut R ultramicrotome (Reichert-Jung GmbH, Heidelberg, Germany). For light microscopy, semithin sections (0.8–1.0 μm thick) were stained with 1% (w: v) Toluidine blue O in 1% (w: v) $Na_2B_4O_7$. Ultra-thin sections (70–100 nm thick) were stained for TEM with 2% (w: v) uranyl acetate followed by 2% (w : v) lead citrate. Hitachi H-600 (Hitachi Scientific Instruments, Mountain View, CA, USA) and JEOL JEM-1200 EX (JEOL USA, Inc., Peabody, MA, USA) TEM were used for observation and photography. To distinguish between silica cells and cork cells in *Spartina* (both are 'short cells' and appear similar under the microscope), paradermal sections were made of the abaxial and adaxial leafepidermis from all species. Sections were stained with 0.1% (w: v) Sudan IV, which stains cork cells red. Scanning electron microscopy: The first fully-expanded, mature leaves from each species were fixed in 5% (v : v) glutaraldehyde in 0.1 M phosphate buffer (pH 7.0) and then dehydrated in an ethanol series. Following this, leaves were cryofractured under liquid nitrogen with forceps. Specimens were then immersed in steps of increasing hexamethyldisilazane (HMDS) concentration (Oshel, 1997). Samples were mounted on a brass plate and were sputter-coated with gold and platinum for 3 min in a SPI Sputter (SPI Supplies, West Chester, PA, USA). Specimens were examined in an ISI-SX30 scanning electron microscope (International Scientific Instruments, Prahran, Victoria, Australia) at 15 kV.

Yousaf et al., [59] considered the leaf epidermal anatomy of selected *allium* species, Family Alliaceae from Pakistan. Leaf epidermal anatomy of the selected *Allium* species had showed variation in size and shape of stomatal cells, stomatal cavity, micro and macro hairs, trichomes, silica bodies and long cells. Leaf epidermal anatomy had proved a significant tool for the resolution of taxonomic confusions of the *Allium* species. Leaves from living and dried specimens were used for anatomical studies. Dried leaves were put in boiling water for a few minutes to soften the leaf until they became unfolded and were been ready for epidermal scrapping. Fresh leaves used directly for anatomical studies. Leaf samples had been prepared according to the modified method of Cotton [60] who followed Clarke's [61] technique. The fresh or dried leaves were kept in a tube filled with 88% lactic acid and placed in hot boiling water bath for about 50-60 minutes. Lactic acid is used to soften the tissues of leaf due to which peeling off was made possible. *Allium* L., species has unifacial leaf, the leaf was placed on tile and flooded with 88% cold lactic acid. The epidermis was cut across the leaf and scrapped away together with the mesophyll cells until only the epidermal layer of the leaf remained on the tile. A sharp scalpel blade was used for that purpose. The epidermis was placed outside uppermost, and mounted in clean 88% lactic acid. The photographs of these mounted materials were taken by a camera (35mm.) mounted on the microscope.

REFERENCES

[1] Ayşegül, G., Gülsen, K. (2012). The Leaf Anatomy of Some *Erica* Taxa Native To Turkey. *Turkey Journal of Botany*, 36: Pp 253-262.

[2] Márcia do R. D., Maria do C. D. (2004). Characters of the Leaf and Stem Morpho-Anatomy of *Alternanthera brasiliana* (L.) O. Kuntze, Amaranthaceae. *Brazilian Journal of Pharmaceutical Sciences*, 4 (1).

[3] Johansen, D.A. (1940). Plant microtechnique. Tata McGraw-Hill Book Company. Bombay: 523.

[4] Rosaria p., Paolode R., Olgade C., Paolo C. (2013). Leaf and stem anatomy in eight *Hypericum* species (Clusiaceae). *Acta Botanica Croat*, 72(2).

[5] Sass, J. E. (1951). *Botanical microtechnique* (2nd. ed.) Ames, Iowa State College Press, 228 p.

[6] Berlyn G.P., Miksche J.P. (1976). *Botanical microtechnique and cytochemistry*. Ames: Iowa State University, p. 121, 276.

[7] Foster, A.S. (1949). *Practical plant anatomy*. 2. ed. Princeton: D. Van Nostrand, p. 218.

[8] Pompelli, M.F., Antunes, W.C., Ferreira, D.T.R.G., Cavalcante, P.P.G.S., Wanderley-ilho, H.C.L., Endres, L. (2012). Allometric Models for Non-Destructive Leaf Area Estimation of the *Jatrophacurcas*. *Biomass Bioenergy* 36, 77–85

[9] Reynold, S. (1963). The Use of Lead Citrate at High pH As an Electron-Opaque Stain in Electron Microscopy. *Journal of Cell Biology*. 17: 217-208

[10] Rodríguez, A., Rodríguez, B., Carrasco, M.N. (2012). High Prevalence of Parental Delivery of Plastic Debris in Cory's Shearwaters (Calonectris diomedea). Mar. Pollut. Bull. 64, 2219–2223.

[11] Dzomeku B. M. (2012).Leaf Anatomical Variation in Relation to Stress Tolerance among Some Woody Species on the Accra Plains of Ghana. *Journal of Plant Development*, 19: Pp13-22.

[12] Rakic, P., Ayoub, A.E., Breunig, J.J., and Dominguez, M.H. (2009). Decision by Division: Making Cortical Maps. *Trends Neurosci*. 32, 291–301.

[13] Nwachukwu C.U. and Mbagwu F.N. (2007). Anatomical Features of the Roots and Leaves of *Hibiscus Rosa Sinensis* and *Abelmoschus Esculenta*. (Fabaceae:Caesalpinoideae. *Feddes Repertorium*, 108 3-4: Pp 15 1 - I58.

[14] Ronald A. B., Aaron M. B., Stephen D. D., Benita M. R. (2003). Leaf Biomechanics, Morphology, and Anatomy of the Deciduous Mesophyte *Prunus Serrulata* (Rosaceae) Evergreen Sclerophyllous Shrub *Heteromeles Arbutifolia* (Rosaceae). *American Journal of Botany*, 90(1): Pp 72–77.

[15] Pereira, D.C., Cláudia Franca Barros and Fabio Rubio Scarano, 2009. *In situ* variation in leaf anatomy and morphology of *Andira legalis* (Leguminosae) in two neighbouring but contrasting light environments in a Brazilian sandy coastal plain. *Acta Botanica Brasilica*. 23(1): 267-273.

[16] Lakušić, B. D. Lakušić, V., Slavkovska, V., and Branka S. (2007). Morpho-Anatomical Differentiation of the Balkan Endemic Species, *teucrium arduini* l. (lamiaceae). *Archive of Biological Science Belgrade*, 59 (4), 369-381.

[17] Zhuang Li, Chen, Y., W. Li, Wang, Z. (2011). Anatomical and Morphological Characteristics Of *Populus Euphratica* in the Lower Reaches of Tarim River under Extreme Drought Environment, *Journal Of Arid Land*, 3(4): Pp 261−267.

[18] Souza, I. Moscheta, S. and Oliveira. J. H. G. (2004). Comparative Morphology and Anatomy of the leaf and stem of *Peperomia Dahlstedtii* C.Dc.,*Ottonia Martiana* Miq. and *Piper Diospyrifolium* Kunth (Piperaceae). Gayana Botanical, 61(1): 6-17.

[19] Dnyansagar, V.R. (1958). Embryological studies in the Leguminosae VIII. *Acacia auriculaeformis* A. Cunn., *Adenanthera pavonina* Linn.,*Calliandra grandiflora* Benth. L. loydia, 21:1-25.

[20] Cutler, D.F. (1978). Applied Plant Anatomy. Longman, London: 103.

[21] Ayşegül Güvenç, M. Mesud Hürkul, Ali Erdem, 2011. The Leaf Anatomy of Naturally Distributed *Juniperus* L. (Cupressaceae) Species in Turkey.*Turkey Journal of Botany*, 35: Pp 251-260.

[22] Marques A.R., Garcia, Q.S., Rezende J.L.P. and Fernandes, G.W. (2000). Variations in Leaf Characteristics of Two Species of *Miconia* in the Brazilian Cerrado under Different Light Intensities.*Tropical Ecology* 41(1): 47-60.

[23] Yunus D., Aye, G. and Kozuharova, E. (2008). A Study of the Anatomical Characteristics of *Vitex agnus-castus* (*Verbenaceae*). *Phytologia Balcanica,* 14 (1): 97 –101.

[24] Davis, P.H. (1982). Flora of Turkey and the East Aegean Islands. Vol. 7. Edinburgh Univ. Press, Edinburgh.

[25] Nwachukwu C.U. and Mbagwu F.N. (2008). Anatomical Features of the Roots and Leaves of *Hibiscus rosasinensis* and *Abelmoschus esculenta. Life Science Journal*. 5(1): 68 – 71.

[26] Hussein, M.M., Abo-Leila, B.H., Metwally, S.A. and Leithy, S.Z. (2012). Anatomical Structure of Jatropha Leaves Affected by Proline and Salinity Conditions. *Journal of Applied Sciences Research*, 8(1): 491-496.

[27] Corgen, J.N. and Widmayer, F.B. (1971). The Effect of Gibberellic Acid on Flower Differentiation Date of Bloom and Flower Hardiness of Poach. *Journal of American Society Science*, 96: 54-57.

[28] Souza L.A, Moscheta, I.S. and Oliveira, J.H.G. (2004). Comparative Morphology and Anatomy of the Leaf And Stem Of *Peperomia Dahlstedtii* C.Dc.,*Ottonia Martiana* Miq. and *Piper Diospyrifolium* Kunth (Piperaceae). *Gayana Botanica*, 61(1): Pp 6-17.

[29] Rizzini, C.T. (1977). Sistematizaçãoterminológica da folha.Rodriguésia 29: 103-125.

[30] Abdal Khalil, H.P.S. Siti. M., Alwani and. Omar M.A.K (2006). Chemical Composition, Anatomy, Lignin Distribution and Cell Wall Structure of Malaysian Plant Waste Fiber. *BioResources* 1(2):220-232.

[31] Hoyam O. A. and Kordofani, M. A. (2012). Morphological and anatomical studies of *Cordia abyssinica* R.Br. and *Cordia sinensis* Lam. (Boraginaceae) in Sudan. *Journal of Chemistry Pharmaceutical Research*, 4(10):4582-4588.

[32] Aguiar, M.O. and Preisinger, H. (2000). Traits of Leaf Anatomy of *Croton lanjouwensis* Jablonski (Euphorbiaceae) in Different Strata of the Plant.

[33] Muhammad S., Dodd P.B. and Sohail, L. (2010). Anatomical Studies of Stems, Roots and Leaves of Selected Citrus Rootstock Varieties in Relation to their Vigour.*Journal of Horticulture of Science and Technology*, 4 (2): 87-94.

[34] Purvis M.J., Collier D.C., Walls D. 1966. Laboratory techniques in botany. 2nd ed. Butterworths, London: 439.

[35] Naima U. L., Olfa B., Fethia H.S., Jaime A.T.D.S., Salah R., Richard H. (2011) Anatomical Adaptations in Vegetative Structure of Apricot Tree (Prunusarmeniaca L.) cv. 'Amor El Euch'grown under Water Stress. *Fruit, Vegetable and Cereal science and Biotechnology*. 5 (2):46-51.

[36] Sayantan T. and Mondal, A. K. (2012). Comparative (Quantitative and Qualitative) Studies of Stomata f Selected Six Medicinally Viable Species Of *Cassia L.*International. *Journal of Lifesciences and Pharma research,* 1 (3).

[37] Gilani,S.S., Khan, M.A., Shinwari, Z.K. and Yousaf, Z. (2002). Leaf Epidermal Anatomy of Selected *Dzgztarza* Species, Tribe Paniceae, Family Poaceae of Pakistan. Pakistan Journal ofBotany, 257-273.

[38] Rahaman, A.A., Oyedotun, R.A. Oladele, F.A. (2011). Diagnostic Significance of Leaf Epidermal Features in the Family Cucurbitaceae. *Insight Botany,* 1(2): Pp 22-27.

[39] Ghazalah Y., Khan, M.A., Nighat S., and Muhammad Q. H. (2010). Taxonomic Significance of Leaf Epidermal Anatomy of Selected *Persicaria* Mill. Species of

Family Polygonaceae from Pakistan. *African Journal of Biotechnology*, Vol. 9(25), Pp. 3759-3768.

[40] Lalita K., Paliwal, G.S. (1972). Studies on the Leaf Anatomy of *Euphorbia* V. Epidermis. *Department of Botany University of Delhi*-110007, Vol. 40(1).

[41] Nels R.L., Albert R. Czlapinski, J.D., Curtis R.F. and Horner, H.T. (2006). Oil Bodies in Leaf Mesophyll Cells of Angiosperms: Overview and a Selected Survey. *American Journal of Botany*, 93(12): Pp 1731–1739.

[42] Anjum P., Rubina A. And Rabab F. (2007). Stomatal Types of Some Dicots within Flora of Karachi, Pakistan. *Pakistan Journal of Botany*, 39(4): 1017-1023.

[43] Sheteol, A.0. and Ayodele, U.A.E. (1997).Epidermal Morphology of the Genus *Dialium Althaea L. (Malvaceae)* Species Growing in Turkey. *Hacettepe University Journal of the Faculty of Pharmacy*, 28(2): Pp 133-148.

[44] Kehinde A.A. and Ariwaodo, J.O. (2012). Comparative Foliar Epidermal Studies of Genus *Pericopsis* (*Papilionaceae*) in Nigeria. *Phytologia Balcanica*, 18 (1): Pp 37 – 41.

[45] Ikechukwu O. Agbawa, B. and Okoli, E. (2006). Leaf Epidermal Microphology in the Systematic of *Abrus (Papilionaceae)* in Parts of Tropical West Africa. *Asian Journal of Plant Science*, 5(1):Pp 42-49.

[46] Farooq A., Khan, M.A., Mushtaq A., Muhammad Z., Zafar I. and Gabe M. (2008). Techniques Listologiques, Masson et Cie., Paris: 11-13.

[47] Oladele, F.A. (1990). Leaf Epidermal Features in *Vernonia amygdalina* and *Vernonia cinerea. Nigerian Journal of Botany*, Vol. 3: Pp 71-77.

[48] Inamdar, J.A. and Shenoy, K. N. (1980). Leaf Architecture in Some Convolvolaceae. *Phyton* (Austria), 21(1): 115-125.

[49] Ramadoss K., Pannem V. and Nesepogu C. (2012). Morpho Anatomical Studies of Leaves of Abutilon indicum (linn.), sweet. Asian Pacific *Journal of Tropical Biomedicine*: 464-469.

[50] Pimentel M. and Sahuquillo, E. (2003).Study of the Leaf Anatomy of the Iberian Species of Genus *Anthoxanthum (Poaceae). Bocconea* 16(2): 683-689.

[51] Ruiz, M.S., Rodicio, M.C. and Corujo, A. (1985): Cuadernode practicas de Citologiae HistologiaVegeta! Y AnimaI. - Santiago de Composte!a.

[52] Devesa, J.A. (1992): Anatomia foliar y palinologia de lasgramineasextremefias. MonografiasBotanicas. - Badajoz. in the C_4 genus *Spartina*(Poaceae). *New Phytologist*, 184: 216–233.

[53] Ellis, R.P. (1976): A Procedure for Standardizing Comparative Leaf Anatomy in the *Poaceae,* I. The Leaf-Blade As Viewed in Transverse Section. - *Bothalia,* 12(1): 65-109.

[54] Ellis,R.P. (1979). A Procedure for Standardizing Comparative Leaf Anatomy in the *Poaceae,* II. The Epidermis As Seen in Surface View.- *Bothalia* 12(4): 641-671.

[55] TerBraak, C.J.F. and Smilauer, P. (1998): A study on the Anatomical Characteristics of *Vitexagnus-castus* (*Verbenaceae*) *Phytologia Balcanica,* 14 (1): 97 –101.

[56] Stearn, W.T. (1983). Botanical Latin, History, Grammar, Syntax, Terminology and Vocabulary. Redwood Press Ltd., Great Britain.

[57] Manik B. (2013).Anatomical and Histological Study of Stem, Root and Leaf of the Medicinal Plant *Amaranthus spinosus* Linn. *Journal of Pharma Science and Technology*, 2(2):68-71.

[58] Brian R. M., Nuria K. Koteyeva, E. V., Voznesenskaya, J. Thomasson, R. and Gerald E. E. (2009). Diversity in Leaf Anatomy, and Stomatal Distribution and Conductance, between Salt Marsh and Freshwater Species. Methods in Plant Histology. The University of Chicago Press, Chicago: 416.

[59] Yousaf, Z., Shinwar, Z.K., Asghar R. And Parveen, A. (2008). Leaf Epidermal Anatomy of Selected *Allium* Species, Family Alliaceae from Pakistan. *Pak. J. Bot.*, 40(1): 77-90.

[60] Cotton, R.L.A. (1974). *Cytotaxonomy of the genus Vulpia*. Ph.D. Thesis, University of Manchester, USA.

[61] Clark, J. (1960). Preparation of Leaf Epidermis for Topographic Study. Stain Technology, 35: 35-39.

In: Leaf Epidermal Anatomical Characters…
Editors: Z. Yousaf, A. Younas and A. Aftab
ISBN: 978-1-63483-190-1

Chapter 3

IMPORTANT LEAF EPIDERMAL MARKERS FOR SYSTEMATICS OF MEDICINAL PLANT SPECIES

A. Aftab*, Z. Tariq, M. Rashid and M. Khan
Department of Botany,
Lahore College for Women University,
Lahore, Pakistan

ABSTRACT

The epidermis is a system of cells, variable in structure and function that together constitute the covering of the plant body in its primary state. The composition consists of relatively unspecialized cells composing the groundmass of the tissue and more specialized cells dispersed through this mass. The epidermal cells are elongated parallel with the long axis of the part. In leaves, petals, ovaries and ovules the epidermal cells may have wavy vertical (anticlinal) walls.

A stomata (plural: stomata) is a central pore on the lower layer of a leaf, used exchange of gases in plants. Stomata are found on all parts of plants, including the petals, petioles, herbaceous stems and leaves. Leaves are the chief "food making" parts of plants. They make food in the presence of light during a process called photosynthesis. Stomatal apertures typically respond to various changes in light intensity, saturation deficit of water vapor and soil moisture availability. Trichomes are actually the outgrowth of epidermal cells of a plant body.

These are typically referred to as the hairs of plants. Trichomes help a plant to defend itself against herbivores. The Silica bodies are important in plants to give their tissues mechanical stability. They also adapt a plant in favorable manner to resist against drought and defense against insects and herbivores. Pollens and phytoliths are important tools to know about the history of the plant and its environmental conditions. As pollen spores and phytoliths share somewhat similar characters regarding their specific gravity. Calcium oxalate crystals are common in flowering plants, including both dicotyledons and monocotyledons. All these characters in plants showthe greatest diversity and could

* arusa_ravi@yahoo.com

be very effective for the resolution of existing taxonomic problems of medicinal plant species.

INTRODUCTION

The composition of epidermis consists of relatively unspecialized cells composing the groundmass of the tissue and more specialized cells disperse through this mass. The ground cells epidermis vary in depth but are often tubular in shape. In elongated plant parts such as stem, petiole, vein ribs of leaves, and leaves of most monocotyledons, the epidermal cells are elongated parallel with the long axis of the part. In leaves, petals, ovaries and ovules the epidermal cells may have wavy vertical (anticlinal) walls. The epidermal cells have living protoplasts and may store various products of metabolism. They contain plastids but the plastids do not usually develop much, if any, chlorophyll small typical glandular chloroplasts, however may be present. (e.g., spinach)

LEAF EPIDERMAL ANATOMICAL STRUCTURES OF LEAF

Epidermal Cells

The epidermis is a system of cells, variable in structure and function that together constitute the covering of the plant body in its primary state. Many of the structural features of the epidermis can be related to the functions that tissue plays as the layer of cells in contact with external environment of the plant. The epidermis is usually one layer of cells in thickness. In some plants, however the protoderm of leaves divides parallel with the surface and their derivatives of these divisions divide again so that a tissue is referred to as multiple epidermis. In some plants sub epidermal layers resemble those of a multiple epidermis but are derived from the ground tissue. Multiple epidermises, therefore, can be identified as such only by developmental studies. The epidermis may last through the life of a given plant part, or it may be later replaced by another protective tissue, called periderm.

Trichomes

Trichomes are actually the outgrowth of epidermal cells of a plant body. These are typically referred to as the hair of plants. They are present almost everywhere on plants such as leaves, fruits, stem etc. They are thought to be functionless but are seen to perform secretary functions as they excrete water and volatile oils as in pepper mint. They are absent in roots. The plants in which trichomes are absent are coca and hemlock and occasionally present in buchu and henna [1]. Trichome help a plant to defend itself against herbivory by its physical as well as chemical means to discourage it. Leaf trichomes protect a plant against herbivory. The variability of structure and size makes it easier to classify trichomes such as:

Non Glandular Trichomes

The non-glandular hairs of the species or relatively thin, non-cutinized walls. These hair are dead or living, capable of movements in response to external conditions. They are pointed and ridged trichomes [2]. Non-glandular trichomes are leaf hair that offera mechanical barrier to some small sized insects, filter and isolation of leaf surface. At high leaf water shortage, non-glandular trichomes extensively improve water maintenance of the leaf tissue. These trichomes are often present at the base of the leaf.

Glandular Trichomes

Glandular trichomes contain or secrete a mixture of chemicals. Glandular trichomes are particular hairs set up on the surface of about 30% of all vascular plants and are accountable for a important part of a plant's derived chemistry (Figure 1). Glandular trichomes have volatile oils and other secretions that are fashioned by the plants. The glandular hair has three phases of development Persecretory, secretory and postsecretory phase [3]. Some crop plants are better defended against pests due to presence of glandular trichomes. Glandular trichomes secrete such type of secretions which are thought to be the precursors of pesticides. These are further divided into three types.

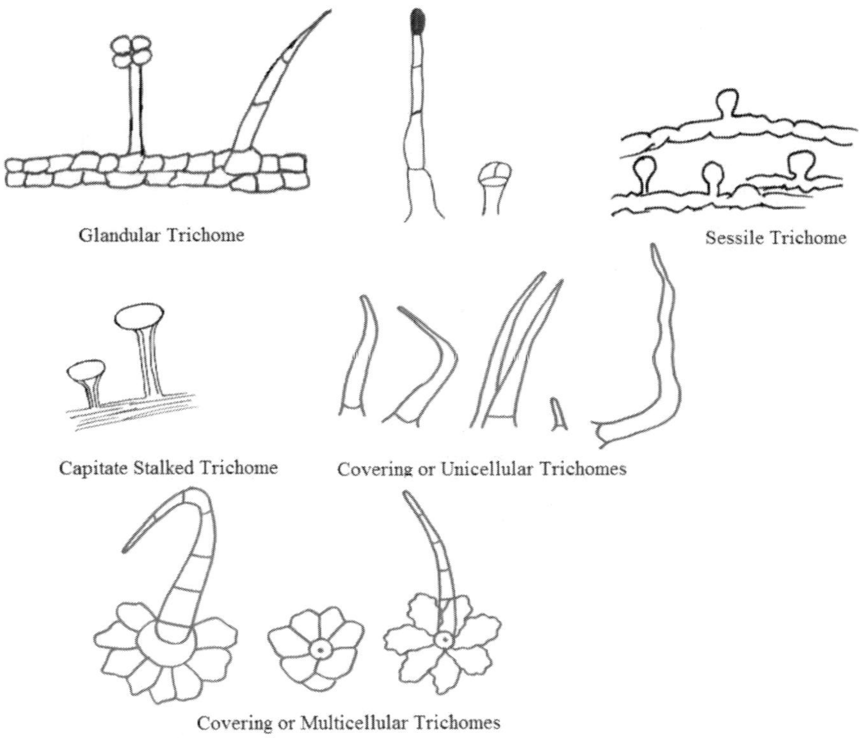

Figure 1. Various Types of Trichomes.

Bulbous

These are the smallest up to 15-20 microns in size. Usually comprised of foot, stalk and head (Figure 1). These can be found anywhere on plant body on the aerial parts of plant. Head cells are the ones producing resins.

CAPITATE-SESSILE

Their size ranges from 25-100 micrometers. These are more concentrated than bulbous glands. These are named as capitate because of globular-shaped head. There is no stalk attached to it if observed under microscope that is why termed as sessile but sometimes a one cell high stalk may be present which is considered to produce cannabinoids in Cannabis and other related compounds (Figure 1).

CAPITATE-STALKED

Their size ranges from 150-500 microns. These usually appear at flowering time. The color of the capitate-stalked trichome changes on maturity (Figure 1).

Hydathodes

It is usually present in angiosperms and secretes water through pores present in the epidermis. When the water is evolved this process is referred to as Guttation. Hydathodes have no mechanism of regulation of their opening as in stomata. They are sub-divided according to number of cells and thickness [1].

Covering Trichomes (Unicellular)

Trichomes consisting of one cell are called unicellular trichomes. Mostly they are found in Lamiaceae family [4, 5, 6, 7].

Covering Trichomes (Multicellular Unbranched)

These types of trichomes include two to three numbers of cells which may be varied in their length and thickness, those containing only two celled trichomes are mostly found in *Ballota kaiseri* and *Ballota undulate* and these are not very common (Figure 1). Whereas three celled trichomes are mostly found in *Ballota saxatailis* [8].

Covering Trichomes (Multicellular Branched)

Trichomes consisting of four to five cells or more are included in multicellular trichomes. These are branched and non-glandular. There are several basal cells and the branching starts readily after epidermal cells develop. The petioles of *Ballota kaiseri* and *Ballota undulate* [8].

Stomata

A stomata (plural: stomata) is a central pore on the lower layer of a leaf, used for exchange of gases in plants. One guard cell is positioned on either side of pore or stomata, and these are thought to be responsible for the opening and closing of the pore.

Carbon dioxide derives inside for photosynthesis, while oxygen and water vapor escape from the stomata.

Stomata are found on all parts of plants, including the petals, petioles, herbaceous stems and leaves. They are formed during the development of these various plant organs so they reflect the environment and its conditions under which they grew.

Types of Stomata

Depending upon the arrangement of guard cells and subsidiary cells stomata may be classified as:

Paracytic or Parallel Celled Stomata

In this the stomata is surrounded by two guard cells and two subsidiary cells which are parallel arranged to each other (Figure 2). e.g., coca and senna leaves.

Diacytic or Cross-Celled Stomata

Diacytic stomata are those in which one or more pairs of subsidiary cells enclose the stoma and the common wall of the subsidiary cells are at right angles to the guard cells. e.g., peppermint, spearmint etc. (Figure 2)

Anisocytic or Unequal Celled Stomata

In this type of stomata three subsidiary cells cover two guard cells; the third guard cell is smaller in size as compared to the others. (Figure 2) e.g., Belladonna, *Datura stramonium*.

Anomocyitic or Irregular Celled Stomata

Varying number of subsidiary cells surround the guard cells e.g., digitalis and lobelia.

Actinocytic Stomata

The subsidiary cells surround the two guard cells in circle. e.g., Araceae, Musaceae [1].

Encylocytic Stomata

These stomata have subsidiary cells forming one or two narrow rings around the guard cells, number of cells four or more, but modified to two or more [9]. (Figure 2)

Laterocytic Stomata

These stomata with more than two subsidiary cells on at least one lateral side of each guard cell pair.

Anticlinal walls between adjacent subsidiary cells radiating from guard cell pairs are equal in thickness to other cell wall.

Staurocytic Stomata

Stomata surrounded by three to five similar subsidiary cells with anticlinal walls arranged crosswise to the guard cells (Figure 2).

The subsidiary cells are more or less radically elongated, but the cell walls at right angles to the long axis of the pore are variously oriented.

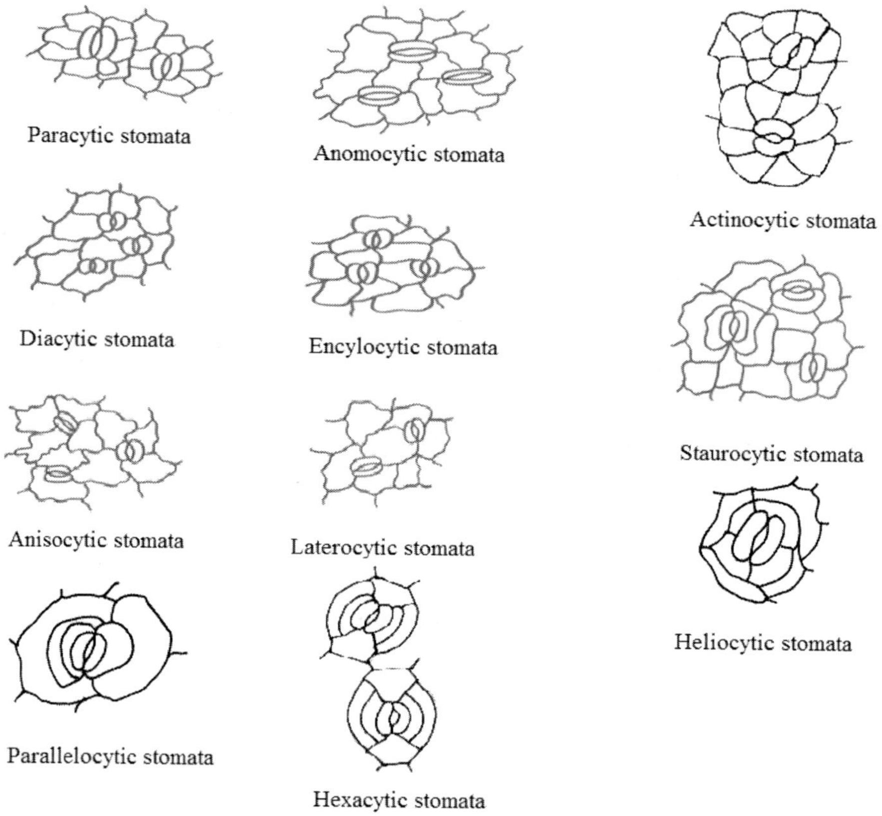

Figure 2. Different Types of Stomata.

Heliocytic

Stomata surrounded by a coil of four or more cells. Equivalent to amphianisocytic [10].

Parallelocytic

This kind of stomata has with alternative complex of three or more C-shaped subsidiary cells of graded sizes parallel to the guard cells.

Hexacytic

Modification of tetracytic type with an additional pair of lateral subsidiary cells [9]. The polar cells may be larger or smaller that the lateral cells (Figure 2).

Stomata Density, Size and Shape

Stomata density means the number of stomata per square millimeter. These density ranges from 100 to 1000 depending on the plant species and the environmental factors. Stomata are more seen on plant surfaces under higher light, lower carbon dioxide concentrations and moist/humid environments. Plants belonging to grasses typically have lower stomata densities than deciduous trees. The size and shape of stomata also vary with different plant species and environmental conditions. For example, grasses have guard cells that resemble slender dumbbells whereas trees and shrubs have guard cells that resemble kidney beans.

Family Cucurbitaceae: The species in which stomata are found on adaxial surface of leaf epidermis include Luffa cyilndrica, Cucumis melo, Lagenaria breviflorus, Lagenaria siceraria, Citrullus colocsynthis, Citrullus lanatus, Luffa acutangula and Momordica foetida. The species on which stomata is present on abaxial surface include Telfaria occidentalis, Momordica charantia, Trichosanthes cucumerina, Coccinia barteri, Coccinia grandis, and Cucumus sativus.

Family Hypoxidaceae: Stomata are present on either surface e.g., Hypoxis obtuse or Hypoxis villosa.

Physiological Function of Stomata

Leaves are the chief "food making" parts of plants. They make food in the presence of light during a process called photosynthesis. In sunlight stomata open, carbon dioxide will enter into the leaf it gets converted to sugars inside the leaf. Simultaneously, water vapor will escape from the leaf through the stomata to the atmosphere in the process of transpiration. If this loss of water remains to continue, they can get deprived of water reserve. This depletion can be dangerous to the physiological function of the plant as it is the most essential solvent in which physiochemical processes of the plant occur. If plants have not adapted to withstand changes in water stress in their respective growth environment, by not regulating the size of their stomata pores in a suitable manner and as a result may fail to survive when water availability decreases or falls.

The single layer of epidermal cells forms the boundary between the atmosphere and the assimilative cells [11]. Another micro morphological structure is stomata which refer to two specialized epidermal cells on a plant body and the aperture which forms between them. In tracheophytes, stomata are integral to water transport in plants because evaporation of water through stomatal pore draws water through the water conducting tissue. Non glandular trichomes present on the surface of leaf related to the regulation of leaf transpiration in stressful conditions. Trichomes also enhancewater retention of the leaf tissue at high leaf water deficit. These are micro morphological trait of the leaf and these morphological structures supporting the function of leaf stoma apparatus.

The Optimal Size of Stomatal Aperture

The Optimization Theory, first proposed by [12] suggested that the mechanism of exchange of gases in a plant is optimum if the photosynthesis goes on at a given average rate of transpiration. This ratio of photosynthesis defines the instantaneous water-use efficiency (WUE).

Stomatal apertures typically in respond to varying changes in light intensity, saturation deficit of water vapor and soil moisture availability. As the size of the stomatal aperture varies, rates of photosynthesis and transpiration will also change because the pore size will provide a resistance to the diffusion of CO_2 inside the leaf and H_2O out of the leaf.

Stomatal Conductance

Plant biologists typically measure stomatal conductance using a specialized instrument called an IRGA (Infra-Red Gas Analyzer).This instrument allows one to clamp a leaf into a chamber and the relative mole fractions of CO_2 and water vapor entering and leaving the chamber are monitored over time.

Oil Bodies

These are such structures in a plant body which contains lipids. These are found in liverworts and vascular plants. In vascular plants these are mostly found in angiosperms. They consist of triacylglycerol surrounded by a layer of phospholipids and oleosin which is proteinaceous in nature [13].

Oil bodies occur in the mesophyll cells of a leaf in many angiosperms. The most important eudicot families in which there is the highest number of species with oil bodies are Rosaceae, Asteraceae, Lamiaceae and Caprifoliaceae. While in the families like Fabaceae, Scorphulariaceae, Apiaceae and Betulaceae few species among them have oil bodies. Among monocot families only a rare of them has these.

Types of Oils

There are two basic types of oils present in plants:

Volatile Oils

Such types of oils which are aromatic and when exposed to air they tend to evaporate. These are composed of terpenes which are low in molecular weight. They gather in glands, cavities or specialized cells. The common examples are: jasmine, rose, spearmint, peppermint, clove etc.

Neutral Oils

The class of oils which are non-aromatic and do not evaporate in air on exposure. They are usually tasteless. Oils of this kind are most ordinary form of storage of lipids. The most common examples are: canola, castor, corn, linseed and olive oil. Functions of oil bodies in leaf mesophyll cells: There are two basic functions of oil bodies or oleosomes: These are actually the intermediate storage products in leaf mesophyll cells during photosynthesis.

Secondly, presence of oil bodies indicates adaptation of the plant in cold temperatures. Pihakaski et al., [14] show that in Diapensia lapponica (Diapensiaceae) which is an evergreen plant of the arctic region has a large lipid body in each of its mesophyll cell during the cold season. There are a large number of plant species which store oils in their seeds to use for germination and even after germination [15, 16, 17]. The size of triacylglycerol in oil bodies is 0.5-2 micrometer in diameter [18].

These oil bodies form large subcellular particles which are called as tapetosomes and their size is approximately 2-3 micrometer in diameter. These are considered to store lipids which act as a pollen coat giving it an aid in water proofing and UV protection. Although in low abundance they are also present in leaves of plants but their structure and function is still unknown [13]. It is also reported that mesocarp of some fruits like olive (*Olea europea*), avocado (*Persea americana*) and oil palm (*Sygarus cocoides*) the lipid particles are present in the size of 10-50 micrometer in diameter. The actual function of these is to attract insects for pollination/seed dispersal [16; 17].

Silica Bodies

The Silica bodies are important in plants to give their tissues mechanical stability, adapt a plant in favorable manner to resist against drought and defense against insects and herbivores [19]. Silica or silicon gets deposited in leaves in the form of phytoliths. These are especially taxonomically important as they are mostly found in Grasses. These are thought to be a distinctive feature of family Poaceae, whose members are specifically monocots. As they have also been derived from fossils they are a subject of importance from paleological and archaeological point of view. They give a lot of information about past environments [20]. So, silica bodies are an important diagnostic factor in grass systematics. These are microscopic bodies which are variable in shapes and sizes.

Classification of Phytoliths

There are four classes of grass-opal phytoliths:

Festucoid

These may be circular, elliptical or rectangular with a crescent variable focus or acicular variable focus and may be oblong. (Figure 3)

Chloridoid

These may be thick or thin with slight elongation.

Panicoid

These may be of cross shape, regular dumbbell, irregular dumbbell or crenate.

Elongate

These are elongated in shape and may be smooth, sinuous, spiny, and spiny with pavement and concave ends [21]. As a common observation grass and angiospermic plants are classified on the basis of their structure. The flowersand seeds are taken into account for taxonomic evaluation. Unlike the fossils of other organisms the fossilized plants or plant parts are rare or not so common in the geological studies.

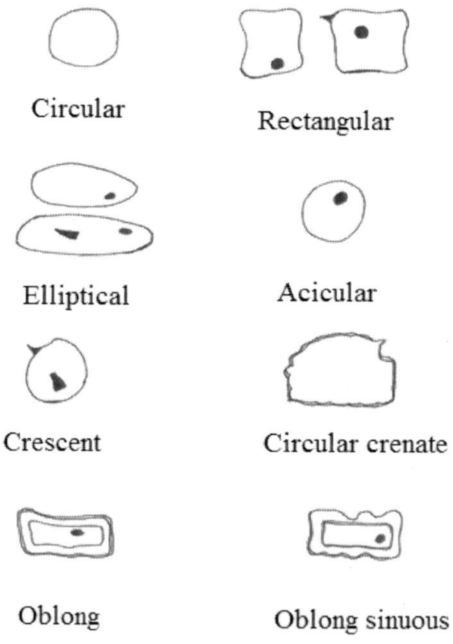

Figure 3. Different Shapes of Phytolith.

Phytoliths have silica deposition and as they are released from epidermal cells of leaves or other parts such as inflorescence or roots in the form of discrete particles when the plant dies. So, phytoliths are of great importance as they are resistant to decay or dissolution and due to their size they can easily be transported along with their pollens or spores as the case may be and if careful handling is done they can be recovered and can give a plenty of information regarding the plant and the environment as well as habitat of that plant.

Classification, as discussed earlier is based on the size, shape and structure of the phytoliths. To describe the micromorphology of 354 genera of grasses [22], emphasized to take into account different aspects like their orientation, size, shape, different types of cells, their location and concentration should all be taken into account while taxonomic classification is being done on the basis of phytoliths.

Several partilcles of atmospheric dust were collected during the journey of Charles Darwin on the H.M.S Beagle. These were recognized by Ehrenberg [23] as grass-opal phytoliths. Twiss et al. [21] also carried out their study on the classification of these phytoliths related to grasses. Their study revealed that according to shape these phytoliths can be grouped into four classes and 26 types.

To summarize them as depicted in the figure:

Class 4: this class has elongated silicified long cells usually 30 to 150 micrometers long.

Class 3, 2 and 1 all have silicified short cells and silica bodies that are only 10-35 micrometers long with some little differences among them [21].

GRASS PHYTOLITHS AS CLIMATIC INDICATORS

Pollens and phytoliths are important tools to know about the history of the plant and its environmental conditions. As pollen spores and phytoliths share somewhat similar characters regarding their specific gravity which is observed to be <2.3 and their size which mostly ranges from 10-150 micrometers. These can travel with the help of wind and water and can occur in soil and sediments together.

The pollen grains are however not a prominent diagnostic feature in classifying grasses and are only grouped in Graminae so far. The panicoid phytoliths fall into the group of C4 grass that pertains to warmer climatic conditions and high moisture content of the plant.

The festucoid phytoliths following C3 pathway are most abundant in northern areas of United States. They can grow well in higher moisture content but only in their growing season. We can say that grasses mostly belong to the area having hot and arid climate near to the tropics of Cancer and Capricorn and they according to Hartley and Slater [24], may have originated from their ancestors in sub-tropical and tropical Africa. So, it can be said that information from the phytoliths can be a valuable tool in the interpretation of the paleoenvironment and palynological studies characterizing woodland communities and grassland communities.

CALCIUM OXALATE

Calcium oxalate crystals are common in flowering plants, including both dicotyledons and monocotyledons. They were first bare by Leeuwenhoek in the 17th century [25] . They have been recognized using light microscopy (LM) and polarization microscopy.

The distinct shapes and short-lived regency of calcium oxalate crystals, especially raphides and styloids, make them readily apparent, particularly in young, vigorously growing tissues, even though smaller, rounded druses are more easily missed. Crystals in general form intracellularly, but extracellular crystals have also been reported, e.g., on the outside surfaces of palisade parenchyma in *Tsuga* leaves [26]. However, since these crystals have a covering of robrous material they are initiated intracellularly [27]. Similarly, in *Welwitschia bainesii* the crystal-encrusted spiculate cells are covered by a sheet-like layer. Some tissues may contain other crystal types in addition to calcium oxalate crystals; in particular silicate crystals are feature of the monocot superorder Commelinanae, many of which have calcium oxalate crystals. Calcium oxalate crystals come into sight in a diversity of shapes which are consistent and repeatable from one generation to the next, representative that the physiological and genetic parameters calculating them are dependable.

There are three main types of calcium oxalate crystal in monocotyledons:

- Raphides
- Styloids
- Druses

These may be treated as three divide characters in cladistic analyses. The different types are sometimes, but not always, mutually select. Three main types of calcium oxalate crystal occur in monocotyledons: raphides, styloids and druses, although intermediates are sometimes recorded and more than one type may be present in a species. The absence of raphides, the most common crystal type in monocots, represents a synapomorphy in some groups, such as Alismatales, Poales and some Liliales.

Raphides

Raphides are bundles of narrow, lengthened needle-shaped crystals usually originate in crystal idioblasts in parenchymatous tissues. Most raphide crystals are four-sided but in some taxa they develop additional into at least six-sided crystals with piercing ends. Grooved raphides are feature of Araceae. Raphides are bundles of narrow, extended needle-shaped crystals, frequently of alike orientation, with pointed ends at maturity. Often (e.g., in Araceae) one end is immediately sharp while the other tapers to a point or is wedge formed. There are varying numbers of crystals in each bundle. They are frequently established in crystal idioblasts in parenchymatous tissues, though there are examples of raphides happening in particular structures, such as aerenchyma, or plant hairs. In *Conanthera campanulata* (Tecophilaeaceae) the uniseriate hairs fringing the tepals contain a bundle of raphides in eac cell and, likewise, raphides have been recorded in trichomes of *Cocos* (Arecaceae) [25]. In some species of *Curculigo* (Hypoxidaceae) numerous small, loosely set crystals are present in leaf epidermal cells, in adding to 'normal' bundles of raphides in mesophyll cells [28].

Although most raphide crystals are four-sided, those of some taxa are at least six-sided, appearing almost elliptical in cross section, for example in *Yucca* [29], *Aga.e*, *Cordyline* [30, 31] and *Ornithogalum* [32].

In these taxa, raphides are initially four-sided with wedge shaped ends but later develop into six- and eight-sided crystals with pointed ends. In *Typha angustifolia* [33], mature crystals are hexagonal towards their ends and octagonal in their central region. During growth, each raphide becomes ensheathed in lamellae and bounded by mucilage (see below). Addition of material to the ends of the raphides extends them from wedge-shaped to sharp, and overlap of lamellae at these pointed ends results in distinctive wing formations [29, 30, 32, 34]. Grooved raphides occur in Araceae. Apparently all raphides in Araceae are grooved, as well as those of *Spirodella* [35] and *Lemna* [36] which are sometimes positioned in a take apart family, Lemnaceae. Grooved raphides consequently characterize a major apomorphy for Araceae. At beginning, the crystals have a rounded appearance but are grooved at maturity.

Styloids

Styloids ('pseudoraphides') are thicker than raphides and usually solitary within a cell. They may have sharp or squared ends, and may be extended or cuboidal. Styloids are quality of some families of Asparagales, notably Iridaceae, where raphides are entirely absent. The existence of styloids is therefore a synapomorphy for some families (e.g., Iridaceae) or groups of families (e.g., Philydraceae, Pontederiaceae and Haemodoraceae), and more comprehensive studies on this aspect may well prove valuable for systematic studies on these groups. In some other families' raphides, styloids and intermediate forms may be present; chiefly in a group of closely associated 'higher' asparagoid families, such as Convallariaceae *sensu lato* and Agavaceae.

Styloids, also called prismatic crystals or 'pseudoraphides', are thicker than raphides and usually solitary within a cell, even though intermediate forms are recorded.

The term styloid encompasses a broad morphological range. Styloids may have pointed ends, or squared ends, and may be lengthened or cuboidal. Some styloids are 'twinned' crystals, e.g., the arrow-shaped styloids of *Iris pseudacorus* [37]. In *Allium*, [38] recorded interpenetrant twin calcium oxalate crystals (styloids) and Frey [25] and Ricci [39] reported a morphological variety of styloids in different species, including normal' rhomboidal crystals, pyramidal crystals with irregular faces, small crystalline granules, and long prisms, sometimes isolated, and sometimes gathered together to form a druse. Large crystals may have adhering small crystals.

In monocot leaves, styloids are usually found either in parenchymatous bundle sheath cells around vascular strands or in crystal idioblasts in adjacent mesophyll tissues, Even though, in *Xanthorrhoea*, the styloids in the leaf are regularly epidermal [40].

In Iridaceae, elongated styloids occur in axially elongated mesophyll cells, and short cuboidal crystals in the shorter cells of the parenchymatous outer bundle sheaths; both types commonly occur in the same leaf. Styloids are characteristic of some families of Asparagales (Lilianae), including some 'higher' asparagoids (Agavaceae, Alliaceae, some Convallariaceae *sensu lato* and Lomandraceae) and some 'lower' asparagoids (Asphodelaceae, Doryanthaceae, and some Hemerocallidaceae *sensu lato*, Lanariaceae, Iridaceae and Xanthorrhoeaceae).

Druses

Druses ('sphñraphides') are crystal conglomerates possibly fashioned around a nucleation site.

They are common in dicotyledons, relatively rare in monocotyledons, where they are almost restricted to some early-branching taxa such as *Acorus*, Araceae and *Tofieldia*. Druses (formerly called 'sphñraphides') are multiple crystals that are thought to have precipitated around a nucleation site to form a crystal conglomerate [34].

The individual part crystals of a druse can also comprise contact twins. Both the monohydrate and dihydrate forms of calcium oxalate can form druses [41, 42].

Druses may have a similar defensive function to that of raphides, as they also have sharp points resulting in considerable irritation if eaten. Druses are common in dicotyledons [25] but relatively rare in monocotyledons, where they are almost entirely restricted to the first-branching taxa, *Acorus*, some Araceae.

Even though there are very few records of druses in the literature, in fact they are quite common in *Acorus*, especially in aerenchymatous tissues. In Araceae they may occur in conjunction with raphides, in *Aglaonema*, *Alocasia*, *Anthurium*, *Colocasia*, *Dieffenbachia*, *Hydrosme*, *Philodendron*, *Pistia*, *Scindapsus*, *Spathiphyllum*, *Symplocarpus*, *Syngonium* and *Zantedechia* [43, 44, 33, 45, 46].

References

[1] Osman, S. B. G. (2012). A textbook of pharmacogonosy by Trichome morphology of Egyptian *Ballota* (Lamiaceae) with emphasis on its Systematic implications. *Pakistan Journal of Botany*, 44(1): 33-46.

[2] Werker, E. (1993). Function of essential oil-secreting glandular hairs in aromatic plants of the Lamiaceae (a review). *Flavour and fragnence* Journal 8: 249-255.

[3] Ascensao, L. and Paris, M. S. (1998) The leaf capitate trichomes of Leonotis leonurus: histochemisrty, ultrastucture and secretion. *Annals of Botany* 81, 263-271.

[4] Abu-Asab, M. S. and P. D. Cantino. 1987. Phylogenetic implications of leaf anatomy in subtribe *Melittidinae Arbor.,* 71: 323-370.

[5] Bonzani, N. E., Costaguta M.and Barboza. G.E. (2007).Anatomical studies in *Mentha* species (Lamiaceae). In: Cantino, P.D. (1990). The phylogenetic significance of stomata and trichomes in the Labiatae and Verbenaceae. *Journal Arnold Arbor*, 68:1-34.

[6] Cantino, P. D. (1990). The phylogenetic significance of stomata and trichomes in the Labiatae and Verbenaceae. *Journal of Arnold Arbor.,* 71: 323-370.

[7] El-Gazzar, A. and Watson. L. (1970). A taxonomic study ofLabiatae and related genera. *New Phytol.,* 69: 451-486.

[8] Osman A. K. (2012). Trichome micro-morphology of Egyptian Ballota (Lamiaceae) with emphasis on its systematic implication. *Pakistan Journal of Botany,* 44(1):33-46.

[9] Van C. W. R. J. (1970). A classification of the stomatal types. *Bot. J. Linn. Soc.* 63: 235-246.

[10] Dilcher, D. L. (1974). Approaches to the identification of angiosperm leaf remains. *The Botanical Review*, 40:1-157.

[11] Bold, H. C. (1980). Morphology of Plants Harper and Brothers, New York.

[12] Cowan, I. R. and Farquhar, G. D. (1977). Stomatal function in relation to leaf metabolism and environment. *Symposium for the Society of Experimental Biology* 31:471-505.

[13] Lersten, N. R., Czlapinski, A. R., Curtis J. D., Freckmann R., Horner H. T. (2006). Oil bodies in leaf mesophyll cells of angiosperms, overview and selected survey. *American journal of Botany*, . 93(12): 1731 – 1739.

[14] Pihakaski K. Pihakaski S. Karunen P. Kallio P. (1987). Seasonal changes in leaf lipids of *Diapensia lapponica*, with special reference to storage lipid bodies. *Nordic Journal of Botany* 7: 281-29.

[15] Napier J. A., Stobart A. K., and Shewry P. R. (1996) The structure and biogenesis of plant oil bodies: the role of the ER membrane and the oleosin class of proteins. *Plant Molecular Biology* 31: 945–956.

[16] Murphy D. J. (2001) The biogenesis and functions of lipid bodies in animals, plants and microorganisms. *Plant Journal* 13: 1–16.

[17] Hsieh, K. and Huang A.H.C. (2004) Endoplasmic reticulum, oleosins, and oils in seeds and tapetum. *Plant Physiol* 136: 3427–3434.

[18] Chien Y. H., Chun I. C., Yao C. L., Yue I. C. (2009). Using and Anthony H.C. Huang. *Plant Physiology*. 150(3): 1192-1203.

[19] Motomura, H., Fujii, T., Suzuki, M. (2005). Silica deposition in abaxial epidermis before the opening of Leaf Blades of Pleioblastus chino (Poaceae, Bambusoideae).

[20] Dolores R. P. and Deborah M. P. (1998). The Silica bodies of Tropical American Grasses: Morphology, Taxonomy, and Implications for Grass Systematics and Fossil Phytolith Identification. *Smithsonian Institution Press Washington,* D.C.

[21] Twiss, P. C., Suess, E., and Smith, R. M. (1969). Morphological classification of grass phytoliths: *Soil Science Society of America, Proceedings,* 33:105-115.

[22] Metcalfe, C. R. (1960). Anatomy of the monocotyledons; I. Gramineae: Clarendon Press, Oxford, 731 p.

[23] Ehrenberg, C. G. (1847). Passatataub und Blutregen: Deutsche Akademie die Wissenschaften der Berlin Abhandung, 269-460.

[24] Hartley, W., and Slater, C. (1960). Studies on the origin, evolution, and distribution of the Gramineae; III. The tribes of the Subfamily Eragrostoideae: *Australian Journal of Botany*, 8:256-276.

[25] Frey A. (1929). Calcium oxalat-Monohydrat und Trihydrat. In: Linsbauer K, ed. *Handbuch der P¯anzenanatomie Vol. 3.* Berlin: Gebru$ der Borntraeger, 82±127.

[26] Gambles, R. L. and Dengler, N. G. (1974). The leaf anatomy of Hemlock, Tsuga canadensis. *Canadian Journal of Botany* ,52:1049-1056.

[27] Horner, H.T. and Franceschi, V.R. (1981). The use of a tissue culture system as an experimental approach to the study of plant crystal cells. *Scanning Electron Microscopy* 3: 245-249.

[28] Rudall P. J., Chase M. W., Cutler D.F., Rusby J., Bruijn, A. (1998*a)*. Anatomical and molecular systematics of Asteliaceae and Hypoxidaceae. *Botanical Journal of the Linnean Society* 127: 1±42.

[29] Eilert G .B. (1974). An ultrastructural study of the development of raphide crystal cells in the roots of Yucca torreyii. PhD Thesis, University of Texas.

[30] Wattendorff J. (1976*b). A third type of raphide crystal in the plant kingdom, six-sided raphides with laminated sheaths in Agave americana L. Planta* 130: 303±311.

[31] Wattendorff J. (1979). Pflanzenliche calcium oxalatekristalle im lichtmikroskop. *Mikrokosmos* 7: 220±224.

[32] Tilton V. R., Horner H. T. (1980). Calcium oxalate raphide crystals and crystalliferous idioblasts in the carpels of *Ornithogalum caudatum. Annals of Botany*, 46: 533±539.

[33] Kausch A. P., Horner H. T. (1981). The relationship of air space formation and calcium oxalate crystal development in young leaves of *Typha angustifolia* L. (Typhaceae). *Scanning Electron Microscopy*, 3: 263±272.

[34] Horner H. T., Kausch A.P., Wagner B.L. (1981). Growth and change in shape of raphide and druse calcium oxalate crystals as a function of intracellular development in *Typha angustifolia* L. (Typhaceae) and *Capsicum annuum* L. (Solanaceae). *Scanning Electron Microscopy* 3: 251±262.

[35] Ledbetter M. C., Porter K. R. (1970). Introduction to the fine structure of plant cells. Berlin, Heidelberg, New York: Springer-Verlag, 148±150.

[36] Arnott H.J. (1966). Studies of calcinations in plants. In: Fleisch H, Blackwood HJJ, Owens M, eds. Third european symposium on calcified tissues. New York: Springer, 152±157.

[37] Kollbeck F, Goldschmidt V. M. and Schrofder R. (1914). Beitrage Kristallographie Mineral 1: 199.

[38] Arnott H.J. (1981). An SEM study of twinning in calcium oxalate crystals of plants. *Scanning Electron Microscopy* 3: 225±234.

[39] Ricci I. (1963). Contributo alla conoscenza tipologica dell'ossalato di calcio nel genere Allium. *Annali di Botanica* 27: 431±450.

[40] Rudall P. J., Chase M. W. (1996). Systematics of Xanthorrhoeaceae *sensu lato:* evidence for polyphyly. *Telopea* 6: 629±647.

[41] Al-Rais A. H., Myers A. and Watson L. (1971). The isolation and properties of oxalate crystals from plants. *Annals of Botany*, 35: 1213±1218.

[42] Franceschi V. R., Horner H. T. (1979). Use of *Psychotria punctata* callus in study of calcium oxalate crystal idioblast formation. *Zeitschrift fu*X *r Pfanzenphysiologie* 92: 61±75.

[43] Gaiser L. O. (1923). Intracellular relations of aggregate crystals in the spadix of Anthurium. *Bulletin of the Torrey Club* 50: 389±398.

[44] Sakai W.S., Hanson M. and Jones R.C. (1972). Raphides with barbs and grooves in *Xanthosoma sagittifolium* (Araceae). *Science,* 178: 314±315.

[45] Sunell L. A. and Healey P. L. (1981). Scanning electron microscopy and energy dispersive X-ray analysis of raphide crystal idioblast in taro. *Scanning Electron Microscopy* 3: 235±244.

[46] Genua J. M. and Hillson C. J. (1985). The occurrence, type and location of calcium oxalate crystals in the leaves of 14 species of Araceae. *Annals of Botany*, 56: 351±361.

In: Leaf Epidermal Anatomical Characters…
Editors: Z. Yousaf, A. Younas and A. Aftab

ISBN: 978-1-63483-190-1

Chapter 4

TAXONOMIC EVALUATION OF FAMILY EUPHORBIACEAE ON THE BASIS OF LEAF EPIDERMAL ANATOMICAL CHARACTERS

A. Younus*

Department of Botany,
Lahore College for Women University,
Lahore, Paistan

ABSTRACT

Present piece of work was conducted to investigate leaf epidermal anatomy of selected plants of Family Euphorbiaceae from district Lahore. The study was conducted in Molecular Taxonomy Lab, Lahore College for Women University, Jail Road, Lahore. Morphology-based taxonomic information of family Euphorbiaceae is contradictory. Therefore, new taxonomic tools were investigated during the present piece of work. Leaf epidermal anatomy is considered as one of the effective techniques used for the generation of more authentic taxonomic information. Important characters identified for the family Euphorbiaceae were, types and shape of cells, cell margins variation, shape and types of trichomes, silica bodies and stomatal cell shapes and type variations. Cells which were pentagonal in shape and elongated with smooth margins were more prominent in *Euphorbia splenden* Bojer ex Hook. and *Jatropha integrima* Jacq. The maximum length of long cell ranged from 196.6μm - 245.2μm and was found in *Euphorbia helioscopia* L. on adaxial side of leaf epidermis. Presence of silica bodies and trichomes can be taken as more taxonomically important characters of Family Euphorbiaceae. Silica bodies were found on abaxial and adaxial leaf surfaces of *Euphorbia splendens* Bojer ex Hook. and silica bodies were found on leaf surface of *Phyllanthus emblica* L. Our data further suggested these epidermal features (stomata, trichome and silica bodies features) are used to delimit the species of family Euphorbiaceae.

* botanistafifa@yahoo.com.

INTRODUCTION

Euphorbiaceae is one of the most cosmopolitan plant families within angiosperms, consisting of about 300 genera and 7500 species [1]. Many members of the family play an important role in supporting tropical rain forest, horticultural value for ornamental, traditional medicine, and food resources. Many of these Euphorbiaceous species have essential medicinal importance. *Acalypha ciliate* is applied as dressing to wound, while *A. ornate*, apart from being a laxative, is also known to have healing application to circumcision wounds. Laboratory experiments on immature female guinea pigs showed that the *Euphorbia hirta* enhanced mammary development and secretion of milk [2].

Although classification of the family Euphorbiaceae has been clearly defined based on morphological characters, however phylogenetic relationships among genera within the family remain unresolved due to remarkable diversification of morphological characters [3]. This is the reason that phylogenetic analysis is used to classify a group of plants based upon micro morphological characters on leaf surfaces as carried out by many researches. They identified that several characters, such as presence/absence of stomata ledge, differentiation of subsidiary cells, shape of epidermal cells and morphology of trichome can provide valuable information in taxonomic studies [4, 5 and 6].

In many studies, these characters appear to be relatively more consistent than those of macro morphological ones [5]. Moreover, stomatal type not only has good diagnostic value for such studies, but it can also be used as indicators for naturally taxonomical similarity [7 and 8]. The diversity of stomata shape in 13 species of Euphorbiaceae was investigated by [3]. He also inferred evolutionary relationships among samples used in the family. The use of data generated from leaf epidermis surface in resolving the taxonomy of taxa has gained much recognition for a long time. Some of the prominent studies that have solved taxonomic problems among the plant taxa with such data are being done by [9, 10, 11, 12, 13, 14, 11, 15 and 16].

Among leaf epidermal features, trichome have several important characters that can be studied with the microscope and can be useful for comparative systematic. Variations in trichome types can provide insight into evolution and relationships within and among species [17 and 18].

In spite of numerous usages to which members of Euphorbiaceae have been put, much research has not been done on the use of leaf epidernmal anatomical features for the delineation and classification of various taxa in this family as a whole. The objective of the present study is to describe and interpret leaf epidermal anatomical features of various genera of family Euphorbiaceae.

OBJECTIVE

The main objective of the present investigation was

1. To explore leaf epidermal characters of some species of family Euphorbiaceae present in Lahore city to solve exiting taxonomic confusions.

HISTORICAL BACKGROUND

Kakkar and Paliwal [19] Studies on leaf anatomy of Euphorbia epidermis, Department of Botany, University of Delhi. A study of epidermis has been 150 species of Euphorbia. It bears straight to variously arched or undulated cell wall. In several species, the cells bear distinct papillae of various shapes and size. A conspicuously thick cuticle is seen in several species, for example in *E. polygonifolia, E. tomentulosa,* and others. Stomata of anomo-, para-, anisocytic type have been observed. Anomocytic type being the most common. Often, more than one type of stomata may occur on the same leaf surface. Stomatal frequencies have been calculated to assess the diagnostic value in several species of the same genus. Unicellular and multicellular, 2-7 celled, uniseriate trichomes, with tapering, blunt and hooked apices occur on the leaves of various species. Rarely, more than one type of trichomes may be encountered on the same leaf surface.

Inamdar, and Gangadhara [20] Studied trichomes of some species of family Euphorbiaceae. Feddes Repertorium, Structure, ontogeny, abundance, classification and taxonomic significance of foliar trichomes in 25 genera and 53 species of the Euphorbiaceae were presented. Seven main types and fourteen sub-types of glandular and eglandular trichomes and stinging emergences were recorded. Inclusion of the genus *Bridelia* under the tribe Phyllantheae and exclusion of *Dalechampia* and *Tragia* from *Acalypheae* and placing in a separate subtribe *Plukenetieae* by Bentham & Hooker seems to be justified on the basis of trichome studies. Species-wise distribution of trichomes showed variation in types among different tribes of Bentham & Hooker.

Gupta, D. and Moulik, B.K. [21] studied morphological nature of the petaloid bract of *E. hetrophlla.* A comparative morphology and anatomical study was undertaken on the foliage leaf, transition leaf and petaloid bract of *E. hetrophlla*, including data on stomata and epidermal cells, vein islet number and veinlet termination number. It was concluded from these studies that the petaloid bract is a morpho-histological modification of the leaf base and the expanded petiolar part.

Ogundipa [22] studied leaf epidermal study of Genus Datura L. of Family Solanaceae. Five West African species datura with narcotic properties were studied. This study was done at the Department of Biological Science, University of Lagos, Nigeria. The most useful anatomical characters for identification were palisade ratio, stomatal density and index, pattern of anti dinall walls, density and types of trichomes, and density and types of crystals. D. candida, D. suaveolens, D. dinnoxia, D. mental and D. stramonium were studied and differentiated on the basis of these anatomical characters.

Geoffrey et al., [23] worked on phylogenetic relationships of Didymocistus and Hymenocardia (Euphorbiaceae) in Annals of the Missouri Botanical Garden. The genus Hymenocardia has been placed either in Euphorbiaceae subfamily *Phyllanthoideae* or in its own family and then associated with the Urticales, particularly the Ulmaceae. Hymenocardia instead has many palynological features found in the *Phyllanthoideae* and should be retained in the Euphorbiaceae in that subfamily. Didymocistus, which has scale-like foliar trichomes and exine sculpturing and ultra structure similar to Hymenocardia, should be transferred from the *Phyllanthoideae-Aporuseae* to a position near Hymenocardia.

Baslar [24] investigated morphological characteristics of the root, stem, leaf, flower, seed and fruit of *Chrozophora tinctoria* (L.) Rafin. (Euphorbiaceae) distributed in West Anatolia, and biometric measurements were taken from samples collected from 42 different localities. This plant is a monoecious, annual herb, 28-63.50 cm. and stellate hairy. Its' leaves are alternate and their shapes vary from rhombic-ovate to ovate-lanceolate. The leaves are bifacial and amphi stomatous. Their stomatas are of the amaryllis and parasitic

Mbagwu et al., [25] worked on leaf epidermal characteristics of four species of the *genus Citrus* (Rutaceae). They collected the material from the different garden of Oweri, Imo state, Nigeria. They studied the four species of *Citrus sinensis*, *C. limon C. aurantifolia* and *C. parandist*. They observed that stomata and trichomes are absent on adaxial surface. Although it was a good piece of work, yet they oversight other anatomical characters like shape of epidermal cells, arrangement of epidermal cells, stomatal cavity and silica bodies. That could be better for identification of species at taxonomic level.

Bonnier et al., [26] described comparative anatomy of the vegetative organs of some *Euphorbia* species from Romania. This article compares the vegetative organs anatomy of 19 *Euphorbia* species from the Romanian flora, focusing on identifying the constant and variable histo-anatomical characters. He worked on the pattern of epidermis cells, the presence or the absence of trichomes, the cortex organization and the repartition mode of secondary conducting tissues. The foliar limb structure differed by the pattern of epidermis cells, the mesophyll organization and the stomatal localization. The anisocytic and anomocytic type of stomata are found in the foliar limb. Anomalous structures occurred in a few taxa.

Perveen et al., [27] worked on stomata types of some dicot within flora of Karachi, Pakistan. For this purpose, they selected the 69 dicot species distuributed in 64 genra and 24 families. They examined the stomatal morphology by using light and scanning electron microscopy. Among these species, six types of stomata were recognized. During study, they ignored the other anatomical characters (shape of cells, size of cell, type of trichome, and so on) at taxonomic level. If they had observed the rest of the characters, then their work would have provided better result.

Rojas [28] studies the anatomy of foliar epidermis of the *Solanum nudum* species, group of *Solanum sect. minate* (*Solanaceae*). The leaf epidermis surfaces of these species were investigated in order to discover and evaluate taxonomically useful characteristics. The study was conducted at venezuele. Variation in epidermal cells, type, dimension and location of stomata presence or absence of stomata location and density of trichomes were evaluated and illustrated. Results showed difference in some anatomical characteristics between S. nudum species group i.e., *S. Aphyodendrom*, S. *companiforme, S. nudum* and *S. tepuiense* and 34 species of *Solanium sect germinate*.

Mbagwu et al., [29] worked on leaf epidermal features of *Canarium schuwenfruther* and *Daeryoda edus* (*Brusiaceae*). They collected these species from different L.G.A of Imo; Ebony and Anambar Ebonyi and Anambra states of Nigeria. They studied the epidermal cells and stomatal cells. Although it was a good piece of work, yet they ignored another anatomical character like arrangement of epidermal cells, different shape of stomata types, variation in trichome and silica bodies, and so on. This could play a vital role in a characterizing on taxonomic tool for identification of species.

Hidayat [3] studied stomatal diversification and phylogenetic analysis of 13 species of Family Euphorbiaceae sensu lato. They investigated the diversity of stomata from 13 species of family Euphorbiaceae. Characters like type of stomata, position of stomata, presence/absence of ledge, and density of both abaxial and adaxial leaf surface were examined. Stomatal characters in the family were quite diverse in the study. Monophyletic nature of Euphorbiaceae has been proven in the study on the basis of phylogenetic analysis using parimioni method. Our data further recommended that the family can be classified into two major groups. However, surprisingly, genus *Phyllanthus* is non-monophyletic.

Idu et al,. [30] made a comparative morphological and anatomical studies on the leaf and stem of some medicinal plants (*jatropha curcas* and *J. tanjorensis*) in the Department of Botany, University of Benin, Nigeria. Morphological and anatomical investigations were carried out on two species of *Jatropha*: *J. curcas* and *J. tanjorensis*. These studies were carried out on the leaf and stem of these plants using the light microscope fitted with a Nikkon digital camera. Trichomes were observed in the leaf section of *J. tanjorensis* whereas these were lacking in *J. curcas*. The lower palisade parenchyma of *J. tanjorensis* consisted of fewer layers but is thicker, compared to that of *J. curcas*, which possessed more but thinner palisade layer.

Aworinde et al., [2] investigated the taxonomic significance of foliar epidermis in some members of Euphorbiaceae Family in Nigeria. A detailed morphological study of the leaf epidermis of the tropical genera of some species *Acalypha*, *Bridelia Euphjorbia*, *Hura*, *Jatropha*, *Manihot* and *Ricinus* in Nigeria is presented. The study revealed several interesting epidermal features, some of which have not previously been reported in the genera. Leaf epidermal characters such as pattern of epidermal cells, types of stomata and presence of trichomes are constant in some species and variable in others and thus of great significance in understanding the relationships between and within species. Leaves are amphistomatic in all species except in *Bridelia ferruginea*, *Euphorbia heterophylla*, *Euhorbia pulcherrima* and *Jatropha gossypifolia* which are hypostimatic. The stomata length, width, density and index also vary in different species.

Haiad B. de Sá et al., [31] studied leaf structure of species from three closely related genera from the tribe *Crotoneae Dumort* (Euphorbiaceae *S.S.*, Malpighiales) Euphorbiaceae *s.s.* is one of the largest angiosperm families, comprising nine major lineages. Molecular analysis indicated a polyphyletic condition in the subfamily *Crotonoideae s.l.*, with four lineages identified, including one termed inaperturate crotonoids, which encompasses the tribe *Crotoneae*. The presence of dorsiventrality, paracytic stomata, collateral vascular bundles, branched non-articulated laticifers and crystal idioblasts in the species of *Croton*. L studies indicate morphological similarities among the genera. The interpretation of the complex trichomes as emergences is a structural novelty for the group. Leaf anatomy provided an excellent source of characters for future studies of the tribe *Crotoneae* (Euphorbiaceae *s.s.*, Malpighiales).

Ahmad et al., [32] evaluated the taxonomic diversity of stomata in dicot flora from (N.W.F.P) in Pakistan. They selected a total 36 dicot species of 34 genera and 20 families. They observed the stomatal types in different species. Among these species they recognized seven (anisocytic, amphianisocytic, axillocytic, anomotetracytic, actinocytic, diacytic and staurocytic) types of stomata. Whereas in amphianisocytic, stomata was the dominant.

Yasmin et al., [33] worked on the micromorphological investigation of foliar anatomy of genera Acononon and Bistorta family *Polygonaceae*. They collected the material from the herbarium of Quaid-e-Azam University, Islamabad. They studied anatomical features such as shape of epidermal stomatal disturibution on abaxial and adaxial surface of leaves and trichomes types under the light microscope. Among these species, they found six types of stomata (paracytic, cyclocytic, staurocytic, anisocytic, anomocytic,and diacytic) and variation among glandular and non glandular trichomes. They emphasized on the anatomical features of these species. They ignored the arrangement of epidermal cell, stomatal variety and variation in trichomes which could provide very important taxonomic tool for identification of species.

Carmen et al., [34] identified the morphology of foliar epidermis in two groups of *Solanum* section Geminata (*Solanaceae*). They selected the eight species from the genus *Solanum*. They studied the quantitative and qualitative anatomical features on abaxial and adaxial epidermis in these species. They observed the large variation through the observation of leaf segment. They revealed the anatomical character.

Shaheen et al., [35] worked on diversity of foliar trichomes and their systematic relevance in the genus *Hibiscus* (*Malvaceae*). They collected the material from different localities of Pakistan, and dried leaves from the herbarium of Quaid-e-Azam University, Islamabad. They examined quanlitative and quantitative morphological characters. They distributed the systematic relevance of genus *Hibiscus*. Glandular capitates trichome was the most dominant on leaf surface. Among these species, they emphasized only trichome anatomical marker. They ignored other anatomical character that could provide better result for systematical studies.

Abdul Rahman and Oladele [36] explained stomatal complex types and epidermis cells in some *Jatropha* Species L.(Euphorbiaceae). Department of Plant Biology, University Ilorin, Nigeria. Epidermis cells and stomata complex types in five West African species of *jatropha* were studied. The essence of this study was to use more fundamental method (anatomical features) to reclassify the members of the genus which is hitherto classified based on their flower colours, a parameter that is not always present in plants. Paracytic stomata occur in all the species with 100% frequency while anisocytic and anomocytic types are also present in *Jatropha. multifida* and *J. podagrica* respectively. Stomatal size is large (60.05 μm) in *J. curcas* and small (26.64 μm) in *J. multifida*. Stomatal density is highest (44 mm^{-2}) in *J. podagrica* and lowest (10 mm^{-2}) in *J. multifida*. Epidermis cells are large (243μm) in size in adaxial surface of *J. neriifolia* and small (56.33 μm) in abaxial surface of *J. podagrica*. Anticlinal cell-wall is either curved or straight in all species studied. Trichome is present in *J. neriifolia*. These epidermal features (stomata and trichome features) are used to delimit the species of *Jatropha* namely, *J. neriifolia, J. gossypifolia, J. multifida, J. curcas, and J. podagrica.*

Senakuni & Chantarnotha [37] Explained morphological survey of foliar trichomes of *Croton* L. (Euphorbiaceae) in Thailand Walai Rukhavej Research Institute, Mahasarakham University, Maha Sarakham 44150, and Thailand. The morphology of foliar trichomes of 23 species of *Croton* in Thailand was studied by scanning electron microscopy and by light microscopy. Seven main trichome types characterized by their morphology were found, namely, stellate, fasciculate, multirate-rosulate, dendritic, lepidote, glandular and simple types. Some trichome subtypes are useful for species identification, particularly stellate rotate trichomes with a central porrect ray found only in *C. tiglium*, appressed stellate trichomes in *C. griffin thii* and glandular trichomes are restricted to *C. caudatus*, *C. crassifolius* and *C. hirtus*.

Table 1. List of Euphorbiaceae species used for anatomical study

S.no	Genus name	Species name
1	*Acalypha*	*Acalypha wilkisiana L.*
2	*Breynia*	*Breynia nivosa J.R and G. Forst*
3	*Codium*	*Codium verigatum A.H.L. Juss.*
4	*Euphorbia*	*Euphorbia helioscopia L.*
5	*Euphorbia*	*Euphorbia hirta L.*
6	*Euphorbia*	*Euphorbia prostrate Ait.*
7	*Euphorbia*	*Euphorbia pulcherrima ex KI*
8	*Euphorbia*	Euphorbia splendens *Bojer ex Hook.*
9	*Excoecaria*	*Excoecaria bicolor L.*
10	*Jatropha*	*Jatropha integrima Jacq.*
11	*Mallotus*	*Mallotus philippinensis (Lam.) Muell.*
12	*Phyllanthus*	*Phyllanthus emblica L.*
13	*Putranjiva*	*Putranjiva roxburji wall.Tent.*
14	*Ricinus*	*Ricinus communis L.*
15	*Sapium*	*Sapium sebiferium Roxb. FI.*
16	*Terwia*	*Terwia nudiflora L.*

STATISTICAL ANALYSIS

Data was evaluated by co-relation matrix and cluster analysis to determine Inter specific relationship. Dendogram was constructed on the basis of unweighted pair group method with arithmetic average (UPGMA).The computer software SPSS v 11.0 was used for this purpose.

RESULTS

16 Species belonging to twelve genus of the family Euphorbiaceae were described with respect to their leaf epidermal characteristics

1. *Acalypha wilkisiana* L.

a. Adaxial Side

Trichomes and silica bodies are absent. Cells are irregular. Margins are double layered and wavy margins. Length of long cell 130.2 µm – 134 µm (Av: 132 µm).Width of long cell 98.4µm – 107.5 µm (Av: 102.95 µm).Length of short cell 21 µm – 26 µm (Av: 23.5 µrn).Width of short cell 18.2 µm – 25 µm (Av: 21.6 µm). Stomata are present.Number of stomata / unit area 5-6.Length of stomatal cavity 40.2µm – 43.9 µm (Av: 41.95 µm).Width of stomatal cavity 7.1µm – 11.4 µm (Av: 9.25 µm). Length of stomatal cell 80.0 µm – 88.5 µm (Av: 84.25 µm). Width of stomatal cell is 47.0 µm - 61 µm (54 µm). Length of long subsidiary cells 111.6µm – 117.9µm (Av: 114.75). Width of long subsidiary cells 18.6µm-21.1µm (Av:19.85). Length of short subsidiary cell 94.3µm – 109.0µm (Av: 101.65). Width of short subsidiary cell 21.1µm – 34.4µm (Av: 27.8µm). (Figure 1)

b. Abaxial Side

Trichomes and silica bodies are absent. Cells are irregular. Margins are double layered and wavy margins. Length of long cell 120.6µm – 136.6 µm (Av: 127µm).Width of long cell 49.8 µm – 133.4µm (Av: 93.2 µm).Length of short cell 64.5 µm – 110.5 µm (Av: 87.5 µm).Width of short cell 28.3µm – 37.6 µm (Av: 32.95 µm). Stomata are present. Number of stomata / unit area is 7-8. Length of stomatal cavity 35.2µm – 37.7 µm (Av: 36.45 µm).Width of stomatal cavity 14.4µm –20.1 µm (Av: 17.25 µm).Length of stomatal cell 62 µm – 65.5 µm (Av: 63.75µm). Width of stomatal cell 35.9µm – 41.8µm (Av: 38.85 µm). Length of long subsidiary cell 80.3µm – 81.9µm (Av: 81.3µm). Width of long cell 17.0µm – 17.1µm (Av: 17.05µm). Length of short subsidiary cells 83.5µm – 90.7µm (Av: 87.1µm). Width of short subsidiary cells 17.7µm – 48.5µm (Av: 19.4µm). (Figure 1)

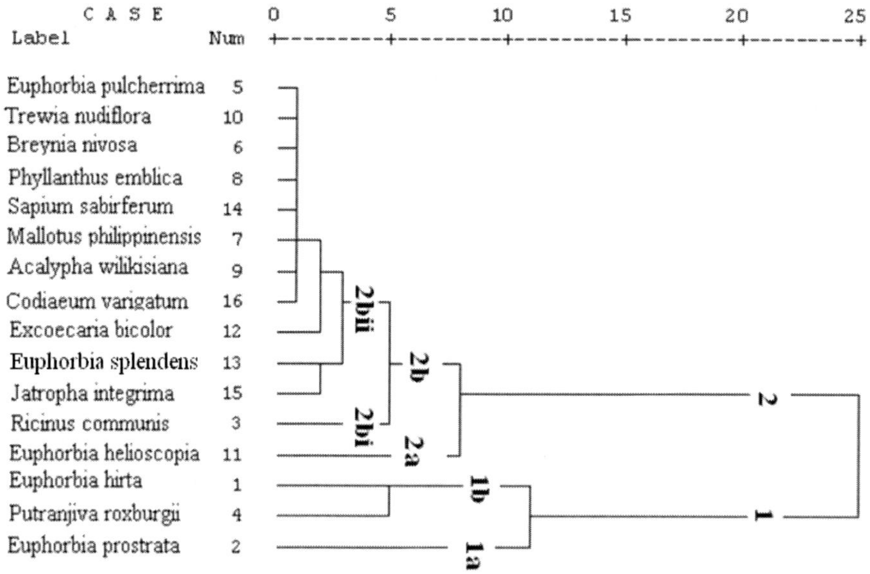

Figure 1. Cluster analysis of species of Family Euphorbiaceae from Lahore district on the basis of leaf epidermal anatomical characters.adaxial surface.

2. *Breynia nivosa* J.R and G. Forst

a. Adaxial Side

Trichome, silica bodies are absent. Cells are tetragonal. Double layered and smooth margins. Length of long cell 84.0µm – 91.2 µm (Av: 87.6µm).Width of long cell 41. µm – 48.3µm (Av: 44.65 µm).Length of short cell 38.5 µm – 51.5 µm (Av: 45.05 µm).Width of short cell 46.4 µm – 51.9 µm (Av: 49.15 µm). Stomata are present. Stomatal type is diacytic. Number of stomata / unit area is 7-8. Length of stomatal cavity 26.0µm – 30.6 µm (Av: 43.3 µm).Width of stomatal cavity 9.2µm – 17.5 µm (Av: 13.35 µm).Length of stomatal cell 63.0 µm – 69.7 µm (Av: 66.35µm). Width of stomatal cell are 31.0µm - 43.4 µm (37.2 µm). Length of long subsidiary cell 68.6µm - 80µm (Av: 75.6µm) Width of long subsidiary cell 30.5µm – 54.6µm (Av: 54.35). Length of short subsidiary cells 71.2µm – 77.9µm (Av: 73.25µm). Width of short subsidiary cells 29.3µm - 44.65µm (Av: 54.35µm). (Figure 2)

b. Abaxial Side

Stomata, silica bodies and trichome are absent.Cells are tetragonal. Double layered and smooth margins. Length of long cell 70.3 μm –72.2 μm (Av: 71.25 μm).Width of long cell 52.4μm – 59.0 μm (Av: 55.7 μm).Length of short cell 44.6 μm – 45.6 μm (Av: 45 μrn).Width of short cell 32.2 μm – 33.9 μm (Av: 33.55 μm). (Figure 2)

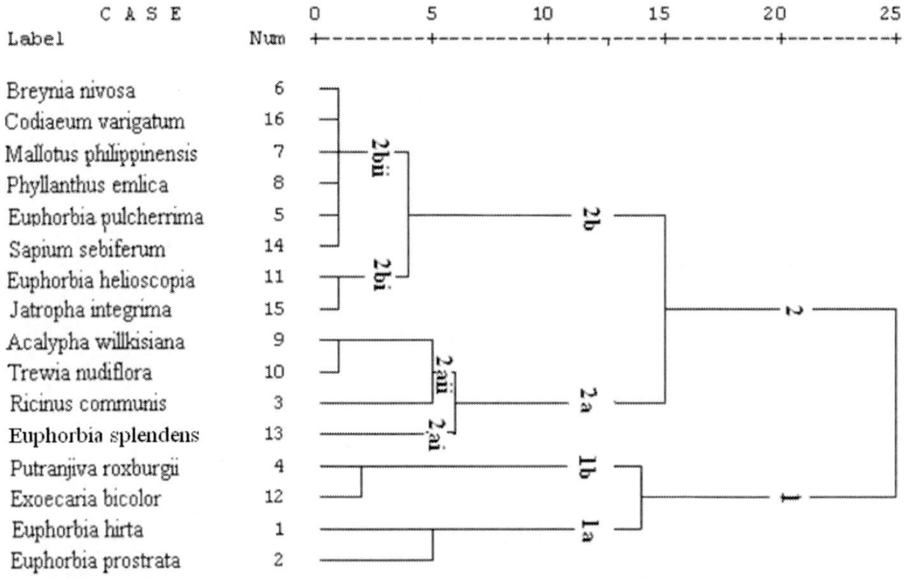

Figure 2. Cluster analysis of species of Family Euphorbiaceae from Lahore district on the basis of leaf epidermal anatomical characters.abaxial surface.

3. *Codium verigatum* A.H.L. Juss.

a. Adaxial Side

Silica bodies, trichomes and stomata are absent..Cells is tetragonal. Cells are double layered and smooth margins. Length of long cell 72.6 μm – 96.3 μm (Av: 84.45μm).Width of long cell 45μm – 53.7μm (Av: 49.35 μm). Length of short cell 43.1μm – 44.7μm (Av: 43.7μrn).Width of short cell 41.7μm – 45.3 μm (Av: 43.5 μm). (Figure 3)

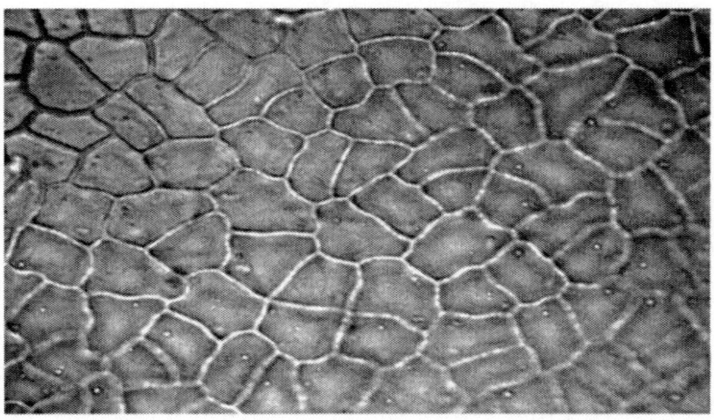

Figure 3. Leaf Epidermal Anatomy of *Codium verigatum* (Adaxial side).

b. Abaxial Side

Trichome and silica bodies are absent. Cells are polygonal. Double layered and smooth margins. Length of long cell 106.2μm – 117 μm (Av: 111.6μm).Width of long cell 40.3 μm – 50.4μm (Av: 45.35 μm).Length of short cell 39.4μm – 47.9 μm (Av: 43.65 μm).Width of short cell 26.2 μm – 30.7μm (Av: 28.45 μm). Stomata are present. Stomatal type is diacytic. Number of stomata / unit area is 6-7. Length of stomatal cavity 36.2μm – 44.4 μm (Av: 43.65 μm).Width of stomatal cavity 13μm – 16 μm (Av: 14.5 μm). Length of stomatal cell 79.6 μm – 95.1μm (Av: 87.35μm). Width of stomatal cell 69μm -77.9 μm (73.15μm).Length of long subsidiary cell 120.1μm – 123.0μm (Av: 121.55μm). Width of subsidiary cell 45.3μm – 47.1μm (Av: 46.3μm). Length of short subsidiary cells 109.2μm – 111.4μm (Av: 110.3μm). Width short of subsidiary cells 40.7μm - 43.4μm(Av: 42.05μm).

4. *Euphorbia helioscopia* L.

a. Adaxial Side

Stomata, silica bodies and trichome are absent. Cells are irregular. Double layered and dentate margined. Length of long cell 194.9 μm – 201.6 μm (Av: 198.25 μm).Width of long cell 79μm – 99.2 μm (Av: 89.1 μm).Length of short cell 71.8 μm – 130.6 μm (Av: 101.2 μrn).Width of short cell 46 μm – 68.6 μm (Av: 57.35 μm). (Figure 4)

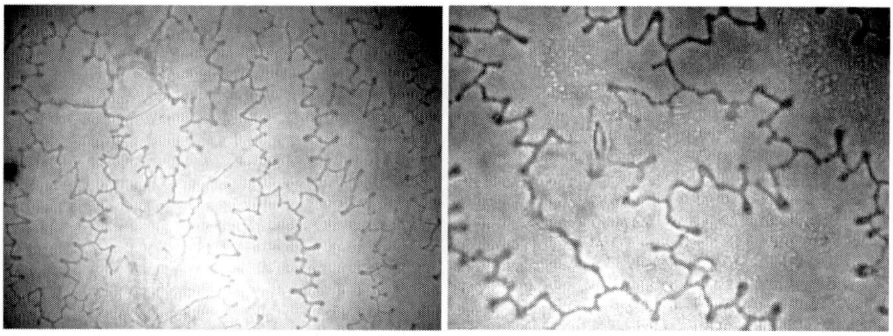

Figure 4. Leaf Epidermal Anatomy of *Euphorbia heliscopia*. Adaxial side (Right side) Abaxial side (Left).

b. Abaxial Side

Trichome and silica bodies are absent. Cells are irregular. Margins are double layered and dentate margins. Length of long cell 196.6 μm – 245.2 μm (Av: 220.9μm).Width of long cell 128.5 μm - 172.8μm (Av: 150.65 μm).Length of short cell 128.3 μm – 138.5 μm (Av: 133.4 μm).Width of short cell 65.9 μm – 83.6 μm (Av: 74.75 μm). Stomata are present. Stomatal type is anomocytic. Numbers of stomata / unit area are 4-5. Length of stomatal cavity 14.2μm – 26.7 μm (Av: 20.45 μm).Width of stomatal cavity 5.8μm – 11.7 μm (Av: 8.75 μm).Length of stomatal cell 36.7 μm – 60.7 μm (Av: 48.7μm). Width of stomatal cell is 22.2μm 50.9 μm (Av: 36.55 μm). Length of long subsidiary cells 179.7μm – 193.1μm (Av: 186.4μm). Width of long subsidiary cells 56.6μm- 72.2μm (Av: 64.4μm). Length of short subsidiary cell 115.3μm – 140.4μm (Av: 127.85μm). Width of short subsidiary cell 75.8μm – 136.9μm (Av: 106.35). (Figure 4)

5. *Euphorbia hirta* L.

a. Adaxial Side

Stomata and trichome are present.Silica bodies are absent. Cells are tetragonal. Double layered and smooth margined. Length of long cell 130.2 µm – 134 µm (Av: 132 µm).Width of long cell 98.4µm – 107.5 µm (Av: 102.95 µm).Length of short cell 21 µm – 26 µm (Av: 23.5 µrn).Width of short cell 18.2 µm – 25 µm (Av: 21.6 µm). Stomata are present. Stomatal type is anisocytic. Number of stomata / unit area 5-6.Length of stomatal cavity 40.2µm – 43.9 µm (Av: 41.95 µm).Width of stomatal cavity 7.1µm – 11.4 µm (Av: 9.25 µm). Length of stomatal cell 80.0 µm – 88.5 µm (Av: 84.25 µm). Width of stomatal cell is 47.0 µm - 61 µm (54 µm). Length of long subsidiary cells 136.6µm – 158.8µm (Av: 147.7µm). Width of long subsidiary cells 37.6µm- 65.0µm (Av: 51.6µm). Length of short subsidiary cell 112µm - 119µm (Av: 115.5µm). Width of short subsidiary cell is 38.1µm – 45.3µm (41.7µm). Trichome is non-glandular and acute tip. Length of trichome 187.0µm-307.5µm (Av: 247.25µm). Width of trichome 30.0µm – 48.6µm (Av: 39.3µm). (Figure 5)

b. Abaxial Side

Trichome and silica bodies are absent. Cells are tetragonal. Double layered and smooth margins. Length of long cell 112.1µm – 186.8 µm (Av: 149.95µm).Width of long cell 48.5 µm - 59.2µm (Av: 89.9 µm). Length of short cell 89.1 µm – 97.9 µm (Av: 93.5 µm).Width of short cell 48.5 µm – 63.3 µm (Av: 55.9 µm). Stomata are present. Stomata type is anisocytic. Number of stomata / unit area is 3-4. Length of stomatal cavity Length of stomatal cavity 26.3µm – 31.1 µm (Av: 28.7 µm).Width of stomatal cavity 5.8µm – 8.1 µm (Av: 6.95 µm).Length of stomatal cell 40.0 µm – 51.0 µm (Av: 45.5µm). Width of stomatal cell is 27µm 34 µm (30.5 µm). Length of long subsidiary cells 54.5µm – 124.9µm (Av: 74.7µm). Width of long subsidiary cells 41.7µm- 45.2µm (Av: 40.45µm). Length of short subsidiary cell 94.9µm – 99.7µm (Av: 112.3µm) Width of short subsidiary cell35.7µm – 48.5µm (Av: 45.1µm). Length of trichome 253.9µ-357.0µm (Av: 352.4µm). Width of trichome 31.8µm-36.8µm (Av: 34.3µm) (Figure 5).

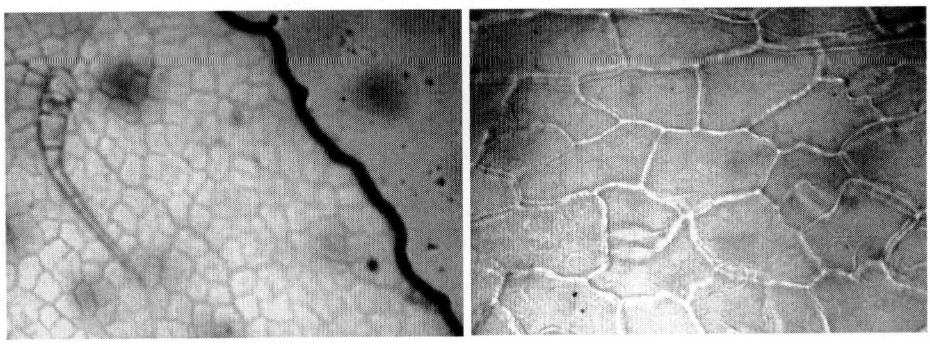

Plate 10. Leaf epidermal anatomy of Euphorbia hirta. (Abaxial side).

Figure 5. Leaf Epidermal Anatomy of *Euphorbia hirta*. Adaxial side (Right) Abaxial side (Left).

6. *Euphorbia prostrate* Ait

a. Adaxial Side

Trichomes and stomata are present. Silica bodies are absent.Cells is tetragonal. Double layered and smooth margins. Length of long cell 58.2 µm – 69.6 µm (Av: 63.9 µm).Width of long cell 46.5µm – 50.5 µm (Av: 48.5 µm).Length of short cell 45.9µm – 46.6 µm (Av: 46.25µrn).Width of short cell 27.2 µm – 33.6µm (Av: 30.4µm). Stomata are present. Number of stomata / unit area is 5-6. Length of stomatal cavity 38.2µm – 43.3 µm (Av: 40.75 µm).Width of stomatal cavity 14.9µm – 15.3µm (Av: 15.1µm). Length of stomatal cell 79.2 µm – 92.7µm (Av: 85.95µm). Width of stomatal cell 59.7 µm - 73.1µm (Av: 66.4µm). Length of long subsidiary cells 154.5µm – 157.4µm (Av: 155.95µm). Width of long subsidiary cells 70.0µm – 127.1 µm(Av:98.55µm) length of short subsidiary cell 55.6µm – 67.5µm(Av: 61.55µm) Width of short subsidiary cell 28.6µm -29.7µm (Av: 29.15µm). Length of trichome 285.6µm – 310.7µm (Av: 298.15µm). Wirth of trichome 27.1µm – 28.2µm(Av: 27.65µm). (Figure 6)

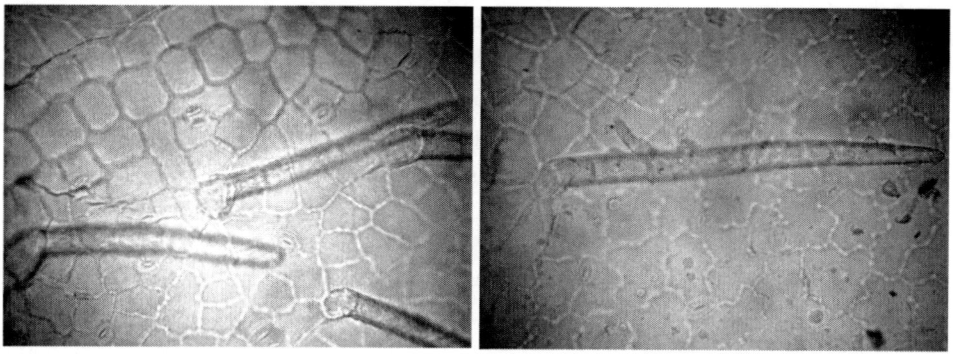

Figure 6. Leaf Epidermal Anatomy of Euphorbia prostrata. Adaxial side (Right) Abaxial side (Left).

b. Abaxial Side

Trichomes are present. Silica bodies are absent. Cells are irregular. Cells are double layered and wavy margins. Length of long cell 79.1µm – 86.0µm (Av: 82.55µm).Width of long cell 69.6µm - 75.0µm (Av: 72.3 µm).Length of short cell 48.3µm – 63.3µm (Av: 55.8µm).Width of short cell 32.0µm – 60.5µm (Av: 46.25µm). Stomata are present. Stomatal type is diacytic. Numbers of stomata / unit area are 6-7. Length of stomatal cavity 40.4µm – 52.4µm (Av: 46.4µm).Width of stomatal cavity 12.8µm – 16.1µm (Av: 14.45µm). Length of stomatal cell 80.7µm – 104.6µm (Av: 92.65µm). Width of stomatal cell 55.6µm - 76.4µm (Av: 66 µm). Length of long subsidiary cells 164.2µm – 226.5µm (Av: 195.35µm). Width of long subsidiary cells 107.0µm – 164.2µm (Av: 135.6µm). Length of short subsidiary cell 67.1µm – 71.7µm (Av: 69.35µm).Width of short subsidiary cell 29.3µm – 38.8µm (34.05µm). Length of trichome 438.9µm – 495.4µm (Av: 467.14µm). Width of trichome is 25.7µm – 29.0µm (Av: 27.35µm) (Figure 6)

7. *Euphorbia pulcherrima* ex KI

a. Adaxial Side

Silica bodies, trichomes and stomata are absent. Cells are irregular. Margins are double layered and wavy. Length of long cell 106.8 μm – 121.2 μm (Av: 114 μm).Width of long cell 63.5μm – 100.7 μm (Av: 82.1 μm).Length of short cell 56.0 μm – 61.7 μm (Av: 58.85 μrn).Width of short cell 76.4 μm – 81.4 μm (Av: 78.9 μm). (Figure 7)

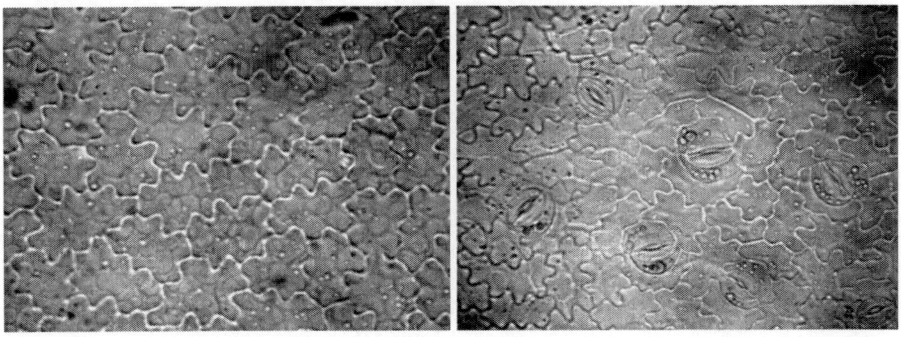

Figure 7. Leaf Epidermal Anatomy of *Euphorbia pulcherrima*. Adaxial side (Right) Abaxial side (Left).

b. Abaxial Side

Trichomes and silica bodies are absent. Cells are irregular. Margins are double layered and wavy. Length of long cell 67.3μm – 94.2 μm (Av: 80.65μm).Width of long cell 55.6 μm – 94.0μm (Av: 74.9 μm).Length of short cell 43.3μm – 65.2 μm (Av: 54.25 μm).Width of short cell 37.5 μm – 50.2 μm (Av: 43.85 μm). Stomata are present. Stomatal type is diacytic are present. Number of stomata / unit area is 5-6. Length of stomatal cavity 27.0μm – 31.8 μm (Av: 29.4 μm).Width of stomatal cavity 8.1μm – 14.0 μm (Av: 9.05μm). Length of stomatal cell 55.5μm – 59.7 μm (Av: 57.6μm). Width of stomata cell is 36.2μm 38.1 μm (37.15 μm). Length of long subsidiary cells 66.2μm – 87.1μm (Av: 65μm). Width of long subsidiary cells 13.9μm – 19.1μm (Av: 16.5μm). Length of short subsidiary cells 52.6μm – 57.6μm (Av: 55.1μm). Width of short subsidiary cells 14.0μm – 24.0μm (Av: 19.0μm). (Figuro 7).

8. *Euphorbia splendens* Bojer ex Hook.

a. Adaxial Side

Trichomes are absent. Cells are pentagonal. Double layered and smooth margined. Length of long cell 133.7 μm – 160.5 μm (Av: 147.1μm).Width of long cell 82.2 μm - 100 μm (Av: 91.1 μm).Length of short cell 77 μm – 94.5 μm (Av: 85.75 μm).Width of short cell 55 μm - 72 μm (Av: 63.5 μm).Stomata are present.stomatal type is diacytic. Number of stomata / unit area 2-3.Length of stomatal cavity 44.9μm – 62.9 μm (Av: 53.9 μm).Width of stomatal cavity 12.6μm – 13.6 μm (Av: 13.1 μm). Length of stomatal cell 89.2 μm – 109.1 μm (Av: 99.15 μm). Width of stomatal cell is 68.1 μm - 83 μm (75.55 μm). Length of long subsidiary cells 108.5μm – 192.0μm (Av: 150.25). Width of long subsidiary cells 39.3μm- 57.5μm (Av: 48.25). Length of short subsidiary cell 153.6μm -174.1 (Av: 163.85μm). Wirth of short subsidiary cell 22.1μm – 69.6μm (Av: 45.85μm). Length of silica bodies' 22.0μm-26.0μm (Av: 26.0μm). Width of silica bodies 18.0μm- 22.4μm (Av: 20.2μm). (Figure 8)

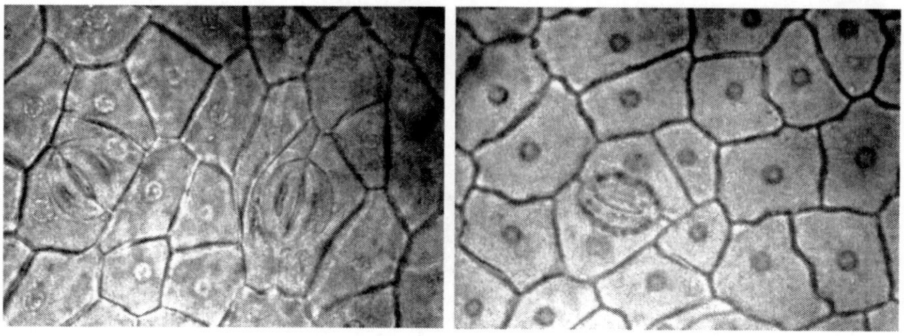

Figure 8. Leaf Epidermal Anatomy of *Euphorbia splendus*. Adaxial side (Right) Abaxial side (Left).

b. Abaxial Side

Trichomes are absent. Cells are tetragonal. Double layered and smooth margins. Length of long cell 89.4μm – 132.4 μm (Av: 110.9 μm).Width of long cell 107.9μm -143.8 μm (Av: 125.85 μm). Length of short cell 75 μm – 98.4 μm (Av: 86.7 μrn).Width of short cell 53.7 μm – 82.9 μm (Av: 68.3 μm).Stomata are present. Stomatal type is diacytic. Number of stomata / unit area is 4-5. Length of stomatal cavity 41.1 μm – 54.2 μrn (Av: 47.65 μm). Width of stomatal cavity 6.4 μm – 14.9 μm (Av: 10.65 μrn).Length of stomatal cell 79.5 μm – 92.3 μm (Av: 85.9 μm).Width of stomatal cell 56.6 μm – 67.4 um (Av: 62 um). Length of long subsidiary cells 127.3μm- 138.1μm (Av: 132.7). Width of subsidiary cells 46.1μm - 54.2μm (Av: 50.15μm). Length of short subsidiary cell 115μm – 144.7μm (Av: 129.85μm). Width of short subsidiary cell 32.0μm – 72.6μm (Av: 52.3μm). Length of silica bodies 19.4μm – 24.2μm (Av: 21.8μm). Width of silica bodies 18.2μm – 26.1μm (Av: 22.15μm). (Figure 8)

9. *Excoecaria bicolor* L.

a. Adaxial Side

Trichome is present. Cells are irregular. Double layered and wavy margined. Length of long cell 109.0μm - 109.4 μm (Av: 109.2 μm). Width of long cell 55.5 μm – 62.3.0 μm (Av: 58.9 μm). Length of short cell 45.3μm – 75.8μm (Av: 60.55μm). Width of short cell 35.4μm – 55.5μm (Av: 45.45μm). Trichome is present and acute tip.Lenght of trichome 242.8μm - 275.6μm (Av: 259.2μm).Width of trichome 35.4μm -63.7μm. (Av: 49.55 μm). (Figure 9).

b. Abaxial Side

Trichomes and silica bodies absent.Cells are irregular. Double layered and wavy margined. Length of long cell 58.1μm – 73.2 μm (Av: 65.65 μm).Width of long cell 42.5 um – 45.2 μm (Av: 87.7 μm). Length of short 39.3μm – 44.9μm (Av: 42.1μm). Width of short cell 23.9μm - 25μm (Av: 24.45μm).stomatal type is diacytic. Stomata are present. Number of stomata / unit area (5-6).Length of stomatal cavity 19.7μm –21.2 μm (Av: 49.5 μm).Width of stomatal cavity 5.0 μm -6.7 μm (Av: 5.85 μm). Length of stomatal cell 38.3 μm –43.7 μm (Av: 41 μm). Width of stomatal cell is 23 μm – 25.6.0 μm (22.7 μm). Length of long subsidiary cell 45.3 µm – 47.1µm (Av: 46.2 µm). Width of long subsidiary cell 13.6µm - 17µm (Av: 15.3 µm). Length of short subsidiary cell is 34.1µm – 38.5µm (Av 36.3µm). Width of sort subsidiary 20µm – 25.6µm (Av: 22.8µm). (Figure 9).

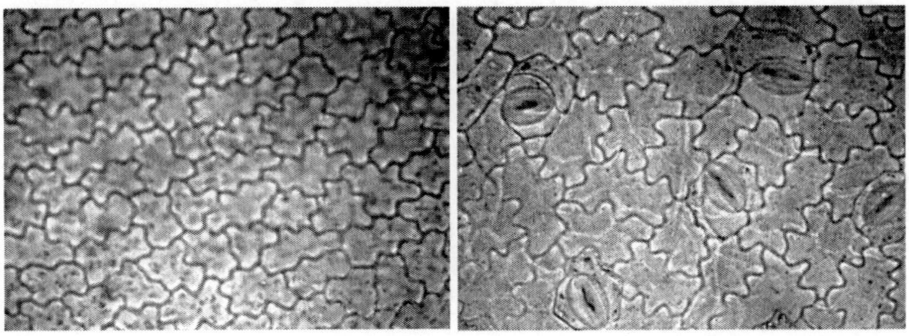

Figure 9. Leaf Epidermal Anatomy of Excoecaria bicolor. Adaxial side (Right) Abaxial side (Left).

10. *Jatropha integrima* Jacq.

a. Adaxial Side

Trichome and silica bodies are absent. Cells are pentagonal. Double layered and smooth margins. Length of long cell 118.6 µm – 161.8 µm (Av: 140.2 µm).Width of long cell 81.9µm – 90.8 µm (Av: 91.6 µm).Length of short cell 69 µm – 70.1 µm (Av: 70.1 µrn).Width of short cell 52 µm -90.8 µm (Av: 71.4 µm).Stomata are present. Stomatal type is diacytic. Number of stomata / unit area is 5-6. Length of stomatal cavity 31.3µm – 40.2 µm (Av: 35.7 µm).Width of stomatal cavity 12.6µm – 14.3 µm (Av: 13.45 µm).Length of stomatal cell 62.1 µm – 67.4 µm (Av: 64.75µm). Width of stomatal cell 47µm - 50.6µm (Av: 48.8 µm). Length of long subsidiary cells 78.5µm – 107.9µm (Av: 93.2µm). Width of long subsidiary cells 38.4µm- 70µm(Av:54.2µm).Length of short subsidiary cell 90.6µm – 92.4 µm (Av: 91.5µm). Width of short subsidiary cell is 73.2µm – 73.4µm (73.3µm). (Figure 10)

b. Abaxial Side

Stomata, Silica bodies and trichome are absent. Cells are hexagonal. Double layered and smooth margined. Length of long cell 161.0 µm – 213.7 µm (Av: 187.35µm).Width of long cell 116 µm - 134.1µm (Av: 125.05 µm).Length of short cell 54.2 µm -83.6 µm (Av: 68.9 µm).Width of short cell 71.3 µm – 81.1 µm (Av: 76.2 µm). (Figure 10)

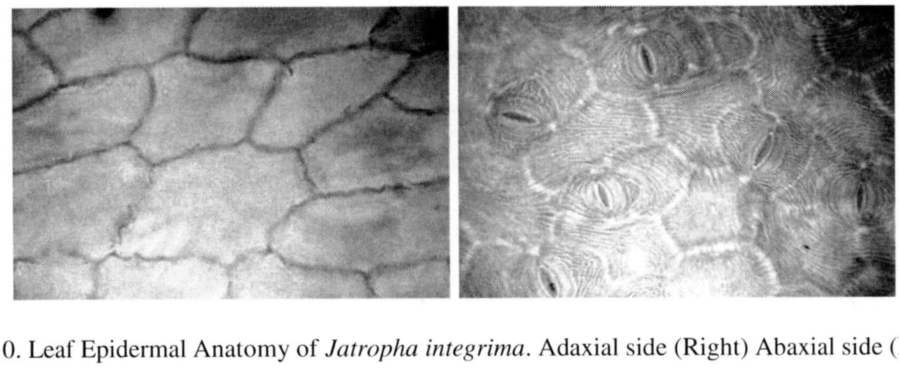

Figure 10. Leaf Epidermal Anatomy of *Jatropha integrima*. Adaxial side (Right) Abaxial side (Left).

11. *Mallotus philippinensis (Lam.)* Muell.

a. Adaxial Side

Stomata, silica bodies and trichome are absent.Cells are irregular. Margins are Double layered and wavy. Length of long cell 52.2 µm – 126.5 µm (Av: 89.35 µm).Width of long cell 26.9µm – 110.4 µm (Av: 78.7 µm).Length of short cell 29.2 µm – 44.8 µm (Av: 37 µrn).Width of short cell 47 µm – 58.3 µm (Av: 52.65 µm). (Figure 11)

b. Abaxial Side

Trichomes are absent. Cells are tetragonal. Double layered margins. Length of long cell 111.6µm – 156.6 µm (Av: 134.1µm).Width of long cell 35 µm – 60.7µm (Av: 47.85 µm).Length of short cell 75.3 µm – 79.3 µm (Av: 77.3 µm).Width of short cell 26.9 µm – 58.3 µm (Av: 42.6 µm). Stomata are diacytic. Number of stomata / unit area is 7-8. Length of stomatal cavity 34.2µm – 41.1 µm (Av: 37.65 µm).Width of stomatal cavity 5.4µm –7.2 µm (Av: 6.3 µm).Length of stomatal cell 78.2µm – 92.7 µm (Av: 85.1µm). Width of stomatal cell is 52.3µm – 56.9 µm (54.6 µm). Length of long subsidiary cell 93.5µm – 105.9µm (Av: 99.7µm). Wirth of long subsidiary cell 26.4µm – 30.5µm (Av: 28.45µm).Length of short subsidiary cells 95.4µm – 100.3µm (Av: 97.85µm). Width of short subsidiary cells 12.8µm-15.1µm (Av: 13.95µm). (Figure 11))

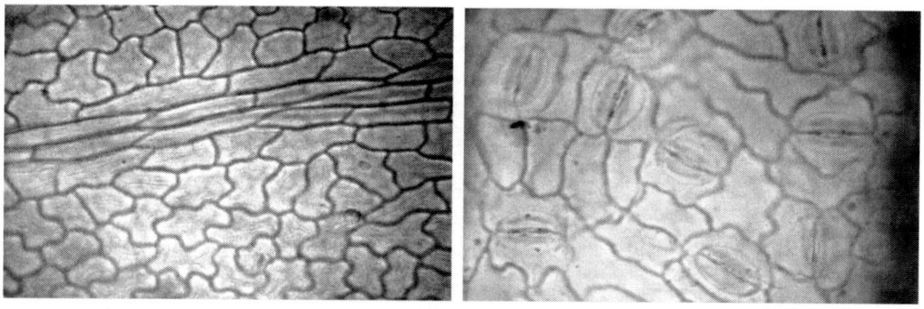

Figure 11. Leaf Epidermal Anatomy of *Mallotus philippinensis.* Adaxial side (Right) Abaxial side (Left).

12. *Phyllanthus emblica* L.

a. Adaxial Side

Trichomes are absent. Silica bodies are present. Cells are irregular. Cells are double layered and wavy margins. Length of long cell 123.2µm – 125.4 µm (Av: 124.3µm).Width of long cell 59.5 µm – 64.9µm (Av: 62.2 µm).Length of short cell 55.4 µm – 74.1 µm (Av: 64.75 µm).Width of short cell 48.1 µm – 50.6 µm (Av: 49.35 µm). Stomata are diacytic. Number of stomata / unit area is 12-13. Length of stomatal cavity 26.5µm – 30.1 µm (Av: 28.3 µm).Width of stomatal cavity 5.7µm – 9.8 µm (Av: 7.75 µm). Length of stomatal cell 53.5µm – 55.3 µm (Av: 54.4µm). Width of stomatal cell is 36.1µm 38.1 µm (37.1 µm). Length of long subsidiary cells 100.8µm – 103.9µm (Av: 102.35µm). Width of long subsidiary cells 53.3µm – 55.3µm (Av: 54.3µm). Length of short subsidiary cells 67.3µm – 69.4µm (Av: 68.35µm). Width of short subsidiary cells 26.3µm – 28.1µm(Av: 27.2µm).Silica bodies are rounded. Length of silica bodies 8.9μm – 12.0μm (Av: 10.45μm) Width of silica bodies 5.5μm - 7.8μm (Av: 8.8μm). (Figure 12)

b. Abaxial Side

Silica bodies, trichomes and stomata are absent..Cells are hexagonal. Margins are double layered and wavy and smooth. Length of long cell 65.4 μm – 119.9 μm (Av: 92.65 μm).Width of long cell 46.1μm – 119.8 μm (Av: 82.95 μm).Length of short cell 47.2 μm – 59 μm (Av: 53.1 μrn).Width of short cell 41 μm – 46.1 μm (Av: 46.05 μm). (Figure 12)

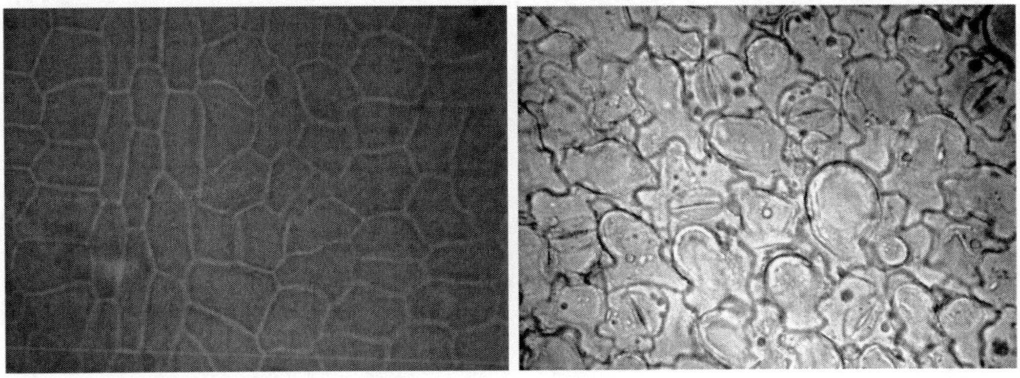

Figure 12. Leaf Epidermal Anatomy of *Phyllanthus emblica.* Adaxial side (Right) Abaxial side (Left).

13. *Putranjiva roxburji* Wall.Tent.

a. Adaxial Side

Stomata and silica bodies are absent. Cells are irregular. Double layered and smooth margins. Length of long cell 39 μm -87.8 μm (Av: 63.4μm).Width of long cell 52.8.0 μm -62 μm (Av: 57.4 μm).Length of short cell 29 μm -42.0 μm (Av: 35.5 μm).Width of short cell 31.6 μm – 43.3 μm (Av: 37.45 μm). Trichomes are non-glandular and trichome tip is acute. Length of trichome 180.4μm-250.3μm (Av: 215.35). Width of trichome 19.6μm- 27.7μm (Av: 23.65μm). (Figure 13)

b. Abaxial Side

Trichomes are present. Silica bodies are absent. Cells are hexagonal. Double layered and smooth margins. Length of long cell 98.6 μm -107.5 μm (Av: 103.05 μm).Width of long cell 37.6μm – 41.4 μm (Av: 39.5 μm).Length of short cell 40.1 μm - 44.3 μm (Av: 42.2 μrn).Width of short cell 22.1 μm -29.1 μm (Av: 25.6 μm).Stomata are present. Stomatal type is diacytic. Number of stomata / unit area are14-15. Length of stomatal cavity 28.3μm – 34.1 μm (Av: 31.2 μm).Width of stomatal cavity 7.2μm -8.1 μm (Av: 7.65 μm).Length of stomatal cell 53.2 μm – 62.4 μm (Av: 57.8μm). Widths of stomata cell are 44.7 μm 52.5 μm (48.6 μm). Length of long subsidiary cells 59.0μm – 77.8μm (Av: 68.4μm). Width of long subsidiary cells 22.5.0μm- 23.3(Av: 22.9μm).length of short subsidiary cells 49.3μm – 57.8μm (Av: 53.55μm). Width of short subsidiary cells 22.0μm -23.3μm (Av: 22.65).

Trichomes are non-glandular. Length of trichome 231.8μm—250.5μm (Av: 241.15). Width of trichome 12.4μm - 26.1μm(Av: 24.8μm). (Figure 13)

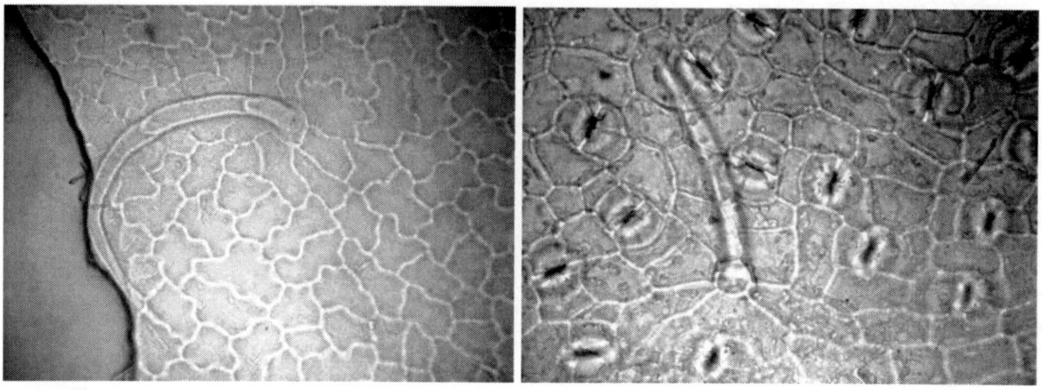

Figure 13. Leaf Epidermal Anatomy of *Putranjiva roxburji.* Adaxial side (Right) Abaxial side (Left).

14. *Ricinus communis* L.

a. Adaxial Side

Trichomes and silica bodies are absent. Cells are irregular and double layered and smooth margins. Length of long cell 72.7 μm -81.8 μm (Av: 77.25 μm).Width of long cell 48.0 μm - 53.9 μm (Av: 50.95 μm).Length of short cell 21.1 μm -49.0 μm (Av: 35.0 μm).Width of short cell 40.5 μm -42.2 μm (Av: 35.05 μm).Stomata are present. Stomatal type is diacytic. Number of stomata / unit area 5-6.Length of stomatal cavity 18.0μm -34.15 μm (Av: 26.075 μm).Width of stomatal cavity 6.7μm -9.4 μm (Av: 8.05 μm).Length of stomatal cell 53.9 μm -59.1 μm (Av: 56.5 μm). Width of stomatal cell is 41.0μm- 44.9μm (42.95 μm). Length of long subsidiary cells 154.5μm – 157.4μm (Av: 155.95μm). Width of long subsidiary cells 70.0μm – 127.1 μm (Av: 98.55μm). length of short subsidiary cell 55.6μm – 67.5μm (Av: 61.55μm) Width of short subsidiary cell 28.6μm -29.7μm (Av: 29.15μm) (Av:24.4). (Figure 14)

b. Abaxial Side

Trichome and silica bodies are absent. Cells are irregular. Double layered and smooth margins. Length of long cell 154.6 μm -164.4 μm (Av: 159.5 μm).Width of long cell 60.0μm - 69.8 μm (Av: 64.9 μm).Length of short cell 52.9 μm -59.6 μm (Av: 56.25 μrn).Width of short cell 19.7 μm -24.2 μm (Av: 21.95 μm).Stomata are present. Stomatal type is diacytic. Number of stomata / unit area is 4-5. Length of stomatal cavity 34.8 μm -58.5 μrn (Av: 46.65 μm). Width of stomatal cavity 8.1 μm - 12.8 μm (Av: 10.45 μrn).Length of stomatal cell 58.4 μm - 81.9 μm (Av: 70.15 μm).Width of stomatal cell 44.1 μm - 98.3 um (Av: 71.2μm). Length of long subsidiary cells 164.2μm – 226.5μm (Av: 195.35μm). Width of long subsidiary cells 107.0μm – 164.2μm (Av: 135.6μm). Length of short subsidiary cell 67.1μm – 71.7μm (Av: 69.35μm) Width of short subsidiary cell 29.3μm – 38.8μm (34.05μm). (Av: 71.2 um). (Figure 14)

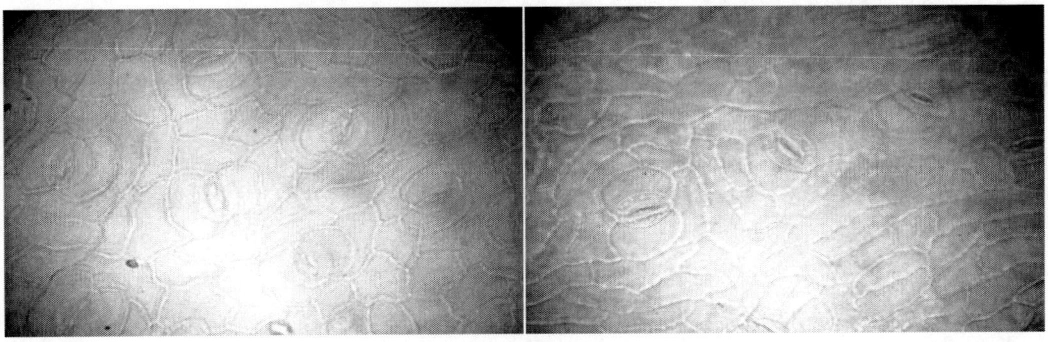

Figure 14. Leaf Epidermal Anatomy of *Ricinys communis*. Adaxial side (Right) Abaxial side (Left).

15. *Sapium sebiferium* Roxb. FI.

a. Adaxial Side

Silica bodies, stomata and trichomes are absent.Cells is irregular. Margins are double layered and wavy margins. Length of long cell 109.0 µm – 113 µm (Av: 111.15 µm).Width of long cell 93.8µm – 103.0 µm (Av: 98.4 µm).Length of short cell 61 µm – 64.6 µm (Av: 62.8 µrn).Width of short cell 54.1 µm – 87.4 µm (Av: 70.75 µm. (Figure 15)

b. Abaxial Side

Trichome and silica bodies are absent. Cells are irregular. Margins are double layered and wavy margins. Length of long cell 114.0µm – 132.7 µm (Av: 123.35µm).Width of long cell 35 µm – 38.6µm (Av: 36.8 µm).Length of short cell 67.5 µm – 81.6 µm (Av: 74.55 µm).Width of short cell 24.2 µm – 54.5 µm (Av: 39.35 µm). Stomata are diacytic. Number of stomata / unit area is 5-6. Length of stomatal cavity 29.7µm –36.3 µm (Av: 33 µm).Width of stomatal cavity 8.0µm – 12.2 µm (Av: 10.1 µm). Length of stomatal cell 56.9 µm – 62 µm (Av: 59.45µm). Width of stomatal cell 34µm – 41.4 µm (Av: 37.7 µm). Length of long subsidiary cells 85.0µm – 87µm (Av: 86µm). Width of long subsidiary cells 35.5µm – 37.7µm(Av: 36.6µm). Length of short subsidiary cell 60.1µm – 62.3µm (Av: 61.2µm). Width of short subsidiary cell 25.6µm – 27.8µm (Av: 26.7µm)(Figure 15).

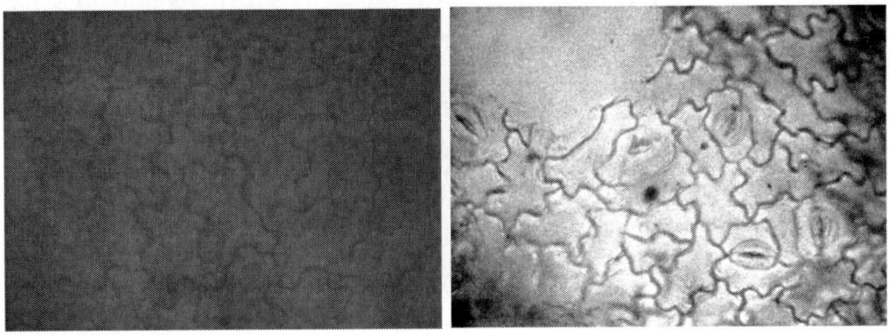

Figure 15. Leaf Epidermal Anatomy of *Sapium sebiferium*. Adaxial side (Right) Abaxial side (Left).

16. *Terwia nudiflora* L.

a. Adaxial Side

Trichomes and silica bodies are absent. Stomata are present. Irregular cells are present. Margins are double layered and wavy smooth. Length of long cell 107.0 µm – 124.8 µm (Av: 115.9 µm).Width of long cell 64.3µm – 70.3 µm (Av: 67.9 µm).Length of short cell 55.7µm – 57.6 µm (Av: 56.65µrn).Width of short cell 33.4 µm – 47.0µm (Av: 40.2µm). Stomata are present. Stomatal type is diacytic are present. Number of stomata / unit area is 1-2. Length of stomatal cavity 49.7µm – 63.5 µm (Av: 56.6µm).Width of stomatal cavity 3.6µm – 6.4µm (Av: 5µm). Length of stomatal cell 90.8 µm – 95.2µm (Av: 93.0µm). Width of stomatal cell 43.4µm - 51.7µm (Av: 47.55µm). Length of long subsidiary cells 116.8µm – 122.5µm (Av: 119.65µm). Width of long subsidiary cells 19.7µm – 26.1 µm (Av: 22.9µm) length of short subsidiary cell 101.0µm – 104.2µm(Av: 102.6µm) Width of short subsidiary cell 45.3µm – 63.7µm (Av: 54.5µm). (Figure 16)

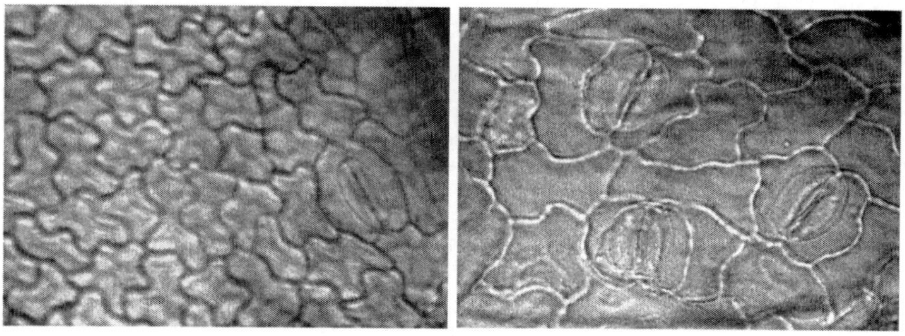

Figure 16. Leaf Epidermal Anatomy of *Terwia nudiflora.* Adaxial side (Right) Abaxial side (Left).

b. Abaxial Side

Trichome and silica bodies are absent. Cells are irregular. Double layered wavy margins. Length of long cell 84.3µm – 128.1µm (Av: 106.2µm).Width of long cell 56.0µm - 58.5µm (Av: 57.25 µm).Length of short cell 45.2µm – 57.2µm (Av: 51.2µm).Width of short cell 25.6µm – 43.0µm (Av: 34.3µm). Stomata are present. Stomatal type is diacytic are present. Number of stomata / unit area is 3-4. Length of stomatal cavity 27.2µm – 39.4µm (Av: 33.3µm).Width of stomatal cavity 9.4µm – 9.8µm (Av: 9.6µm). Length of stomatal cell 61.7µm – 64.9µm (Av: 63.3µm). Width of stomatal cell 41.2µm - 44.1µm (Av: 42.65 µm). Length of long subsidiary cells 70.9µm – 81.9µm (Av: 76.4µm). Width of long subsidiary cells 15.0µm – 23.4µm (Av: 19.2µm). Length of short subsidiary cell 74.1µm – 74.8µm (Av: 74.45µm). Width of short subsidiary cell 19.6µm – 23.3µm (21.45µm). (Figure 16)

DISCUSSION

Many of the species of Family Euphorbiaceae are known for their medicinal importance. The use of data generated from leaf epidermal surfaces in resolving the taxonomic problems of taxa has gained much recognition for a long time [40]. Sufficient research has not been done on the use of anatomical features for the delineation and classification of the various taxa [2]. Leaf epidermis features like stomatal types, density, size, epidermal cell types, epidermal cell size and trichomes, are of taxanomic values in delimiting species of the same genus or genera within a family [36, 41]. It has been observed that anatomical studies showed variation in size/shape of long cell, short cell, trichomes, silica bodies, and stomata [42]. Same observation of anatomical features was made during the present piece of investigation. Cells are rectangular or polygonal in shape and elongated in *Codiaeum verigatum*.

During the present work, the family Euphorbiaceae was found to be characterized by different types of cells that is, irregular, tetragonal, hexagonal, pentagonal, polygonal and their margins were may be smooth, dentate or wavy. The cells were single-and double-layered. The size and shape of organelles were different on abaxial and adaxial surfaces of leaf epidermis because of bifacial nature of leaf, which was also confirmed by the findings of [16]. The maximum length of long cell ranged from 196.6μm - 245.2μm which was found in *Euphorbia helioscopia* on adaxial side of leaf epidermis. Similarly, maximum length of short cell ranged from 128.3μm to 138.4μm, exhibited by the same species which is a unique leaf character and is sufficient to differentiate it from other species of family Euphorbiaceae.

Trichomes are important taxonomic character of Euphorbiaceae. This is considered as identifiable character for species. The length of trichome varies from species to species. According to our findings during this study, trichomes could be used for identification of the species of Euphorbiaceae. In *Euphorbia prostrate,* length of trichome ranged from 438.9μm – 495.4μm on adaxial side of leaf epidermis. Trichomes not only showed variation in length but also in origin. It might be intracellular, from the surface of outer memberane/ inner memberane. Diversity was also found among trichomes as they were prickle-like or hairy. Trichomes with acute or blunt ends were also present. Same observations were made by [37], while they were working to identify the leaf epidermial anatomy of Croton. Trichomes were altogether absent on both abaxial and adaxial leaf surface in *Ricinus communis*, *Euphorbia pulcherrima, Breynia nivosa, Mallotus philippinensis, Phyllanthus emblica, Acalypha wilkisiana, Trewia nudiflora, Euphorbia helioscopia, Sapium sebiferium, Jatropha integrima* and *Codiaeum verigatum*. All trichomes were of non glandular type in family Euphorbiaceae.

In many cases, stomata were found to be more abundant on abaxial than adaxial side of leaf, sometimes totally absent during the present work. Environmental condition and genetic factor strongly influence the morphogenesis of stomata [43]. The ledge of stomata is present in several genera of family Euphorbiaceae. This structure is important for opening and closing of stomata, and in a general outer shape of stomata, [5, 43, 44] occasional differences were observed in the structure of stomata present on foliar organs in a number of plants. Later, [44] revealed that the stomata on the upper and lower surfaces of the leaves may differ in their structure.

Present study shows that stomata were kidney shaped. There were many stomatal types that is, diacytic, anomocytic & anisocytic and so on. Stomata were totally absent on abaxial epidermis of *Putranjiva roxburgi, Euphorbia pulcherrima, Breynia nivosa, Mallotus philippinensis, Phyllanthus emblica, Euphorbia helioscopia, Excoecaria bicolor, Sapium sebiferium, Jatropha integrima* and *Codiaeum verigatum*. Whereas stomata were present on adaxial surface of all 16 species under study. Presence of silica bodies were also considered as one of taxonomically important characters of Euphorbiaceae. This anatomical character was present on both abaxial and adaxial surface of *Euphorbia splendus* and only on adaxial surface of *Phyllanthus emblica*. In Family Euphorbiaceae, silica bodies were regular and round shaped.

From the result of present studies, it was concluded that leaf epidermal anatomical features are important taxonomic markers, and they can be effectively utilized for resolving the persistent taxonomic problems of Family Euphorbiaceae.

Table 2 shows the relationship between the species of Family Euphorbiaceae. Co-relation of this family ranged 0.01 - 1.00. The value of co-relation for *Sapium sebiferum,Phyllanthus emblica* and *Mallotus philippinensis* is 1.00 (Table 2). The species that showed minimum co-relation (*Euphorbia splendus* and *Euphorbia pulcherrima)* is 0.01. The species that showed negative related species were *Ricinus communis* with Excoecaria bicolor and *Putranjiva roxbergii*. The negative relationship ranged from -0.02 and -0.05.

Table 3 shows the relationship between the species of Family Euphorbiaceae. Co-relation of this family ranged from 0.19 to 0.99. The value of co-relation for *Excoecaria bicolor, Euphorbia prostrate, Trewia wilkisiana* and *Mallotus philippinensis was presented in* Table 3. The species that showed minimum co-relation were *Excoecaria bicolor* and *Euphorbia prostrate* co-relation value, 0.19. The species that showed minimum co-relation were *Trewia wilkisana* and *Mallotus philippinensis, co-relation value* is 0.99.

Table 4 shows correlation between characters of the species of Family Euphorbiaceae. Co-relation of this family ranged from 0.02 - 0.99. The value of co-relation for length of stomatal cell and length of stomatal cavity is 0.99 (Table 4). The characters that showed minimum co-relation width of stomatal cell, length of long cell, length of stomatal cell and width of long cell showed a co-relation value of 0.02. The characters that showed negative range that is from -0.48 to -0.01, the value of co-relaton for wirth of stomatal cell and wirth of long cell is -0.01. The characters that showed minimum negative co-relation length of long subsidiary cell and width of short cell.

Table 5 shows correlation between characters of species of Family Euphorbiaceae. Co-relation of this family ranged from 0.01 to 0.96. The value of co-relation for width of trichome and length of trichome is 0.96 (Table 5). The characters that showed minimum co-relation length of stomatal cavity and length of trichome is 0.01. The characters that showed negative co-relation have range from -0.54 to -0.01. The value of co-relation for length of short subsidiary cell and width of trichome is -0.01. The characters that showed minimum negative co-relation length of stomatal cavity and length of long cell are -0.54.

Cluster analysis based on leaf epidermal anatomical characters revealed that 16 species of Family Euphorbiaceae were divided into two main clusters - 1 and 2 at phylogenetic distance 100% (Figure 1). At 46%, genetic linkage group 1 is further divided into 1b and 1a. The species included in 1a are *Euphorbia prostrata* and in 1b are species *Euphorbia hirta* and *Putranjiva roxburgii*. At 35%, genetic linkage group 2 is divided into 2b and 2a. The species included in 2a are *Euphorbia hirta* and *Putranjiva roxbergii*. At 29%, genetic linkage group 2b is further divided into 2bi and 2bii. The specie included in 2bi is *Ricinus communis*. The species included are *Euphorbia splendus, Jatropha integrima, Excoecaria bicolor, Codiaeum verigatum, Acalypha wilkisiana, Mallotus philippinensis, Sapium sebiferium, Phyllanthus emblica, Breynia nivosa, Tewia nudiflora, Euphorbia pulcherrima*. The species are related on the basis of anatomical characters.

Table 2. Co-relation between the species of family Euphorbiaceae based on leaf epidermal anatomical characters. abaxial surface

Species	1	2	3	4	5	6	7	8	9	10	11	12	13	14	15	16
1	1															
2	0.91	1														
3	0.48	0.46	1													
4	0.77	0.77	-0.05	1												
5	0.26	0.04	0.25	0.32	1											
6	0.27	0.05	0.26	0.32	0.98	1										
7	0.29	0.05	0.25	0.32	0.99	0.98	1									
8	0.28	0.05	0.25	0.32	0.98	1.00	0.99	1								
9	0.60	0.35	0.78	0.07	0.51	0.53	0.55	0.54	1							
10	0.56	0.36	0.82	0.03	0.47	0.48	0.47	0.47	0.95	1						
11	0.29	0.06	0.28	0.31	0.95	0.97	0.94	0.95	0.54	0.50	1					
12	0.77	0.75	-0.02	0.99	0.38	0.39	0.38	0.38	0.10	0.07	0.40	1				
13	0.52	0.33	0.80	0.01	0.46	0.48	0.46	0.47	0.92	0.94	0.49	0.04	1			
14	0.27	0.05	0.25	0.32	0.99	0.99	1.00	1.00	0.52	0.47	0.94	0.38	0.46	1		
15	0.31	0.06	0.27	0.32	0.98	0.98	0.99	0.98	0.57	0.49	0.97	0.39	0.48	0.98	1	
16	0.24	0.02	0.24	0.30	0.99	0.99	0.98	0.98	0.51	0.47	0.98	0.38	0.47	0.98	0.99	1

Key: 1: Euphorbia hirta; 2: Euphorbia prostrate; 3: Ricinus communis; 4: Putranjiva roxburgi; 5: Euphorbia pulcherrima; 6: Breynia nivosa; 7: Mallotus philippinensis; 8: Phyllanthus emblica; 9: Acalypha wilkisiana; 10: Trewia nudiflora; 11: Euphorbia helioscopia; 12: Excoecaria bicolor; 13: Euphorbia splendens; 14: Sapium sebiferum; 15: Jatropha integrima; 16: Codiaeum varigatum.

Table 3. Co-relation between the species of family Euphorbiaceae based on leaf epidermal anatomical characters. adaxial surface

Species	1	2	3	4	5	6	7	8	9	10	11	12	13	14	15	16
1	1															
2	0.91	1														
3	0.29	0.35	1													
4	0.97	0.94	0.33	1												
5	0.36	0.23	0.78	0.34	1											
6	0.32	0.25	0.86	0.32	0.90	1										
7	0.37	0.25	0.84	0.38	0.94	0.92	1									
8	0.36	0.27	0.92	0.35	0.94	0.93	0.95	1								
9	0.39	0.21	0.77	0.35	0.97	0.87	0.95	0.93	1							
10	0.37	0.24	0.84	0.38	0.97	0.93	0.99	0.96	0.97	1						
11	0.39	0.25	0.83	0.33	0.91	0.88	0.88	0.95	0.92	0.91	1					
12	0.33	0.19	0.70	0.29	0.94	0.82	0.81	0.84	0.91	0.88	0.84	1				
13	0.34	0.25	0.81	0.31	0.96	0.92	0.91	0.93	0.93	0.94	0.91	0.89	1			
14	0.37	0.26	0.88	0.37	0.92	0.93	0.98	0.98	0.94	0.97	0.92	0.81	0.88	1		
15	0.37	0.23	0.81	0.33	0.94	0.96	0.90	0.95	0.91	0.94	0.94	0.86	0.93	0.93	1	
16	0.35	0.26	0.87	0.31	0.82	0.93	0.88	0.91	0.85	0.89	0.88	0.75	0.87	0.89	0.89	1

Key: 1: *Euphorbia hirta*; 2: *Euphorbia prostrate*; 3: *Ricinus communis*; 4: *Putranjiva roxburgi*; 5: *Euphorbia pulcherrima*; 6: *Breynia nivosa*; 7: *Mallotus philippinensis*; 8: *Phyllanthus emblica*; 9: *Acalypha wilkisiana*; 10: *Trewia nudiflora*; 11: *Euphorbia helioscopia*; 12: *Excoecaria bicolor*; 13: *Euphorbia splendens*; 14: *Sapium sebiferum*; 15: *Jatropha integrima*; 16: *Codiaeum varigatum*.

Table 4. Co-relation between the characters of family Euphorbiaceae based on leaf epidermal anatomical characters. abaxial surface

Characters	1	2	3	4	5	6	7	8	9	10	11	12	13	14	15	16
1	1															
2	0.77	1														
3	0.63	0.26	1													
4	0.45	0.41	0.69	1												
5	-0.29	-0.25	-0.29	-0.44	1											
6	-0.18	-0.18	-0.25	-0.40	0.95	1										
7	0.07	0.03	-0.15	-0.43	0.15	0.09	1									
8	0.18	0.11	0.30	0.07	-0.03	-0.06	0.62	1								
9	0.05	0.02	-0.20	-0.47	0.19	0.12	0.99	0.62	1							
10	0.02	-0.01	-0.17	-0.44	0.23	0.13	0.96	0.69	0.98	1						
11	-0.05	-0.13	-0.23	-0.48	0.21	0.12	0.93	0.59	0.96	0.97	1					
12	-0.25	-0.38	-0.22	-0.41	0.32	0.17	0.66	0.43	0.72	0.79	0.88	1				
13	0.16	0.14	-0.10	-0.36	0.09	0.07	0.96	0.75	0.95	0.94	0.89	0.59	1			
14	0.07	0.00	-0.14	-0.41	0.13	0.09	0.98	0.59	0.97	0.93	0.92	0.66	0.95	1		
15	0.23	0.16	0.42	0.24	-0.15	-0.14	0.44	0.97	0.42	0.50	0.38	0.22	0.61	0.42	1	
16	0.23	0.16	0.42	0.24	-0.15	-0.14	0.44	0.97	0.42	0.50	0.38	0.22	0.61	0.42	1	1

Key: 1: Length of long cell; 2: Width of long cell; 3: Length of short cell; 4: Width of short cell; 5: Length of trichome; 6: Width of trichome; 7: Length of stomatal cavity; 8: Width of stomatal cavity; 9:Length of stomatal cell; 10 Width of stomatal cell; 11: Length of long subsidiary cell; 12: Width of long subsidiary ce; 13: Length of short subsidiary cell; 14: Width of short subsidiary cell; 15: Length of silica bodies; 16: Width of silica bodies.

Table 5. Co-relation between the characters of family Euphorbiaceae based on leaf epidermal anatomical characters. adaxial surface

Characters	1	2	3	4	5	6	7	8	9	10	11	12	13	14	15	16
1	1															
2	0.48	1														
3	0.80	0.74	1													
4	0.43	0.60	0.58	1												
5	-0.14	-0.05	-0.05	-0.02	1											
6	-0.07	-0.06	-0.02	-0.04	0.96	1										
7	-0.54	-0.17	-0.52	-0.32	0.01	-0.11	1									
8	-0.09	-0.15	-0.18	-0.10	-0.07	-0.15	0.31	1								
9	-0.14	-0.06	-0.02	0.14	0.23	0.08	0.38	-0.06	1							
10	0.11	0.07	0.05	0.13	0.23	0.13	0.26	-0.26	0.86	1						
11	0.51	0.38	0.38	0.22	0.22	0.07	0.10	0.06	0.46	0.64	1					
12	0.22	0.04	-0.04	0.04	0.34	0.19	0.37	0.30	0.37	0.52	0.84	1				
13	0.63	0.56	0.70	0.58	-0.04	-0.01	-0.17	0.28	0.10	0.03	0.35	0.13	1			
14	0.64	0.66	0.58	0.75	-0.06	-0.07	-0.26	0.11	-0.15	-0.04	0.44	0.28	0.62	1		
15	-0.05	0.36	0.17	0.31	-0.16	-0.17	0.19	-0.10	0.31	0.27	0.14	0.00	0.37	0.09	1	
16	-0.05	0.38	0.18	0.32	-0.16	-0.16	0.22	-0.09	0.33	0.28	0.15	0.00	0.39	0.10	1.00	1

Key: 1: Length of long cell; 2: Width of long cell; 3: Length of short cell; 4: Width of short cell; 5: Length of trichome; 6: Width of trichome; 7: Length of stomatal cavity; 8: Width of stomatal cavity; 9: Length of stomatal cell; 10: Width of stomatal cell; 11: Length of long subsidiary cell; 12: Width of long subsidiary ce; 13: Length of short subsidiary cell; 14: Width of short subsidiary cell; 15: Length of silica bodies; 16: Width of silica bodies.

Cluster analysis based on leaf epidermal anatomical characters revealed that 16 species of Family Euphorbiaceae were divided into two main clusters, 1 and 2 at phylogenetic distance 100% (Figure 2). At 53%, genetic linkage group 1 is further divided into 1b and 1a. The species included in 1a are *Euphorbia prostrata* and *Euphorbia hirta*. In 1b, included species are *Excoecaria bicolor* and *Putranjiva roxburgii*. At 60%, genetic linkage group 2 is divided into 2b and 2a. At 38%, genetic linkage group 2a is divided into 2ai and 2aii.The specie included in 2ai was *Euphorbia splendus*. The species included in 2aii are *Ricinus communis, Tewia nudiflora* and *Acalypha wilkisiana*. At 38%, genetic linkage group 2b is further divided into 2bi and 2bii. The species included in 2bi were *Jatropha integrima* and *Euphorbia helioscopia*. The species were included in 2bii *Breynia nivosa, Codiaeum verigatum, Mallotus philippinensis, Phyllanthus emblica, Breynia nivosa, Euphorbia pulcherrima* and *Sapium sebiferium*. The species are related on the basis of anatomical features.

CONCLUSION

During the present work, the family Euphorbiaceae was found to be characterized by different types of cells, that is, irregular, tetragonal, hexagonal, pentagonal, polygonal and their margins smooth, dentate or wavy. The maximum length of long cell ranged from 196.6μm - 245.2μm which was found in *Euphorbia helioscopia* on adaxial side of leaf epidermis. Similarly, maximum length of short cell ranged from 128.3μm – 138.4μm, exhibited by the same species which is a unique leaf character and is sufficient to differentiate it from other species of Family Euphorbiaceae. Stomata were totally absent on abaxial epidermis of *Putranjiva roburgi, Euphorbia pulcherrima, Breynia nivosa, Mallotus philippinensis, Phyllanthus emblica, Euphorbia helioscopia, Excoecaria bicolor, Sapium sebiferium, Jatropha integrima* and *Codiaeum verigatum*. Presence of silica bodies are also considered as one of taxonomically important characters of Euphorbiaceae. This anatomical character was present on both abaxial and adaxial surfaces of *Euphorbia splendens* and only on adaxial surface of *Phyllanthus emblica*. In Family Euphorbiaceae silica bodies were regular and round shaped. Trichomes were altogether absent on both abaxial and adaxial leaf surfaces in *Ricinus communis, Euphorbia pulcherrima, Breynia nivosa, Mallotus philippinensis, Phyllanthus emblica, Acalypha wilkisiana, Trewia nudiflora, Euphorbia helioscopia, Sapium sebiferium, Jatropha integrima* and *Codiaeum verigatum*.

REFERENCES

[1] Jones, B.S. and Luchsinger. E.A. (1987). Plant Systematics. 2nd ed. New York: McGraw-Hill Book Company.

[2] Aworinde, D.O., Nwoye, D.U., Olagoke. A.A. (2009).Taxonomic significance of foliar epidermis in some members of Euphorbiaceae Family in Nigeria. *Journal of Botanical Research,* 4:17-28.

[3] Hidayat,T. and Kusdianti, O. (2009). Stomata Diversification and Phylogenenetic Analysis of 13 Species of Family *Euphorbiaceae sensu lato* Biodiversitas, *Journal of Botanical Research,* 10:19-22.

[4] Wu, Q.and. Culter. D.F (1985). Taxonomic, Evolution and Ecological Implication of the Leaf Anatomy of Rhizomatous Iris Species. *Botanical Journal of the Linnean Society,* 90: 253-303.

[5] Yukawa, T., Ando, T., Karasawa, K. and Hashimoto., K. (1992). Existence of Two Stomatal Shapes in the *Genus Dendrobium* (Orchidaceae) and Its' Systematic Significance. *American Journal of Botany*, 79: 946-952.

[6] Sudarsono, R. and Budiwati, D. (2005). Taksonomi Tumbuhan Tinggi Malang: Universitias Negeri Malang.

[7] Fahn, A. (1991). Anatomi Tumbuhan. Yogyakarta: Gadjah Mada University Press.

[8] Sharma, O.P. (1993). Plant Taxonomy. New Delhi: Tata MGraw-Hill Publishing Company Limited.

[9] Ogundipipie, J.A and Olatunji, J.A. (1991). Anatomy of Foliar Epidermis of the *Solanum nudum* Species Group of *Solanum* sect germinate (solanacece*). Acta Hort.* (ISHS), 745: 287-294.

[10] Olowokudejo, J. D. (1993). Comparative Epidermal Morphology of West African Species of *Jatropha L.(*Euphorbiaceae) *Botanical Journal of the Linnean Society,* 3: 139-15

[11] Das, S. (2002). The Stomata and Glandular Hairs in Some Indian Mangroves. Acta Botanical Croatica, 61:199-205.

[12] Illoh, H.C. (1995). Foliar Epidermis and Petiole Anatomy of Four Species of *Celosia* L. in Nigeria. *Feddes Repertorium,* 106: 15-23.

[13] Isawumi, M.A. (1996). Trichomes Types and their Taxanomic Value at the Sectional and Species Level in West African Vernonia (Compositae). Compositae Systematics. Vol.1. Hind, D.J.N and H. Beentje(Eds.). Royal Botanic Gardens, Kew, pp: 29-39.

[14] Croxdale, J. L.(2000). Stomatal patterning in Angiosperms. *American Journal of Botany,* 87: 1069-1080.

[15] Ayodele, A.E. and Olowokudejo, J.D. (2006). The Family of Polygonaceae in West Africa Taxonomic Significance of Leaf Epidermal Characters. South African Journal of Botany, 72:442-459.

[16] Webster, G. L., Aguilar, D. A. and Smith, B. A. (1983). Systematic Distribution of Foliar Trichomes Types in Croton (Euphorbiaceae). *Botanical Journal of the Linnean Scoiety,* 121:41-57.

[17] Payne, W. W. (1978). A Glossary of Plant Hair Terminology. *Brittonia,* 30: 239-255.

[18] Theoblald, W. L., Krahulik, J. L. and Rollins. R. C. (1979). Trichomes Description and Classification. In: Metcalfe C.R. and Chalk, L. (eds.), Anatomy of dicotyledons, 2nd edition: 40-53. Clarendon Press, Oxford.

[19] Kakkar, L. and Paliwal. G. S. (1972). Studies on the leaf anatomy of Euphorbia v. Epidermis. *Journal of Experimental Botany,* 40: 102-950.

[20] Inamdar, J. A. and Gangadhara. M. (1977). Studies on the Trichomes of Some Euphorbiaceae. *Feddes Repertorium*, 88: 103–111.

[21] Gupta, D. and Moulik, B.K. (1990). Studies on Morphological Nature of the Petaloid Bract of *E. hetrophlla* bulletine of the Botanical Society of Bengal., 44 (1-2): 97-101.

[22] Ogundipa, O.T. and Olatunji. O. A. (1992). The Leaf Anatomy of the Species of *Cochlospermum Kunth.* in West Africa. *Feddes Repertorium*, 102: 173-187.

[23] Geoffrey, A. L., Michael G. Simpson A. (1994). Phylogenetic Relationships of *Didymocistus* and *Hymenocardia (Euphorbiaceae). Journal of Experimental Botany,* 81: 239-244

[24] Baslar. S. (2000). An Investigation on *Chrozophora tinctoria (L.)* Rafin. Distributed in West Anatolia. *Turkey Journal of Botany,* 24: 103-112

[25] Mbagwu, E. (2007). Study of the Leaf Epidermal Characteristics of Four Species of Genus *Citrus (*Rutaceae). *Botanical. Journal of the Linnean society,* 5: 149-45.

[26] Bonnier,G. A and Sablon. L. D. (1905). Comparative Anatomy of the Vegetative Organs of Some Euphorbia Species (Euphorbiaceae) juss. *Romania Journal of Biology*, 51-52: 39-47.

[27] Perveen, N. S., Murthy, K. S. R. and Pullaiah. T. (2002). Leaf Epidermal Characters in Cortolaria spp. Papilionodeae from Eastern. *Ghart. Phytomorphology*. 50: 205-212.

[28] Rojos, C. B. (2007). Anatomy of Foliar Epidermis of the *Solanum nudum* Species Group of *solanum* sect germinate (solanacece*). Acta Hort.* (ISHS), 745: 287-294.

[29] Mbagwu E. (2008). Leaf Epidermal Features of *Canarium schuwenfruther* and *Daeryoda edus* (*Brusiaceae*). *Boanical Journal of the Linnean society,* 3: 139-15.

[30] Idu, M., Timothy, O., Onyibe, H. I. and Comor. A. O. (2009). Comparative Morphological and Anatomical Studies on the Leaf and Stem of Some Medicinal Plants *Jatropha curcas L. and Jatropha tanjorensis J.L.* Ellis and Saroja (Euphorbiaceae). *Ethnobotanical leaflet*, 13: 1232-39.

[31] Haiad B. de Sá,. Serpa-Ribeiro, A.C.C C., Barbosa,N., Pizzini, D. de O. Leal, L. de Senna-Valle and de Santiago-Fernandes. L.D.R. (2009). Leaf Structure of Species from Three Closely Related Genera from Tribe *Crotoneae* Dumort. (Euphorbiaceae *s.s.*, Malpighiales) Prop. Bot., 283 (3-4): 179-202.

[32] Ahmad, K., Khan., M. A. Shaheen M. A., and Nazir. A. (2010). Evaluate Taxonomic Diversity of Stomata in Dicot Flora from (N.W.F.P). *Pakistan Journal of Botany,* 12: 115-119.

[33] Yasmin S., Salman D. and Waqas. S. (2009). The Micromorphological Investigation of Foliar Anatomy of Genra *Acononon* and Family *Polyonaceae Botanical Journal of the Linnean. Society, 40: 33-21.*

[34] Carmen S. D., Cutler, W. D , Kustiandi. Y. G. (2009). The Morphology of Foliar Epidermis in Two Groups of *Solanum* Section *Germinata* (Solanaceae) Annals of Botany, 60: 422 - 430.

[35] Shaheen, W., Younas, A.R., Asad. B. (2009). Diversity of Foliar Trichome and Their Systematic Relevance in Genus *Hibiscus* (Malvaceae). *Journal of Botany,* 22: 23-43.

[36] Abdulrahman, A. A. and Oladele. F. A. (2010). Stomatal Complex Types and Epidermis Cells in Some *Jatropha* Species L. (Euphorbiaceae) *Nigerian Journal of Pure & Applied Science,* 23: 2160 - 2163.

[37] Senakuni, C. and Chantarnotha. P. (2010). A Morphological Survey of Foliar Trichomes of *Croton* L. (Euphorbiaceae) in Thailand. *Botany Journal of the Linnean Society,* 111: 299-423.

[38] Hardin, J. W. (1990). Variation Patterns and Recognition of Varieties of *Tilia americana Sys. Journal of Botany,* 15: 33-48.

[39] Oladele, D. A. (2000). Anatomical Studies of Family Euphorbiaceae. *Botany of Journal of the Linnean Society*., 6: 159-23.

[40] Metacalfe, C. R. (1960). Anatomy of the Dicotyledons. 2nd Edn., Clarendon Press, Oxford.

[41] Willmer, C.M. (1983). Stomata Londen: Longman.

[42] Copeland, J. (1902). Comparative Morpho-histological Studies of the Tribe *Eragrostideae* (Poaceae) from West Africa. *Journal of Economic and Taxonomic Botany,* 19: 41-58.

[43] Stober. 1917. Stomatal Diversification in Family Euphorbiaceae. *Botanical Journal of the Linnean Society,* 141: 399-436.

[44] Loftfield. (1921). Some Findings in Leaf Characters of Cassava Varieties. *Japan Agricultural Research Quarterly* 17: 67-72.

In: Leaf Epidermal Anatomical Characters ...
Editors: Z. Yousaf, A. Younas and A. Aftab

ISBN: 978-1-63483-190-1

Chapter 5

TAXONOMIC EVALUATION OF FAMILY MORACEAE ON THE BASIS OF LEAF EPIDERMAL ANATOMICAL CHARACTERS

H. A. Rehman [*] ***and A. Ramzan***
Department of Botany,
Lahore College for Women University,
Lahore, Pakistan

ABSTRACT

Present research work was carried to investigate the Leaf Epidermal Anatomy of economically important species of Family Moraceae. Information of taxonomy based on morphology of family Moraceae is contradictory. Therefore new taxonomic tools were investigated.

The study of leaf epidermal anatomy is considered as one of the effective techniques used for the generation of more authentic taxonomic information. Valuable characters identified for the family Moraceae were, types and shapes of cells, cell membrane variation, shape and types of trichomes, oil droplets, silica bodies and stomatal cell. The average length of organelles of leaf epidemis showed more accurate picture of the difference between the species than considering their full range of length and width. The comparative lengths of glandular and non-glandular trichome are considered as an important taxonomic marker for dicot flora.

Five different types of stomata were observed (anisocytic, anomocytic, dicytic, cyclocytic and staurocytic). Staurocytic stomata were most dominant in family Moraceae. Stomata were found on both abaxial and adaxial sides of *Ficus religiosa* leaves. But adaxial side bear stomata in cluster form.

Round and irregular shape of silica bodies were also found. Irregular silica bodies were found in *Ficus macrophylla* and round shape silica bodies were seen in *Ficus racemosa, Ficus nerifolia,* and *Ficus triangularis.* Leaf epidermal anatomy was found to be an important tool for the identification and characterization of family Moraceae and it

---

[*] sweet_pinklibra@hotmail.com

can be effectively utilized for solving the persistent taxonomic problems of family Moraceae.

INTRODUCTION

Most of the species of Family Moraceae showed various epidermal structures, like different types of trichomes, calcium deposition structures and hydothodes are distributed on leaves. The trichomes are generally categorized into glandular, non-glandular and stinging trichome. The stinging trichomes have been studied by many investigators in different angiospermic families [1, 2, 3, 4, 5 and 6]. The foliar epidermis is one of the most noteworthy taxonomic characters from the biosystematics point of view and the taxonomic studies of number of families are made on the basis of leaf epidermis [7, 8, 9 and 10].As the leaf epidermal characters of family Moraceae were not investigated for the taxonomic purposes. Therefore the aim of present study is to search out the solution of existing taxonomic problems of some species of family Moraceaewhich overlap in most of their morphological characters and to clarify the relationship of the critical taxa by utilization of leaf epidermal characters.

HISTORICAL BACKGROUND

Gangadhara and Inamdar [11] worked on the taxonomic significance of trichomes and stomata in the *Urticales*. They studied structure and ontogeny of stomata and trichomes of 23 species and 3 varieties of the *Urticales*. They found different types of stomata i.e., anomocytic, more rarely paracytic; anisocytic and sometimes helicocytic and transitorial types are found in*Urticaceae* and *Dorstenia*, rarely in *Artocarpus*. The ontogeny of anomocytic and actinocytic stomata is perigenous, of paracytic either mesogenous or perigenous, of anisocytic either mesogenous or mesoperigenous, and of helicocytic and transitional types mesogenous. Among trichomes they observed non-glandular unicellular (wide spread), bicellular or uniseriate filiform (*Cannabis*); glandular capitate with uni or bicellular (*Moraceae, Ulmaceae, Cannabaceae*), uniseriate filiform (*Ulmaceae*) or multiseriate stalk (*Cannabis*); sunken glands (*Artocarpus*); uniseriate glandular with uniseriate stalk (*Celtis*), and stinging emergences (*Urticaceae*). They concluded that the *Urticales* represented a natural order with four families *Ulmaceae, Moraceae, Urticaceae* and *Cannabaceae* which are distinct but interrelated with each other.

Lackey [12] worked on leaflet anatomy and taxonomic relationship of *Phaseoleae* (Leguminous: sub family papilionoideae) . In department of Botany, Smithsonian Institution, Washington. Author studied the leaflet anatomy of the tribe phaseolese by light microscopy of cleaned material and by scanning electron microscopy of leaflet surfaces. Anatomical results compared with recently revised of sub tribal classification of the tribe. Taxonomically important characters that were found Bulbous based and vesicular glands are restricted to the sub tribe *canjaninae* hooked hairs are present in three generic groups. Glandular hairs were particularly present, branched hairs were also present.

Ogundipe [13] worked on leaf epidermal anatomy of genus Datura L.of family *Solanaceae*. Five West African species of *Datura* with narcotic properties were studied. The most useful anatomical characters for identification were palisade ratio, stomatal density and types of crystals. *D.candida, D.suaveolens, D.dinnoxia, D.metal and D.stramonium* were studied and differentiated on the basis of these anatomical characters.

YiFu et al., [14] worked on comparative study on the stinging trichomes and some related epidermal structures in the leaves of *Dendrocnide meyeniana*, *Girardinia diversifolia*, and *Urtica thunbergiana.* They studied the morphology and distribution of the stinging trichomes and the related epidermal structures in the mature leaves. They observed stinging, glandular and non-glandular trichomes on the adaxial and abaxial sides of leaf.. Among these studied species, the stinging trichome of *Dendrocnide meyeniana was* the shortest, but the density on the adaxial surface was the highest. In *Girardinia diversifolia*, the stinging trichomes were the longest, but the density is the lowest. The silicon and calcium were the major elements of stinging cells in all studied plants. In *Dendrocnide meyeniana*, the average silicon/calcium ratio was the highest, while the lowest silicon/calcium ratio was found in the stinging cells located on the veins of abaxial surface in *Urtica thunbergiana*. In *Girardinia diversifolia*, the amount and surrounding area of pedestal cells were the highest while in *Dendrocnide meyeniana*, however, were the lowest.

Kanademir [15] studied the morphological, anatomical and karyological characters of endemic *Salvia hypergia* Fich and Mey (Lamiaceae). They investigated the endemic plant. From this work they observed that these species have perennial root system, rectangular herbaceous stem and simple leaves with glandular and non-glandular hairs on both sides of leaves. Anatomically, the internal morphological properties of root, stem, leaf, and petiole were observed.

Liscousky and Cosa [16] worked on comparative anatomy of leaf and stem in species of *cestreae* G. Don (Solanaceae) from Argentina. Leaf and stem anatomy of thirteen native and cultivated species of *lestrum* L. and *sessa vestioides* were studied. The best distinguished leaf characters among species were secondary vein Architecture, marginal ultimate venation and stomata in hairs. Cutical characteristics, and presence of fibres adjacent to external and internal phloem in the middle vein, scleried in the middle vein, scleried presence and abundance in mesophyll and middle vein.

Carpenter et al., [17] worked out leaf cuticular morphology link planteaceae and proteaceae. They collected the specimen from the Foliar museum of natural history, university of Floride. For this purpose, they used the light and scanning electron microscopy for the investigation of leaf cuticular morphology of extent species of plantantaceae, they observed the anatomical characters types of stomata, epidermal cells on abaxial and adaxial surface. They studied the morphology linking plantanaceae and proteaceae. Although it is good piece of work but they just emphasized on the particular characters among these species and ignored all other anatomical characters like shape and size of epidermal cells and arrangement of cells, variation in types of stomata cells and silica bodies etc. This could be proven very important taxonomic tool for identification of species.

Ozdemir and Atan, [18] investigated morphology and anatomy of some endemic *alkana* species. Two *alkana* species were examined for morphological and anatomical differences. Aim of this study was to provide information of morphological and anatomical properties of *alkana friedinii, Rech, Fie* and *A.cardifalia L.Koch* .It has been observed that *A.froedinii* have some different characteristics from *A.cardiofolia* such as numerous setiform and glandular hairs.

Klimko et al., [19] worked on the morphological variability of the leaf epidermis in selected taxa of the genus *Ficus* L. (*Moraceae*) and its taxonomic implication. They collected the material from the botanical garden, Adam Mickiewicz University, Pozan. They selected the leaf epidermal of 36 taxa (Species, subspecies, varieties and cultivars) of the genus *Ficus* (subgenera *Ficus, Synoecia, Sycamores, pharmacosyces* and *Urostigma)* that was observed under light microscope and simple electron microscope. They investigated a number of micro morphological characters at taxonomic level. They optimized the plant growth under cultivation condition to minimize the modifying effect of environmental condition on the epidermal leaf traits. They emphasised on leaf epidermal characters and ignored other anatomical characters. If they observed these anatomical characters that could be helpful in producing better result.

Perveen et al., [20] worked on stomatal types of some dicots within flora of Karachi, Pakistan. For this purpose they selected 69 dicot species distributed in 64 genera and 24 families. They examined the stomatal morphology by using light and scanning electron microscopy. Among these species, six types of stomata were recognized. During study they ignored the other anatomical characters (shape of the cells, size of the cells, types of trichome etc) at taxonomical level. If they observed the rest of characters then their work could be provided better results.

Munsif et al., [21] worked on leaf epidermal anatomy of Genera *Lantana verbena* and *vitex* family verbenaceae from Pakistan. They studied abaxial and adaxial leaf epidermal anatomy of eight species of verbenaceae. They used leaf epidermal anatomy as a tool to identify Genera *Lantana,* and *Verbana* and *Vitex* of *verbenaceae.* Anatomical characters such as stomata types and size, epidermal cells shape and size and trichome shapes were used as anatomic tools. *Lantana camara* , *L. Indica, Verbena bonariensis, Ver. Officinalis, Ver.bipinnatifida, Ver. Hybrid, Vitex negundo* and *V.trifolia* were differentiated on the basis of trichome shapes.

Rojas, [22] studied anatomy of the foliar epidermis of the *Solanum nudum* species, group of Solanum sect.germinata (Solanaceae). The leaf epidermis surfaces of these species were investigated in order to discover and evaluate the taxonomically useful characteristics. Study was conducted at venezuele. Variation in epidermal cells type, dimension and location of stomata presence or absence, stomata location and density of trichome was evaluate and illustrated. Results showed difference in some anatomical characteristics between *S.nudum* species group i.e., *S.Aphyodendrom, S.campaniforme, S.nudum,* and *S. Tepuiense* and 34 species of *Solanium sect germinate.*

Jiang et al., [23] worked on the leaf epidermis of two species of *flos lonicera.* Anatomical study was done by scanning electron microscope. Leaf epidermal surface of these species were studied to examine stomata, trichome a dermal cell. Leaf epidermal anatomical characters of the two species were studied, and successfully differentiated on the basis of stomatal location, epidermal cells and trichome density and shape.

Yousaf et al., [24] studied leaf epidermal anatomy of selected *Allium* species anf the study was conducted at department of botany in University of Arid Agriculture Rawalpandi, Pakistan. Leaf epidermal anatomy of a selected *Allium* species showed variation in size and shape of stomatal cells, stomatal cavity, micro and macro hairs, trichomes, silica bodies and long cells. Leaf epidermal anatomy is helpful for the resolution of taxonomic confusion.

Allium consanguineum had a great diver's leaf epidermal anatomy. This species had a largest stomatal cells (6-14μm) and silica bodies (6-14μm). Presence of micro hairs is an important distinguished character for *Allium carolianum*, the length of the micro hairs varies from 150-200μm, dumb bell shaped of silica bodies were observed in six different species viz., *A. dolichastylum, A. borszczewii, A. micranthum, A. consanguiem, A. stocksian, and A.stoliczki.* trichomes were present in the *A. lamondae, A. miserbile, A .congicallum, A. gilli* and *A. dalichostylum.*

Yasmin et al., [25] worked out the micro morphological investigation of foliar anatomy of the genera *Acononon* and *Bistorta* family *Polygonaceae.* They collected the material from the herbarium of Quaid-e-Azam University Islamabad. They studied anatomical features such as pattern of epidermal stomatal distribution on the abaxial and adaxial surface of leaves and trichomes types under the light microscope. Among the species they found six types of stomata (paracytic, cyclocytic, staurocytic, anisocytic, anomocytic, and dicytic) and variation among glandular and non-glandular trichome. They just emphasized on the anatomical features of these species and ignored the arrangement of epidermal cell, stomal variety and variation in trichome which could provide very important taxonomic tool for identification of species.

Azmati et al., [26] worked on a viable alternative mechanism in adapting the plants to heavy metal environment in Department of Chemistry, Department of Botany and Department of Biochemistry Jinnah University for Women, Nazimabad, and Karachi, Pakistan. This study was undertaken to test the hypothesis that variations in leaf anatomy and morphology reflect their adaptability to the environmental stress. A self defense mechanism system related with the trichomes on the surface of leaves for the detoxification of Pb was observed in *Phaseolus mungo* and *Lens culinaris* through light microscopy in leaves. The presence of trichomes and increased number of stomata in the adaxial (upper) leaf surface of both species seems to constitute an important morphological mechanism for survival that allows this species to maintain good photosystem II efficiency during the stress. Foliar morphological variability in *Phaseolus mungo* and *Lens culinaris* may be considered an adaptive advantage that enables leaves to develop and function in habitats marked by strong variations of Pb toxicity with solar radiation, air, temperature and humidity. These hairs may constitute a shield against Pb pollution and act as a physical barrier for the protection of plants and can act as a biomonitor of environmental contamination and biogeochemical indicator of Pb. Enhancement in the hairs on leaves at high dose of Pb may be related with reducing evaporation of water from the surface of leaves in stress condition. The protection provided by the trichomes could afford advantages under stress conditions, especially during leaf development. The importance of increased number of stomata in relation with the absorption of CO_2 with increase in enzymatic activity, creatine, glucose and reducing sugars for both species under the metal stress were examined.

Ahmed et al., [27] worked on taxonomic diversity of stomata in dicot flora of a district tank (N.W.F.P) in Pakistan. They observed 36 dicot species distributed in 34 genera and 20 families were investigated for stomatal diversity. Seven types of stomata were found in which amphianisocytic was the dominant, one found in 12 species while staurocytic and diacytic were found in 7 and 6 species respectively. In 7 species two and in one specie three different types of stomata were found. During study they ignored the other anatomical characters (shape of the cells, size of the cells, types of trichomes etc) at taxonomical level. If they observed the rest of characters then their work could be provided better results.

Shaheen et al., [28] worked on diversity of foliar trichome and their systematic relevance in the genus *Hibiscus* (Malvaceae). They collected the material from the different localities of Pakistan and dried leaves from the herbarium of Quaid-e-Azam University Islamabad. They examined qualitative and quantitative morphological characters. They distributed the systematic relevance of both glandular and non-glandular foliar trichome in species of genus *Hibiscus*. Glandular capitate trichome was the most dominant on leaf surface. Among of these species they just emphasized only on trichome anatomical marker.

Kadiri and Ajayi [29] worked on Phyto-anatomical characteristics of the West African. *Musanga cercropioides* M. Smithii R. Br. (Moraceae) Dept. of Botany & Microbiology Dept. of Pharmacognosy, College of Medicine, University of Lagos, Nigeria to investigated the two complimentary studies of anatomy and phytochemistry used in pharmacognostic drug research on *Musanga cercropiodes* by using light microscopy and phytochemical methods. Diagnostic anatomical features of the plant include bulbous trichome bases, and flaky strand-like waxes. They also observed the unique features of the plant i.e., hypostomatic leaf, anomocytic stomata, abaxially restricted simple unicellular and non glandular trichomes and crescentiform and hairy petiole. The bioactive compounds which were present in both leaf and bark were alkaloids, flavonoids, tannins, free and bound anthraquinone, saponin and cardiac glycosides but anthocyanosides and cyanogenic glycosides were absent. *Musanga cercropioides* is a popular plant in folkloric medication in West Africa.

Table 1. List of species of Family Moraceae used for leaf epidermal anatomical study

S.NO	Genus Names	Species Names
1	*Artocarpus*	*Artocarpus integrifolia* Spreng
2		*Artocarpus lakoocha* Roxb.
3	*Benjamina*	*Benjamina comosa* L.
4	*Broussonatia*	*Broussonatia papyrifera* (L.)Vent
5	*Ficus*	*Ficus bengalensis* L.
6		*Ficus benjamina* L.
7		*Ficus elastic* Roxb
8		*Ficus glomerata* Roxb
9		*Ficus infactoria* Roxb
10		*Ficus lyrata* Warb.
11		*Ficus macrophylla* Roxb
12		*Ficus nerifolia* J.E.Sm.
13		*Ficus pumila* L.
14		*Ficus racemosa* L.
15		*Ficus religiosa* L.
16		*Ficus retusa* L.
17		*Ficus triangularis* Roxb.
18		*Ficus virens* Dryad
19	*Morus*	*Morus alba* L.
20		*Morus nigra* L.

Rahaman et al., [30] worked out the stomatal complex types, stomatal density and stomatal index in some species of *Dioscorea.* They collected the fresh material from the Grain Research Unit (GRU) of the International Institute of Tropical Agriculture; Ibadan. They observed the three stomatal types like paracytic, anisocytic and tetracytic stomata. However they just emphasized on the stamatal types and they ignored other important anatomical characters like cell arrangement, types of the cells, various types of trichomes and silica bodies etc. That could be proved for identification of species at taxonomic level.

Shaheed and Illoh, [31] worked out taxonomic study of some species in *Cassinae* (Leguminaceae) using leaf epidermal characters. They selected the ten species from the genus *Cassia* L. They collected the specimen from the different localities within the South Western part of Nigeria. The aim was to establish some useful diagnostic features for all the species. They worked on the trichomes as well as the length of epidermal cells and types of cell wall. Although it is good piece of work yet they emphasized on the some specific features and they ignored the arrangement of the epidermal cells, subsidiary cell and silica bodies etc which could be proved better for identification of species. Ahmed et al., [32] worked out taxonomic diversity in epidermal cell of some sub-tropical plant species. For this purpose they selected 40 species belonging to 38 genera of 22 families.

The shape and type of large epidermal cells among these varies from straight and polygonal. They concluded that large variation in epidermal cells configuration at different taxonomic level. But they ignored other anatomical characters that could provide an intergenic and intragenic variation in many species. There is a need of more comprehensive work on the shape and type of leaf epidermal cells for their further details of taxonomic characters.

RESULTS

20 species belonging to five genus of family Moraceae were described with respect to their leaf epidermal characteristic as follows:

1. Artocarpusintegrifolia Spreng

a. Adaxial Side

Trichomes, stomata and Silica bodies were absent. Oil droplets were present. Numbers of oil droplets per unit area were 2. Length of oil droplets 19.1 µm – 24.4 µm (21.75 µm). Width of oil droplets 13 µm - 13.6 µm (Av: 13.3 µm). Polygonal cells haddouble layer and wavy margins. Length of long cells 89.5 µm – 111.4 µm (Av: 100.45 µm). Width of long cells 35.5 µm – 36.2 µm (Av: 35.85 µm). Length of short cells 21.5 µm – 25 µm (Av: 23.25 µm). Width of short cells 18 µm – 21.9 µm (Av: 19.95 µm). (Figure 1)

b. Abaxial Side

Trichomes, oil droplets and silica bodies were absent.Polygonal cells had double layer and wavy margins. Length of long cells 82.2 µm – 100.5 µm (Av: 91.35 µm). Width of long cells 36.9 µm – 52.4 µm (Av: 44.65 µm). Length of short cells 25.9 µm – 53 µm (Av: 39.45

μm). Width of short cells 19 μm – 39 μm (Av: 29 μm). Stomata were present. Type of stomata was anomocytic. Numbers of stomata per unit area were 11-12. Length of stoamtal cavity 46.6 μm – 54.1 μm (Av: 50.35 μm). Width of stamatal cavity 18.4 μm – 20.1 μm (Av: 19.25 μm). Length of guard cells 71.9 μm – 83.1 μm (Av: 77.5 μm). Width of guard cells 50.2 μm – 55 μm (Av: 52 μm). Subsidiary cells were present. Numbers of subsidiary cellswere 5 – 6. Length of long cells 88.1 μm – 92.7 μm (Av: 90.4 μm). Width of long cells 27.8 μm – 44.4 μm (Av: 36.1 μm). Length of short cells 40.3 μm – 46.9 μm (Av: 43.6 μm). Width of short cells 37.2 μm –69.1 μm (Av: 53.15 μm). (Figure 1)

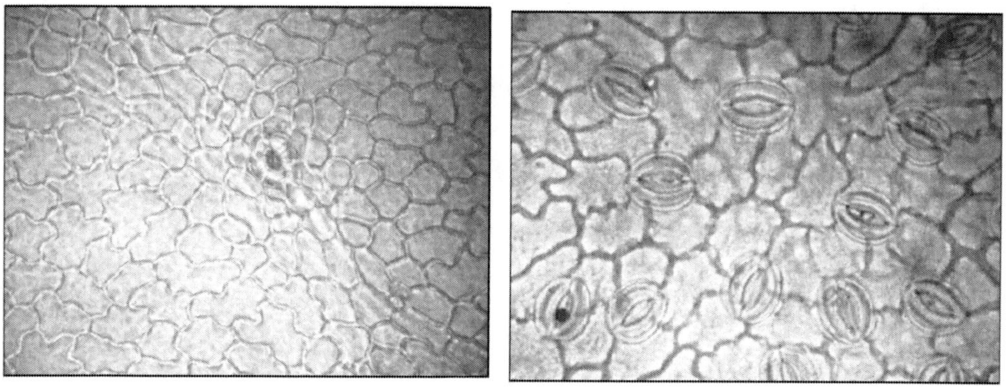

Figure 1. Leaf Epidermal Anatomy of Artocarpus integrifolia. Adaxial side (Right) Abaxial side (Left).

2. Artocarpuslakoocha(Leaf), Roxb

a. Adaxial Side

Trichomes, oil droplets, stomata and silica bodies were absent.Polygonal cells had double layer and smooth margins. Length of long cells 108.2 μm -132.2 μm (Av: 120.2 μm). Width of long cells 36.1 μm – 46.0 μm (Av: 41.05 μm). Length of short cells 69.6 μm – 82.5 μm (Av: 76.05 μm). Width of short cells 36.0 μm - 45 μm (Av: 40.5 μm). (Figure 2)

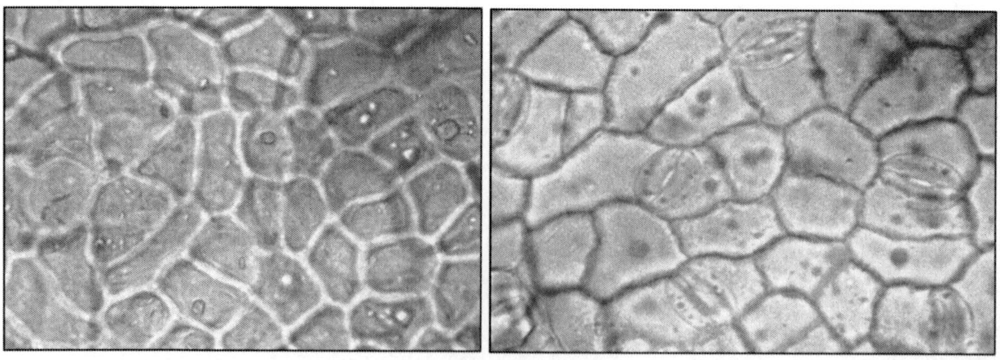

Figure 2. Leaf Epidermal Anatomy of Artocarpus lakoocha. Adaxial side (Right) Abaxial side (Left).

b. Abaxial Side

Trichomes, oil droplets and silica bodies were absent.Polygonal cells had double layer and smooth margins. Length of long cells 133.2 μm – 154.4 μm (Av: 143.8 μm).Width of long cells 80.6 μm – 82.1 μm (Av: 81.35 μm). Length of short cells 58.7 μm – 71.6 μm (Av: 65.15 μm). Width of short cells 38.1 μm – 73.2 μm (Av: 55.65 μm). Stomata were present. Type of stomata was diacytic. Numbers of stomata per unit area were 7. Length of stomatal cavity 31.1 μm – 51.1 μm (Av: 41.1 μm). Width of stomatal cavity 7.8 μm – 13.9 μm (Av: 10.85 μm). Length of guard cells 74.6 μm – 87.3 μm (Av: 80.95 μm).Width of guard cells 51.1 μm – 59.2 μm (Av: 55.35 μm). Subsidiary cells were present. Length of long subsidiary cells 55.2 μm – 111.6 μm (Av: 83.4 μm). Width of long subsidiary cells 79.6 μm – 87.5 μm (Av: 83.55 μm). Length of short cells 22.5 μm – 36 μm (Av: 29.25 μm). Width of short cells 77.8 μm – 94.6 μm (Av: 86.2 μm). (Figure 2)

3. Benjamina Comosa Linn

a. Adaxial Side

Trichomes, oil droplets and silica bodies were absent.Hexagonal cellshad double layer andsmoothmargins. Length of long cells 93 μm – 102.2 μm (Av: 97.6 μm). Width of long cells 39.1 μm – 39.8 μm (Av: 39.45μm). Length of short cells 52.8 μm – 62 μm (Av: 57.40 μm). Width of short cells 31.1 μm –33 μm (Av: 32.05 μm).

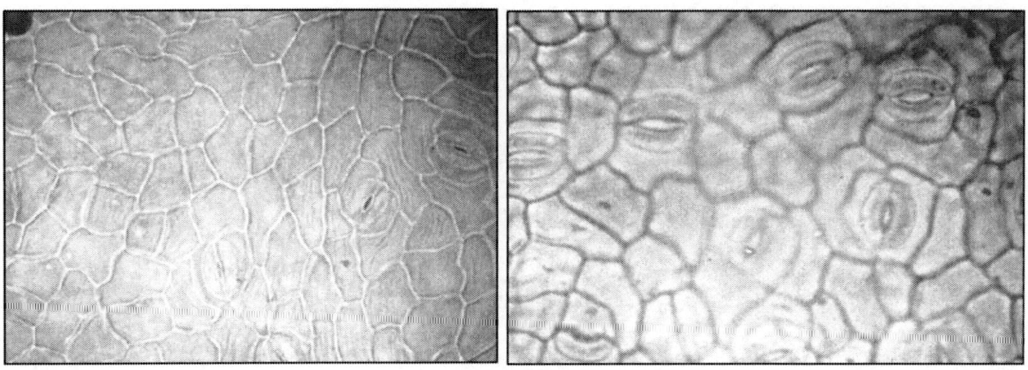

Figure 3. Leaf Epidermal Anatomy of Benjamina comosa. Adaxial side (Right) Abaxial side (Left).

Stomata were present. Type of stomata was diacytic. Numbers of stomata per unit area were 4-5. Length of stoamtal cavity 33.9 μm – 20.1 μm (Av: 27 μm). Width of stamatal cavity 6.7 μm – 10 μm (Av: 8.35 μm). Length of guard cells 52.5 μm – 463.6 μm (Av: 58.05 μm). Width of guard cells 35.5 μm – 49.7 μm (Av: 42.60 μm). Subsidiary cells were present. Number of subsidiary cells were 2. Length of long cells 66.8 μm – 71.7 μm (Av: 69.25 μm). Width of long cells 28 μm – 41.2 μm (Av: 34.60 μm). Length of short cells 37 μm – 48 μm (Av: 42.50 μm). Width of short cells 16.1 μm – 22.7 μm (Av: 19.4 μm). (Figure 3)

b. Abaxial Side

Trichomes, oil droplets and silica bodies were absent. Hexagonal cells had double layer and smooth margins. Length of long cells 83.2 μm – 105.5 μm (Av: 94.35 μm). Width of long cells 31.1 μm – 35.8 μm (Av: 33.45 μm). Length of short cells 21.9 μm – 31.9 μm (Av: 26.9 μm). Width of short cells 21.3 μm – 29 μm (Av: 25.15 μm). Stomata were present. Type of stomata was diacytic. Numbers of stomata per unit area were 13-14. Length of stoamtal cavity 28.6 μm – 37.6 μm (Av: 33.1 μm). Width of stamatal cavity 6.9 μm – 8.6 μm (Av: 7.65 μm). Length of guard cells 57.3 μm – 61.6 μm (Av: 59.45 μm). Width of guard cells 38.5 μm – 39.8 μm (Av: 39.15 μm). Subsidiary cells were present. Numbers of subsidiary cells were 2. Length of long cells 71.0 μm – 73.6 μm (Av: 72.3 μm). Width of long cells 17.2 μm – 20.8 μm (Av: 19 μm). Length of short cells 50.6 μm – 52.3 μm (Av: 51.45 μm). Width of short cells 13.5 μm – 17.2 μm (Av: 15.35 μm). (Figure 3)

4. Broussonatia Papyrifera (L.) Vent

a. Adaxial Side

Silica bodies and stomata were absent. Non-glandular trichomes were present. Number of trichomes per unit area was 1. Length of trichomes 118.8 μm – 122.2 μm (Av: 120.5 μm). Width of trichomes 28 μm – 31.2 μm (Av: 29.6 μm). Oil droplets were present. Numbers of oil droplets per unit area were 2. Length of oil droplets 33.0 μm – 41.6 μm (Av: 37.3 μm). Width of oil droplets 32.8 μm – 39.0 μm (Av: 35.9 μm). Polygonal cellshad double layer and smooth margins. Length of long cells 51.0 μm – 59.2 μm (Av: 55.1 μm). Width of long cells 25.5 μm –26.2 μm (Av: 25.85 μm). Length of short cells 27.9 μm – 28.6 μm (Av: 28.25 μm). Width of short cells 20.9 μm – 28.4 μm (Av: 24.65μm). (Figure 4)

Figure 4. Leaf Epidermal Anatomy of Broussonatia papyrifera. Adaxial side (Right) Abaxial side (Left).

b. Abaxial Side

Oil droplets and silica bodies were absent. Glandular trichomes were present. Numbers of trichomes per unit area were 3. Length of trichomes 55.1 μm – 64.5μm (Av: 59.8 μm). Width of trichomes 37.0μm – 41.3 μm (Av: 39.15 μm). Polygonal cells haddouble layer and smooth margins. Length of long cells 66.9 μm – 67.8 μm (Av: 67.35 μm). Width of long cells 29.1

μm – 42.7 μm (Av: 35.9 μm). Length of short cells 10.2 μm – 19.0.μm (Av: 14.6 μm). Width of short cells 13.0 μm – 23.2 μm (Av: 18.1 μm). Stomata were present. Numbers of stomata per unit area were 15. Type of stomata wasstaurocytic. Length of stoamtal cavity 20.0 μm – 30.7 μm (Av: 25.3μm). Width of stamatal cavity 10.3 μm – 13.5 μm (Av: 11.75 μm). Length of guard cells 41.6 μm – 43.3 μm (Av: 42.45 μm). Width of guard cells 28.3 μm – 29.7 μm (Av: 29 μm). Subsidiary cells were present. Numbers of subsidiary cellswere 4-5. Length of subsidiary cells16.6 μm – 70.3 μm (Av: 65.95 μm). Width of subsidiary cells 21.2 μm – 29.5 μm (Av: 25.45 μm). (Figure 4)

5. Ficus Bengalensis L.

a. Adaxial Side

Stomata and silica bodies were absent. Non-glandular trichomes were present. It is originated from the central cell and was uniserate with acute ended. Length of cells 79.8 μm – 160.6 μm (Av: 120.2 μm). Width of cells 21.6 μm – 26.9 μm (Av: 24.25 μm). Oil droplets were present. Numbers of oil droplets per unit area was 1. Length of oil droplets31 μm- 32 μm (Av: 31.5 μm). Width of oil droplets 27 μm – 29.4 μm (Av: 28.2 μm). Polygonal cellshad double layerand rough margins. Length of long cells 43.1 μm – 53.1 μm (Av: 48.1 μm). Width of long cells 15 μm – 26 μm (Av: 20.5 μm). Length of short cells 22.8 μm – 23.4 μm (Av: 23.1 μm). Width of short cells18.9 μm - 21 μm (Av: 19.95 μm). (Figure 5)

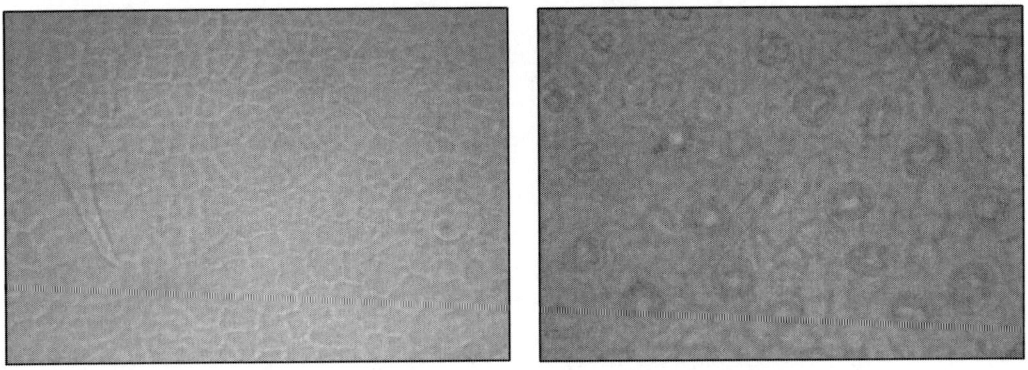

Figure 5. Leaf Epidermal Anatomy of Ficus bengalensis. Adaxial side (Right) Abaxial side (Left).

b. Abaxial Side

Trichome,oil droplets and silica bodies were absent. Polygonal cells had double layer andrough margins. Length of long cells 47.1 μm – 47.3 μm (Av: 47.2 μm). Width of long cells 15 μm – 26 μm (Av: 20.5 μm). Length of short cells 27.2 μm – 27.9 μm (Av: 27.5 μm). Width of short cells 17.1 μm - 19 μm (Av: 18.05 μm). Stomata were present. Type of stomata was cyclocytic. Numbers of stomata per unit area were 16-17. Length of stomatal cavity 58.0 μm – 88.1 μm (Av: 73.05 μm). Width of stomatal cavity 46.7 μm – 70.6 μm (Av: 58.65 μm). Length of guard cells 35.4 μm 50.8 μm (Av: 43.1 μm). Width of guard cells 33.6 μm – 42.7 μm (Av: 38.15 μm).

Subsidiary cells were present. Numbers of subsidiary cell were 4. Length of long cells43.1 μm – 47.4 μm (Av: 45.25 μm). Width of long cells12.0 μm – 15.8 μm (Av: 13.6 μm). Length of short cells 29.2 μm – 44.2 μm (Av: 36.7 μm). Width of short cells13.6 μm – 14.4 μm (Av: 14 μm).

6. Ficus Benjamina L.

a. Adaxial Side

Trichomes, oil droplets stomata and silica bodies were absent. Polygonalcells haddouble layer and rough margins. Length of long cells 54.7 μm -74.4 μm (Av: 64.55 μm). Width of long cells 16.3 μm – 23.9 μm (Av: 20.1 μm).Length of short cells 22.5 μm – 27.5 μm (Av: 25.0 μm).Width of short cells23.1 μm - 24.2 μm (Av: 23.65 μm). (Figure 6)

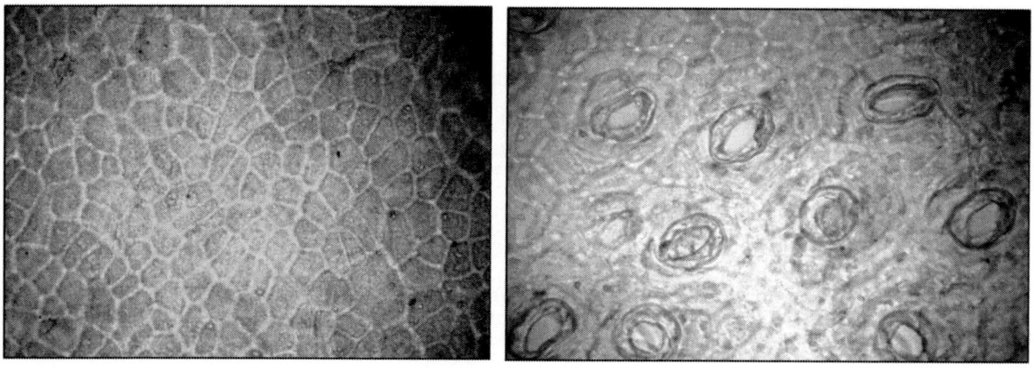

Figure 6. Leaf Epidermal Anatomy of Ficus elastic. Adaxial side (Right) Abaxial side (Left).

b. Abaxial Side

Oil droplets and silica bodies were absent. Glandularrtrichomes were present. Shape of trichome was multiseriate. Numbers of cells were7 – 11. Length of central cells 40.8 μm – 65.8 μm (Av: 53.3 μm). Width of central cells 42.2 μm – 65.0 μm (Av: 53.6 μm). Length of long cells 64.6 μm- 78.5 μm (Av: 71.55 μm). Width of long cells 28.9 μm – 34.9 μm (Av: 31.9 μm). Length of short cells 34.1 μm – 38.3 μm (Av: 36.2 μm). Width of short cells29.7 μm – 41.5 μm (Av: 35.6 μm).Polygonal cells had double layer and rough margins. Length of long cells 81.0 μm – 106.2 μm (Av: 93.6 μm).Width of long cells 24.1 μm – 35.0 μm (Av: 29.55 μm). Length of short cells 29.4 μm - 39.8 μm (Av: 34.6 μm). Width of short cells 27.5 μm – 28.2 μm (Av: 27.85 μm).Stomata were present.Type of stomata was cyclocytic. Numbers of stomata per unit area were 2-3. Length of stomatal cavity 43.9 μm – 52.4 μm (Av: 48.15 μm). Width of stomatal cavity 14.2 μm – 34.4 μm (Av: 24.3 μm). Length of guard cells 105.8 μm – 118.0 μm (Av: 111.9 μm).Width of guard cells 96.6 μm – 102.0 μm (Av: 99.3 μm). Length of subsidiary cells 56 μm – 67.2 μm (Av: 61.6 μm). Width of subsidiary cells 45.3 μm – 46.6 μm (Av: 45.95 μm). Length of short cells 45.8 μm – 52.6 μm (Av: 49.2 μm). Width of short cells 23.4 μm – 24 μm (Av: 23.7 μm).

7. Ficus Elastic, Roxb

a. Adaxial Side

Trichomes, oil dropletsand stomata were absent. Hexagonal cells haddouble layer and rough margins. Length of long cells 58.2 µm - 68.8 µm (Av: 63.5 µm). Width of long cells 26.9 µm – 29.2 µm (Av: 28.05 µm). Length of short cells 26.2 µm – 31.1 µm (Av: 28.65 µm). Width of short cells 21.2 µm - 22 µm (Av: 43.2 µm). Silica bodies were present. Numbers of silica bodies per unit area were 4 – 5. Silica bodies were round. Length of silica bodies 21.2 µm – 30.5 µm (Av: 25.85 µm). Width of silica bodies 18.1 µm – 26.6 µm (Av: 22.35 µm). (Figure 6)

b. Abaxial Side

Trichomes, oil droplets andsilica bodies were absent.Hexagonal cells haddouble layer and rough margins. Length of long cells 59.9 µm – 68.9 µm (Av: 64.2 µm). Width of long cells 24.7 µm – 28.8 µm (Av: 26.75 µm). Length of short cells 38.1 µm – 44.2 µm (Av: 41.15 µm). Width of short cells 21.8µm – 29.1 µm (Av: 25.45 µm). Stomata were present. Type of stomata was cyclocytic. Numbers of stomata per unit area were 8-9. Length of stoamtal cavity 31.4 µm – 51.9 µm (Av: 41.65 µm). Width of stamatal cavity 20.9 µm – 21.1 µm (Av: 21 µm). Length of guard cell 62.3 µm – 73.2 µm (Av: 67.75 µm). Width of guard cells 53.8 µm – 57 µm (Av: 55.4 µm). Subsidiary cells were present. Numbers of subsidiary cells were 4 – 5. Length of long cells 56.9 µm – 73.8 µm (Av: 65.35 µm). Width of long cells 22.8 µm - 41 µm (Av: 31.9 µm). Length of short cells 32.8 µm – 44 µm (Av: 38.4 µm). Width of short cells 22.1 µm – 26.2 µm (Av: 24.15 µm). (Figure 6)

8. Ficus glomerata

a. Adaxial Side

Trichomes, stomata and silica bodies were absent. Oil droplets were present. Numbers of oil droplets per unit area were 2. Length of oil droplets 19 µm –24.0 µm (Av: 21.5 µm). Width of oil droplets 18.1 µm – 22 µm (Av: 20.05 µm). Hexagonal cells had double layer and wavy margined. Length of long cells 83.4 µm – 97.3 µm (Av: 90.35 µm). Width of long cells 57.8 µm – 61.6 µm (Av: 59.7 µm). Length of short cells 41.0 µm – 55.9 µm (Av: 48.45 µm). Width of short cells 38.0 µm – 42.5 µm (Av:40.25 µm). (Figure 7)

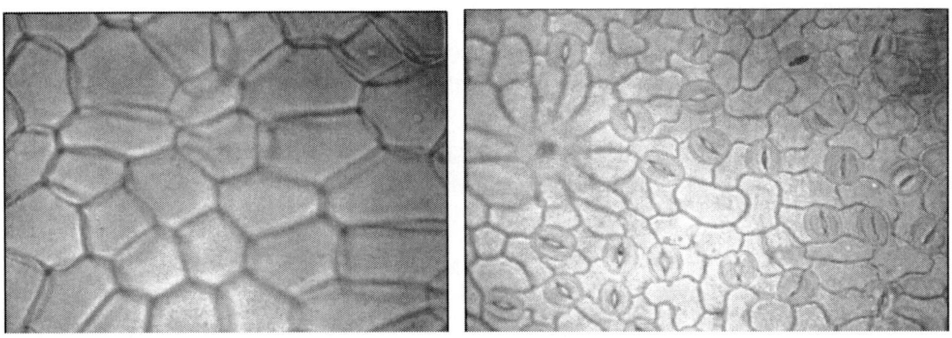

Figure 7. Leaf Epidermal Anatomy of Ficus glomerata. Adaxial side (Right) Abaxial side (Left).

b. Abaxial Side

Oil droplets and silica bodies were absent.Glandular trichomeswere present.Length of long cells 78 µm – 84.2 µm (Av: 81.1 µm). Width of long cells 29.2 µm – 45.3 µm (Av: 37.25 µm). Length of short cells 39 µm – 58.9 µm (Av: 48.95 µm). Width of short cells 29.1 µm – 31.8 µm (Av: 30.45 µm). Polygonal cells had double layer with smooth margins. Length of long cells 62.0 µm – 74.1 µm (Av: 68.05 µm). Width of long cells 29.0 µm – 44.7 µm (Av: 36.85 µm). Length of short cells 38.0 µm – 49.7 µm (Av: 43.85 µm). Width of short cells 32.0 µm – 34.1 µm (Av: 33.05µm). Stomata were present. Shape of stomata was staurocytic. Numbers of stomata per unit area were 27-28. Length of stoamtal cavity 22.0 µm – 32.3 µm (Av: 27.15 µm). Width of stamatal cavity 9.0 µm – 9.8 µm (Av:9.4 µm). Length of guard cells 45.1 µm – 50.0 µm (Av: 47.55 µm). Width of guard cells 36.3 µm – 41.1 µm (Av: 38.7 µm). Subsidiary cells were present. Numbers of subsidiary cells were 4 – 5. Length of long cells 64.9 µm – 68.5 µm (Av: 66.7 µm). Width of long cells 24.2 µm – 32.2 µm (Av: 28.2 µm). Length of short cells 25.7 µm – 34.4 µm (Av: 30.05 µm). Width of short cells 31.4 µm – 39.1 µm (Av: 35.25 µm). (Figure 7)

9. Ficus Infactoria Roxb

a. Adaxial Side

Trichomes, oil droplets, stomata and silica bodies were absent. Polygonal cells haddouble layer and smooth margins. Length of long cells 87.6 µm – 88.1 µm (Av: 87.85 µm). Width of long cells 48.2 µm – 62.6 µm (Av: 55.4 µm). Length of short cells 44.0 µm – 60.0 µm (Av: 52 µm). Width of short cells 39.8 µm – 43.0 µm (Av: 41.4 µm). (Figure 8)

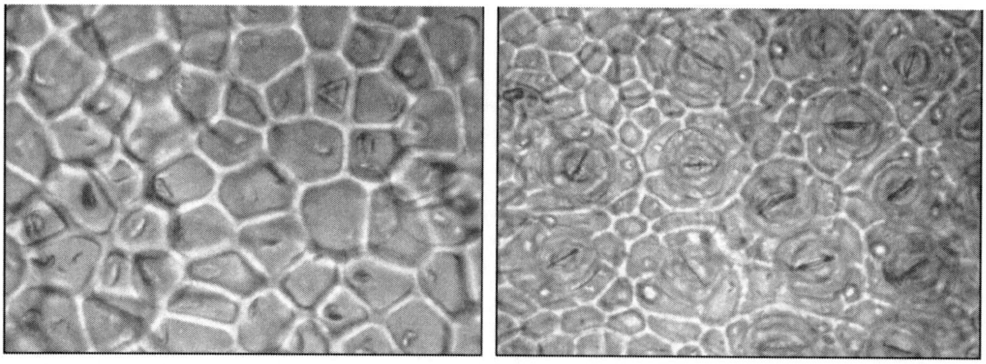

Figure 8. Leaf Epidermal Anatomy of Ficus infactoria. Adaxial side (Right) Abaxial side (Left).

b. Abaxial Side

Trichomes, oil droplets and silica bodies were absent. Polygonal cells had double layer and smooth margins. Length of long cells 43.8 µm – 56.6 µm (Av: 50.2 µm). Width of long cells 18 µm – 27.8 µm (Av: 22.8 µm). Length of short cells 28.3 µm – 33.7 µm (Av: 31.0 µm). Width of short cells 21.3 µm – 24.3 µm (Av: 22.8 µm). Stomata were present. Type of stomata was staurocytic. Numbers of stomata per unit area were 14-15. Length of stoamtal cavity 28 µm – 45.0 µm (Av: 36.5 µm). Width of stamatal cavity 5.8 µm – 9.0 µm (Av: 7.4 µm). Length of guard cells 55.1 µm – 64.0 µm (Av: 59.55 µm). Width of guard cells 63.0 µm

– 78.3 μm (Av: 70.65 μm). Subsidiary cells were present. Numbers of subsidiary cells were 4 – 5. Length of long cells 61.4 μm – 88.4 μm (Av: 74.9 μm). Width of long cells 31.2 μm – 35.4 μm (Av: 33.3 μm). Length of short cells 30.2 μm – 37.2 μm (Av: 33.7 μm). Width of short cells 19.8 μm – 30.1 μm (Av: 24.95 μm). (Figure 8)

10. Ficus Lyrata Warb

a. Adaxial Side

Trichomes, oil droplets,stomata and silica bodies were absent.Polygonal cells haddouble layer and smooth margins. Length of long cells 57.7 μm – 61.7 μm (Av: 59.7 μm). Width of long cells 17.5 μm – 21.5 μm (Av: 19.5 μm). Length of short cells 20 μm – 27 μm (Av: 23.5 μm). Width of short cells 19.8 μm – 26 μm (Av: 22.9 μm). (Figure: 9)

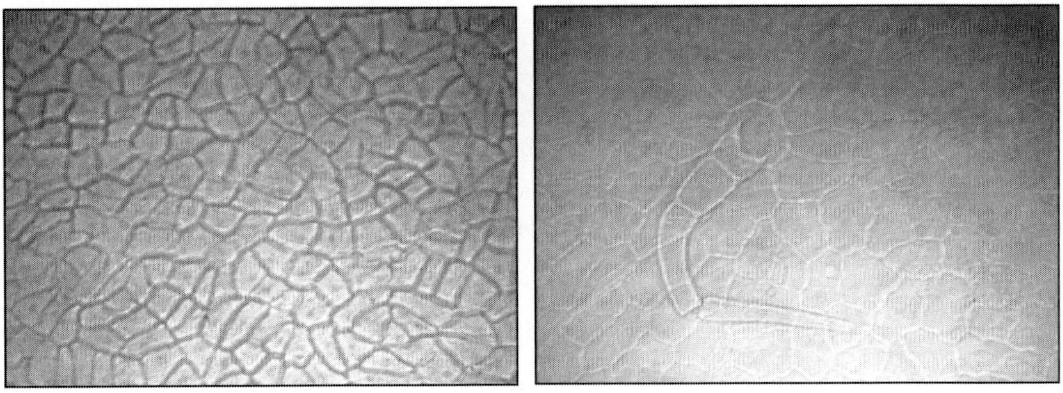

Figure 9. Leaf Epidermal Anatomy of Ficus lycata. Adaxial side (Right) Abaxial side (Left).

b. Abaxial Side

Oil droplets and silica bodies were absent. Non-glandular trichomes were present.Tip of trichomes was acute. Shape of trichomes wasmultiseriate.Length of trichomes 160.65 μm – 177.55 μm (Av: 169.1 μm). Width of trichomes 39.9 μm – 41.1 μm (Av: 40.5 μm).Polygonal cells haddouble layer and wavy margins. Length of long cells 80.5 μm – 85.0 μm (Av: 82.75 μm). Width of long cells 33.2 μm – 36.1 μm (Av: 34.65 μm). Length of short cells 35.1 μm – 41.9 μm (Av: 38.5 μm). Width of short cells 27.3 μm – 31.9 μm (Av: 29.6 μm). Stomata were present. Type of stomata was anisocytic. Numbers of stomata per unit area were 10. Length of stoamtal cavity 13.2 μm – 17.1 μm (Av: 15.15 μm). Width of stamatal cavity 4.0 μm – 5.0 μm (Av: 4.5 μm). Length of guard cells 26.2 μm – 29.2 μm (Av: 27.7 μm). Width of guard cells 12.1 μm – 22.0 μm (Av: 17.05 μm). Subsidiary cells were present. Numbers of subsidiary cells were 3. Length of long cells 58.9 μm – 66.1μm (Av: 62.5 μm).

Width of long cells 38 μm – 41.2 μm (Av: 39.6 μm). Length of short cells 29.2 μm – 51.4 μm (Av: 40.3 μm). Width of short cells 23.1 μm – 32.8μm (Av: 27.95 μm). (Figure 9)

11. Ficus Macrophylla, Roxb

a. Adaxial Side

Trichomes and Stomata were absent. Oil droplets were present. Length of oil droplets cells 39.8 μm –41.0 μm (Av: 40.4 μm). Width of oil droplets cells 32.6 μm – 38.2 μm (Av: 35.4 μm). Polygonal cells had double layer and rough margins. Length of long cells 66.9 μm – 91.3 μm (Av: 79.1 μm). Width of long cells 23.6 μm – 25.5 μm (Av: 24.55 μm). Length of short cells 25.3 μm – 35 μm (Av: 30.15 μm).

Width of short cells 21.0 μm – 24.1 μm (Av: 22.55 μm). Silica bodies were present and irregular in shape. Length of silica bodies 18.9 μm– 20.2 μm (Av: 19.55 μm). Width of silica bodies 15.6 μm – 17.9 μm (Av: 16.75 μm).

b. Abaxial Side

Trichomes and silica bodies were absent. Oil droplets were present. No. of oil droplets were 2. Length of long cells 22.2 μm – 30.8 μm (Av: 26.5 μm). Width of long cells 26.2 μm – 37.6 μm (Av: 31.8 μm). Polygonal cells had double layer and rough margins. Length of long cells 43.0 μm – 58.8 μm (Av: 50.9 μm). Width of long cells 20 μm – 21.9 μm (Av: 20.95 μm). Length of short cells 17.8 μm – 24.2 μm (Av: 21 μm). Width of short cells 13.9 μm – 17.9 μm (Av: 15.9μm).

Stomata were present. Shape of stomata was staurocytic. Numbers of stomata per unit area were 10-11. Length of stoamtal cavity 20.8 μm – 23.3 μm (Av: 22.05 μm). Width of stamatal cavity 11 μm – 11.5 μm (Av:11.25 μm). Length of guard cells 38.1 μm – 40.0 μm (Av: 30.05 μm). Width of guard cells 20.9 μm – 31 μm (Av: 25.95 μm). Subsidiary cells were present. Numbers of subsidiary cells were 5. Length of long cells 40 μm – 52.3 μm (Av: 46.15 μm).

Width of long cells 17 μm – 21.9 μm (Av: 19.45 μm). Length of short cells 27.8 μm – 28.5 μm (Av: 28.15 μm). Width of short cells 20 μm – 36.1 μm (Av: 28.05 μm).

12. Ficus Nerifolia

a. Adaxial Side

Trichomes, oil droplets, and stomata were absent. Polygonal cells had double layer and smooth margins. Length of long cells 51.5 μm – 51.8 μm (Av: 51.65 μm). Width of long cells 18 μm – 19.1 μm (Av: 18.55 μm). Length of short cells 14.9 μm – 19.2 μm (Av: 17.05 μm). Width of short cells 16.3 μm – 19 μm (Av: 17.65 μm). Round shape silica bodies were present. Length of silica bodies 22.6 μm - 24.2 μm (Av: 23.4 μm). Width of silica bodies 20 μm – 24.3 μm (Av: 22.15 μm). (Figure 10)

b. Abaxial Side

Oil droplets and silica bodies were absent. Non-glandular trichomes were present. Numbers of trichomes per unit area were 2. Tip of trichomes was acute. Length of trichomes 164.5 μm – 240.4 μm (Av: 202.45 μm). Width of trichome 26.8 μm – 36 μm (Av: 31.4 μm). Polygonal cells haddouble layer and smooth margins. Length of long cells 47.9 μm – 57.3 μm (Av: 52.6 μm). Width of long cells 14 μm – 17.7 μm (Av: 15.85 μm). Length of short cells

23.8 μm – 28.6 μm (Av: 26.2 μm). Width of short cells 12.6 μm – 14.8 μm (Av: 13.7 μm). Stomata were present. Numbers of stomata per unit area were 19. Shape of stomata was cyclocytic. Length of stoamtal cavity 13.3 μm – 15.3 μm (Av: 14.3μm). Width of stamatal cavity 13 μm – 13.9 μm (Av: 13.45 μm). Length of guard cells 32.3 μm – 33.6 μm (Av: 32.95 μm). Width of guard cells 33.5 μm – 33.7 μm (Av: 33.6 μm). Subsidiary cells were present. Numbers of subsidiary cells were 4. Length ofsubsidiary cells42.3μm – 47.2 μm (Av: 44.75 μm). Width of subsidiary cells 13.6 μm – 15 μm (Av: 14.3 μm). (Figure 10)

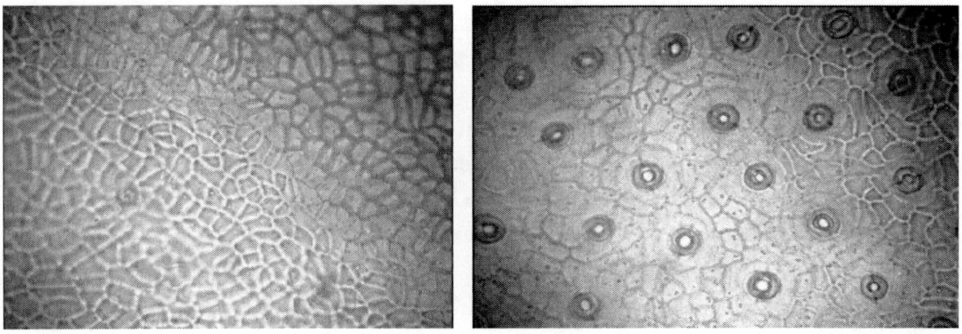

Figure 10. Leaf Epidermal Anatomy of Ficus nerifolia. Adaxial side (Right) Abaxial side (Left).

13. Ficus Pumila L.

a. Adaxial Side

Trichomes, silica bodies and stomata were absent. Oil droplets were present. Length of oil droplets 13.0 μm – 18.0 μm (Av: 15.5 μm). Width of oil droplets 11.3 μm – 13.9 μm (Av: 12.9 μm). Pentagonal cells had double layerand smooth margins. Length of long cells 55.7 μm – 62.2 μm (Av: 58.95 μm). Width of long cells 31.6 μm – 33.7 μm (Av: 12.9 μm). Length of short cells 18 μm – 22 μm (Av: 20 μm). Width of short cells 19 μm – 20 μm (Av: 19.5 μm). (Figure 11)

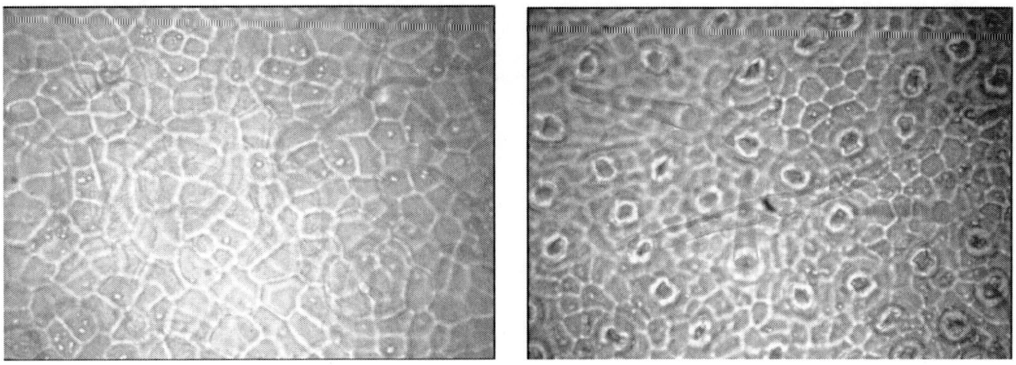

Figure 11. Leaf Epidermal Anatomy of Ficus pumila. Adaxial side (Right) Abaxial side (Left).

b. Abaxial Side

Oil droplets and silica bodies were absent. Non-glandular trichomes were present. Numbers of trichomes per unit area were 4. Length of trichomes 112 µm – 312.1 µm (Av: 212.05 µm). Width of trichomes 20 µm – 28.2 µm (Av: 24.1 µm). Pentagonal cells had double layer and smooth margin. Length of long cells 36.1 µm – 37.6 µm (Av: 36.85 µm). Width of long cells 18.4 µm – 25.7 µm (Av: 22.05 µm). Length of short cells 23.3 µm – 24 µm (Av: 23.65 µm). Width of short cells 20.1 µm – 20.6 µm (Av: 20.35 µm). Stomata were present. Numbersof stomata per unit area were 31. Type of stomata was diacytic. Length of stoamtal cavity 23.3 µm – 29.2 µm (Av: 26.25 µm). Width of stamatal cavity 12.7 µm – 19 µm (Av: 15.85 µm). Length of guard cells 40.4 µm – 45.5 µm (Av: 42.95 µm). Width of guard cells 32 µm – 32.4 µm (Av: 32.2 µm).

Subsidiary cells were present. Numbers of subsidiary cells were 2. Length of long cells 32.6 µm – 36.4 µm (Av: 34.5 µm). Width of long cells 17 µm – 21.5 µm (Av: 19.25 µm).

Length of short cells 14.1 µm – 24.1 µm (Av: 19.1 µm). Width of smaller cells 220 µm – 22.4 µm (Av:21.2µm). (Figure 11)

14. Ficus Racemosa L.

a. Adaxial Side

Oil dropletsand stomata were absent. Non- glandular trichomes were present. Numbers of trichomes per unit area were 2. Tip of trichomes was acute. Length of trichomes 161 µm – 164.2 µm (Av: 162.6 µm). Width of trichomes 29.7 µm – 31.9 µm (Av: 30.8 µm).

Polygonal cellshaddouble layer and smooth margins. Length of long cells 74.2 µm – 80.2 µm (Av: 77.2 µm). Width of long cells 38.6 µm – 39.3 µm (Av: 38.95 µm).

Length of short cells 28 µm – 50.8 µm (Av: 39.4 µm). Width of short cells 32 µm – 40.2 µm (Av: 36.1 µm). Round shape silica bodies were present. Length of silica bodies 19.0 µm - 32.1 µm (Av: 25.55 µm). Width of silica bodies 18.8 µm – 21 µm (Av: 19.9 µm). (Figure 12)

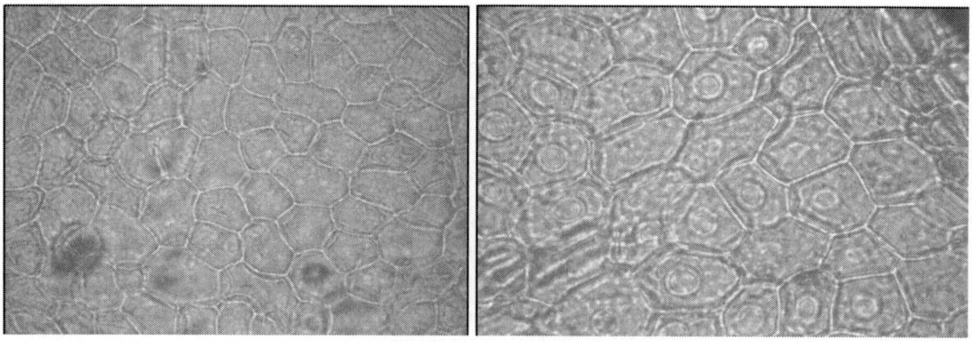

Figure 12. Leaf Epidermal Anatomy of Ficus racemosa. Adaxial side (Right) Abaxial side (Left).

b. Abaxial Side

Oil droplets were absent. Glandular trichomes were present. Numbers of trichomes per unit area were 20 -21. Length of trichomes 23.7 µm – 41.8 µm (Av: 32.75 µm). Width of trichomes 17 µm – 39.6 µm (Av: 28.3 µm). Polygonal cells had double layer and smooth margins. Length of long cells 106.8 µm – 110.9 µm (Av: 108.85 µm). Width of long cells

49.7 μm – 73.8 μm (Av: 61.75 μm). Length of short cells 53 μm – 76.2 μm (Av: 64.6 μm). Width of short cells 46.2 μm – 47 μm (Av: 46.6 μm). Stomata were present. Numbers of stomata per unit area were 31. Type of stomata was anomocytic. Length of stoamtal cavity 20.6 μm – 39.3 μm (Av: 29.95μm). Width of stamatal cavity 11.2 μm – 14.6 μm (Av: 12.9 μm). Length of guard cells 39.4 μm – 44.5 μm (Av: 41.95 μm). Width of guard cells 32 μm – 34.4 μm (Av: 33.2 μm). Subsidiary cells were present. Numbers of subsidiary cells were 6. Length ofsubsidiary cells 44.1μm – 59.1 μm (Av: 51.8 μm). Width of subsidiary cells 15.8 μm – 22.8 μm (Av: 19.3 μm). Round shape silica bodies were present. Length of silica bodies 27 μm – 67.9 μm (Av: 47.45 μm). Width of silica bodies 13 μm – 43 μm (Av: 28.0 μm). (Figure 12)

15. Ficus Religiosa(Leaf), L.

a. Adaxial Side

Oil droplets and silica bodies were absent.Non-glandular trichomeswere present. Length of trichomes 50.9 μm – 134.0 μm (Av: 92.4 μm). Width of trichomes 18.0 μm – 73.3 μm (Av: 45.6μm). Hexagonal cells had single layer and smooth margins. Length of long cells 106.3 μm – 116.8 μm (Av: 111.55 μm). Width of long cells 57 μm – 63.6 μm (Av: 60.3 μm). Length of short cells 47 μm – 54 μm (Av: 50.5 μm). Width of short cells 37.6 μm – 44 μm (Av: 40.8 μm). Staurocytic stomata were present in cluster form.Numbers of stomata per unit area were 22.Length of stomatal cavity 17.5 μm – 27.9 μm (Av: 22.7 μm). Width of stamatal cavity 9.2 μm – 13.6 μm (Av: 11.4 μm). Length of guard cells 34 μm – 35 μm (Av: 34.5 μm). Width of guard cells 22.1 μm – 36.1 μm (Av: 29.1 μm). Numbers of subsidiary cellswere 4 - 6. Length of long cells 41 μm – 42.2 μm (Av: 41.6 μm). Width of long cells 27.7μm – 35.5 μm (Av: 31.6μm). Length of short cells 20.2 μm – 26.5 μm (Av: 23.35 μm). Width of short cells 29.4 μm – 34.4 μm (Av: 31.9 μm). (Figure 13)

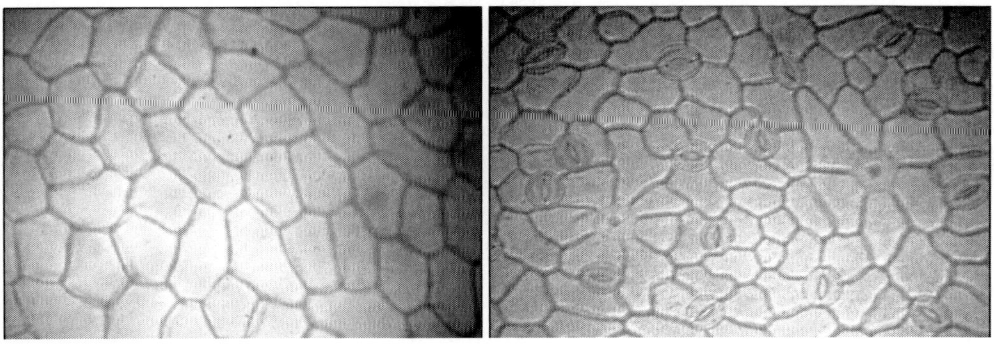

Figure 13. Leaf Epidermal Anatomy of Ficus religiosa. Adaxial side (Right) Abaxial side (Left).

b. Abaxial Side

Oil droplets andsilica bodies were absent.Glandular trichomes were present.Length of long cells 43 μm – 82.7 μm (Av: 62.85 μm). Width of long cells 43.4 μm – 54.4 μm (Av: 48.9 μm). Length of small cells 50.5 μm – 103.7 μm (Av: 77.1 μm). Width of small cells 32.2 μm - 62.8 μm (Av: 47.5 μm).Hexagonal cells haddouble layer and smooth margins. Length of long cells71.7 μm – 79.4 μm (Av: 75.55 μm). Width of long cells 32.2 μm – 41.2 μm (Av:

36.7 μm). Length of short cells 26 μm – 37.2 μm (Av: 31.6 μm). Width of short cells22.7 μm – 28.8 μm (Av: 25.75 μm). Staurocytic stomata were present.Numbers of stomata per unit area were16-17. Length of stoamtal cavity 30.3 μm – 33.5 μm (Av: 31.9 μm). Width of stamatal cavity 9 μm – 11.2 μm (Av: 10.1 μm). Length of guard cells 45.1 μm – 47.5 μm (Av: 46.3 μm). Width of guard cells 30.6 μm –39 μm(Av: 34.8 μm). Numbers of subsidiary cells were 2 – 4. Length of long cells 62.8 μm – 83.5 μm (Av: 73.15 μm). Width of long cell 17 μm – 35.4 μm (Av: 26.2 μm). Length of smaller cell 49.8 μm – 56.6 μm (Av: 53.2 μm). Width of smaller cell 26.4 μm – 49.2 μm (Av: 37.8 μm). (Figure 13)

16. Ficus Retusa L.

a. Adaxial Side

Trichomes, stomata and silica bodies were absent. Oil droplets were present.Length of oil droplets 24.7 μm – 29.5 μm (Av: 27.1 μm). Width of oil droplets 27.7 μm – 30.4 μm (Av: 29.05 μm). Pentagonal cells had double layer and smooth margins. Length of long cells 58.7 μm – 71.0 μm (Av: 64.85 μm). Width of long cells 26.0 μm – 32 μm (Av: 29 μm). Length of short cells 22.4 μm – 24 μm (Av: 23.2 μm). Width of short cells 21.9 μm – 24 μm (Av: 22.95 μm).

b. Abaxial Side

Trichomes and silica bodies were absent. Oil droplets were present. Length of oil droplets 22.5 μm – 34.9 μm (Av: 28.7 μm). Width of oil droplets 24.8 μm - 51 μm (Av: 37.9 μm). Pentagonal cells had double layer and smooth margins. Length of long cells 110 μm – 130.7 μm (Av: 120.35 μm). Width of long cells 29.2 μm – 49.0 μm (Av: 39.1 μm). Length of short cells 53.3 μm – 64.4 μm (Av: 58.85 μm). Width of short cells 36.7 μm – 49.8 μm (Av: 43.25 μm). Stomata were present. Type of stomata was staurocytic.Numbers of stomata per unit area were 3 -4. Length of stoamtal cavity 79.4 μm – 83.2 μm (Av: 81.3 μm). Width of stamatal cavity 29.8 μm – 33.2 μm (Av: 31.5 μm). Length of guard cells 94.3 μm – 94.4 μm (Av: 94.35 μm). Width of guard cells 74.8 μm – 85 μm (Av: 79.9 μm). Subsidiary cells were present. Numbers of subsidiary cells were 4 – 5. Length of long cells96.7 μm – 117 μm (Av: 106.85 μm). Width of long cells 43μm – 53 μm (Av: 48μm). Length of short cells 60.3 μm – 76.3 μm (Av: 68.3 μm). Width of short cells 46.4 μm – 53 μm (Av: 49.7 μm).

17. Ficus Triangularis (Leaf), Roxb

a. Adaxial Side

Stomata were absent. Non-glandular trichomes were present. Numbers of trichomes per unit area were 2. Tip of trichome was acute. Length of trichomes 307.1 μm – 334.2 μm (Av: 320.65 μm). Width of trichomes 19.8 μm – 22.2 μm (Av: 21 μm). Oil droplets were present. Numbers of oil droplets per unit area were 2. Round shape silica bodies were present. Numbers of silica bodies per unit area were 5. Length of silica bodies 16.8 μm – 17.0 μm (Av: 16.9 μm). Width of silica bodies 11.4 μm – 15.3 μm (Av: 13.3 μm). Polygonal cellshad double layer and smooth margins. Length of long cells 112.4 μm – 112.9 μm (Av: 112.65

μm). Width of long cells 68.4 μm –77.4 μm (Av: 72.9 μm). Length of short cells 35.6 μm – 44.7 μm (Av: 40.15 μm). Width of short cells 26.4 μm – 31.4 μm (Av: 28.9 μm). (Figure 14)

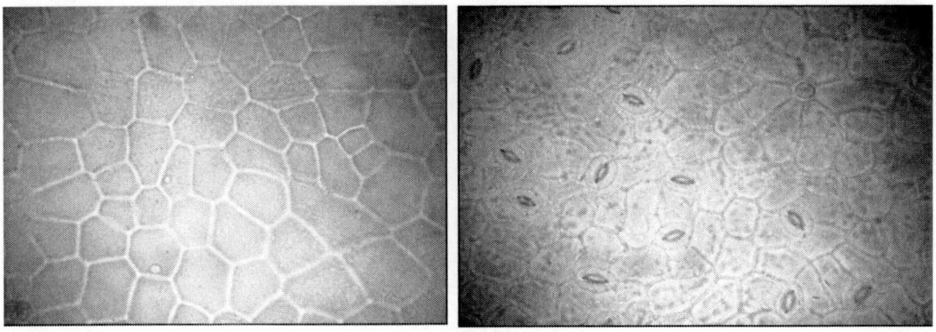

Figure 14. Leaf Epidermal Anatomy of Ficus triangularia. Adaxial side (Right) Abaxial side (Left).

b. Abaxial Side

Trichomes and silica bodies were absent. Oil droplets were present. Numbersof oil droplets per unit area were 2. Length of oil droplets 20.2 μm – 21.4 μm (20.8 μm). Width of oil droplets 16.9 μm – 18.2 μm (Av: 17.55 μm). Polygonal cells haddouble layer and smooth margins. Length of long cells 71.6 μm – 79.6 μm (Av: 75.35 μm). Width of long cells 28 μm – 31.1 μm (Av: 29.55 μm). Length of short cells 35.1 μm – 37.0μm (Av: 36.05 μm). Width of short cells 30.3 μm – 33 μm (Av: 31.65 μm). Stomata were present. Numbers of stomata per unit area were 13. Type of stomata was staurocytic. Length of stoamtal cavity 25.8 μm – 28.8 μm (Av: 27.3μm). Width of stamatal cavity 12.5 μm – 13.9 μm (Av: 13.2 μm). Length of guard cells 37 μm – 50.3 μm (Av: 43.65 μm). Width of guard cells 29.8 μm – 30.4 μm (Av: 30.1 μm). Subsidiary cells were present. Numbers of subsidiary cellswere 5-6. Length ofsubsidiary cells56.2 μm – 57.1 μm (Av: 56.65 μm). Width of subsidiary cells 35.7 μm – 74.1 μm (Av: 54.9 μm). (Figure 14)

18. Ficus Virens, Dryand

a. Adaxial Side

Trichomes, oil droplets,stomata and silica bodies were absent. Hexagonal cellshaddouble layer and smooth margins. Length of long cells 120.1μm – 123.2 μm (Av: 121.65 μm). Width of long cells 67.7 μm – 84.7μm (Av: 76.2 μm). Length of short cells42.1 μm – 45.8 μm (Av: 43.95 μm). Width of short cells 33.3 μm – 34 μm (Av: 33.65 μm). (Figure 15)

b. Abaxial Side

Trichomes, oil droplets and silica bodies were absent. Hexagonal cells had double layered and smooth margins. Length of long cells 96.2μm – 101.2 μm (Av: 98.7 μm). Width of long cells 40.6 μm – 43.6 μm (Av: 42.1 μm). Length of short cells 42.3 μm – 50.6 μm (Av: 46.45 μm). Width of short cells 30.3 μm – 45.9 μm (Av: 38.1 μm). Stomata were present. Type of stomata was cyclocytic. Numbers of stomata per unit area were 6. Length of stoamtal cavity 36.2 μm – 72.8 μm (Av: 54.5 μm). Width of stamatal cavity 17.5 μm – 20.9 μm (Av: 19.2 μm). Length of guard cells 54.1 μm – 103 μm (Av: 78.55 μm). Width of guard cells 34.9 μm

– 76.2 μm (Av: 55.55 μm). Subsidiary cells were present. Numbers of subsidiary cells were 6 – 7. Length of long cells62 μm – 75.9μm (Av: 68.95 μm). Width of long cells 26.8μm – 32.4 μm (Av: 29.6 μm). Length of short cells46.1μm – 46.8 μm (Av: 46.45 μm). Width of short cells 29.1 μm – 47.4 μm (Av: 38.25 μm). (Figure 15)

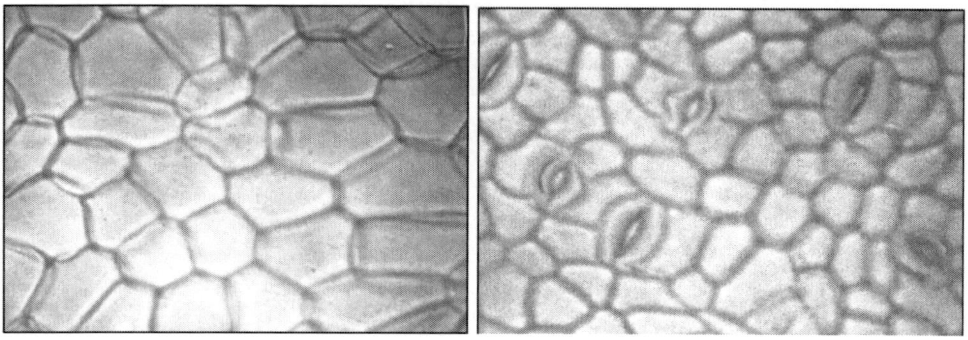

Figure 15. Leaf Epidermal Anatomy of Ficus virens. Adaxial side (Right) Abaxial side (Left).

19. Morus Alba

a. Adaxial Side

Oil droplets, silica bodies and stomata were absent. Glandular trichomes were present. Numbers of trichomes per unit area were 2. Length of trichomes 212 μm – 301.4 μm (Av: 256.7 μm). Width of trichomes 22.4 μm – 29 μm (Av: 25.7 μm). Hexagonal cellshaddouble layer and smooth margins. Length of long cells 127.3 μm – 141.9 μm (Av: 134.6 μm). Width of long cells 79.6 μm –84.2 μm (Av: 81.9 μm). Length of short cells62 μm – 74.8 μm (Av: 68.4 μm). Width of short cells 47.4 μm – 56 μm (Av: 51.7 μm). (Figure 16)

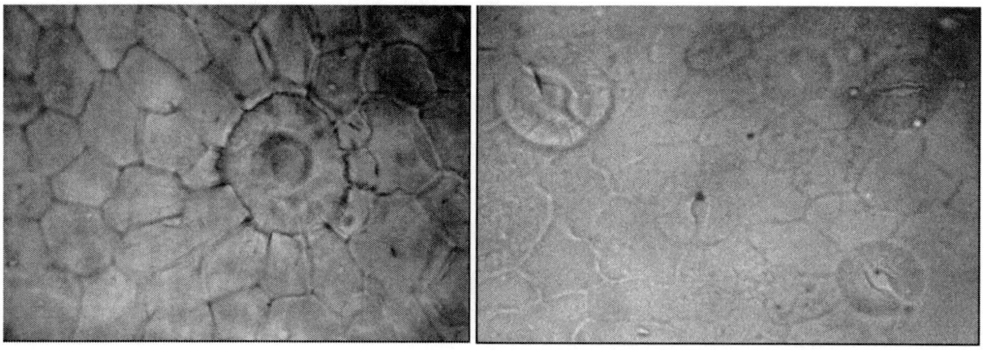

Figure 16. Leaf Epidermal Anatomy of Morus alba. Adaxial side (Right) Abaxial side (Left).

b. Abaxial Side

Oil droplets and silica bodies were absent. Non-glandular trichomeswere present.Tip of trichomes was acute. Numbers of trichomesper unit area is were4. Length of trichomes 287.8 μm – 334 μm (310.9 μm).Width of trichomes 25.5 μm – 27.8 μm (Av: 26.65 μm). Hexagonal cells haddouble layer and smooth margins. Length of long cells 73.1 μm – 94 μm (Av: 83.55 μm). Width of long cells 44.6 μm – 63.6 μm (Av: 54.1 μm). Length of short cells 36.5 μm –

48.7μm (Av: 42.6 μm). Width of short cells 28.2 μm – 40.3 μm (Av: 34.25 μm). Stomata were present. Numbers of stomata per unit area were 6. Type of stomata was staurocytic. Length of stoamtal cavity 58.6 μm – 59.7 μm (Av: 59.15μm). Width of stamatal cavity 21.2 μm – 28.6 μm (Av: 24.9 μm). Length of guard cells 90.6 μm – 111.9 μm (Av: 101.25 μm). Width of guard cells 84.6 μm – 111.7 μm (Av: 98.15 μm). Subsidiary cells were present. Numbers of subsidiary cells were 5-6. Length of subsidiary cells99.7 μm – 100.3 μm (Av: 100 μm). Width of subsidiary cells 31.1 μm – 48.1 μm (Av: 39.6 μm). (Figure 16)

20. Morus Nigra L.

a. Adaxial Side

Trichomes, oil droplets, silica bodies and stomata were absent. Glandular trichomes were present. Numbers of trichomes per unit area were 2. . Length of trichomes 112.4 μm – 120.4 μm (Av: 116.4 μm). Width of trichomes 75.6 μm – 80.9 μm (Av: 28.25 μm). Hexagonal cellshaddouble layer and smooth margins. Length of long cells 176 μm – 232.7 μm (Av: 204.35 μm). Width of long cells 112.4 μm –112.9 μm (Av: 112.65 μm). Length of short cells120.4 μm – 121.8 μm (Av: 121.1 μm). Width of short cells 75.1 μm – 87.1 μm (Av: 81.1 μm). (Figure 17)

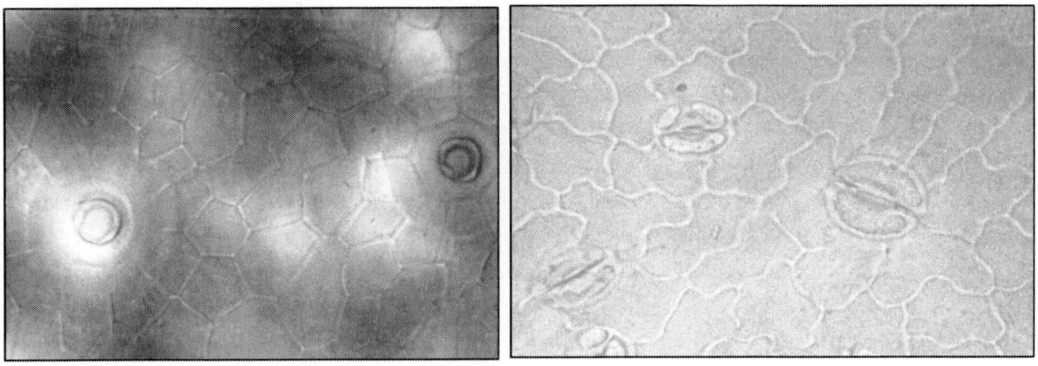

Figure 17. Leaf Epidermal Anatomy of Morus nigra. Adaxial side (Right) Abaxial side (Left).

b. Abaxial Side

Trichomes, oil droplets and silica bodies were absent.Polygonal cells haddouble layer and smooth margins. Length of long cells 129.6 μm – 137.1 μm (Av: 133.39 μm). Width of long cells 57.4 μm – 66.3 μm (Av: 61.85 μm). Length of short cells 69 μm – 81.μm (Av: 75 μm). Width of short cells 70.3 μm – 87.5 μm (Av: 78.9 μm). Stomata were present. Numbers of stomata per unit area were 3. Type of stomata was staurocytic. Length of stoamtal cavity 34.1 μm –44.7 μm (Av: 39.4μm). Width of stamatal cavity 17.1 μm – 17.7 μm (Av: 17.4 μm). Length of guard cells 88.3 μm – 97.1 μm (Av: 92.7 μm). Width of guard cells68.6 μm – 88.5 μm (Av: 78.55 μm). Subsidiary cells were present. Numbers of subsidiary cells were 5-7. Length of subsidiary cells116.3 μm – 120.9 μm (Av: 118.6 μm). Width of subsidiary cells 56.0 μm – 72.0 μm (Av: 64 μm). (Figure 17)

DISCUSSION

Leaf Epidermal Characters of Family Moraceae

The role of Anatomical data in traditional taxonomy has been recognized since the variation within the species, genera or a family is usually reflected in anatomical features [33]. Leaf epidermal anatomical markers like shape of stomata, types of trichome and other characters are useful taxonomical tools. The present study dealt with the diversity in the types and shape of leaf epidermal characters in family Moraceae. Although studies on grass morphology and wood anatomy of the plants are available yet this identification criteria would be incomplete without foliar epidermis morphology [34].

The functions of epidermis are water regulation, protection and defence against other organisms [35]. The occurrence of trichome is common in family Urticaceae- and the different trichome may contribute some function (water retention) described by Mauseth [35]. Trichome is highly variable appendages of the epidermis. There are different types of trichome like glandular, non-glandular, blunt hairs and scales.. Trichome structure vary widely at intra and inter specific level [36]. They were sometimes remarkably uniform and may be used for taxonomic purposes. Same observations were made by Metcalfe and Chalk [37] for dicotyledons. In tree species trichomes were found on both sides (abaxial and adaxial) of leaves.

Family Moaceae is characterised by glandular trichome, originated from intracellular region. This type of trichome was found on adaxial side of *Morus alba and Morus nigra* leaves. Glandular trichome was present on abaxial side of *Broussonatia papyrifera, Ficus benjamina, Ficus glomerata, Ficus religiosa,* and *Ficus racemosa* leaves. Glandular trichomes and their distribution play an important role for the identification and evolutionary studies of plant species. Shaheen et al., [28] used the diversity and distribution of glandular and non-glandular trichome within the genus *Sida* L. (Malvaceae). They observed sellate, peltate, long stalked, short stalked and capitates trichomes. However Shaheen et al., did not study other anatomical markers like cells types; arrangement of epidermal cells various types of stomata and silica body etc.

Origination of non-glandular trichom from cell wallis an important anatomical feature, used for the delimitation of various taxa of dichotyledons. Non-glandular trichomes were found on adaxial side of leaves collected from *Broussonatia papyrifera, Ficus bengalensis, Ficus religiosa, Ficus racemosa,* and *Ficus riangularis* leaves. Non-glandular trichomes were seen on abaxial sides of leaves in *Ficus lyrata, Ficus pumila, Ficus nerifolia,* and *Morus alba.*

Trichomes were not found on adaxial sides of *Ficus benjamina, Ficus glomerata, Ficus lyrata, Ficus pumila* and *Ficus nerifolia.* Trichomes were not present on abaxial side of *Ficus bengalensis, Morus alba,* and *Ficus triangularis.*

Different types of stomata studied by Stresburger firstly [38], followed by Vesque [39],recognized four broad categories of stomata based on the presence and arrangement of accessory cells as well as their mode of development. Stomatal patterning was related to ordered placement of stomata on the leaf surface. The pattering process involved the selection of undifferentiated cells to become stomata [39]. On the basis of arrangement the guard epidermal cells neighbouring the guard cells, more than 25 main types of stomata in dicot have been recognized [37].

Five different types of stomata eanomocytic, anisocytic, dicytic, cyclocytic and staurocytic were found in family Moraceae. Ahmed et al., [27] recognized seven types of stomata i.e., anisocytic, amphianisocytic, axillocytic, anomotetracytic, actinocytic, dicytic, and staurocytic. They emphasis on one anatomical marker that is stomata but they ignored other leaf anatomical characters. Diacytic stomata were found on both abaxial and adaxial side of *Benjamina comosa* leaves. Diacytic stomata were present on abaxial side of *Artocarpus lakoocha,* and *Ficus pumila*. Staurocytic stomata were found on both abaxial and adaxial sides of *Ficus religiosa* leaves. Staurocytic stomata were present on abaxial side of *Broussonatia papyrifera, Ficus glomerata, Ficus infactoria, Ficus retusa, Ficus macrophylla, Ficus triangularis* and *Morus alba* leaves. Anomocytic stomata were found on abaxial side of *Artocarpus integrifolia,* and *Ficus racemosa.* Anisocytic stomata were present on abaxial side of *Ficus lyrata.* Cyclocytic stomata were seen on abaxial side of *Ficus bengalensis, Ficus benjamina, Ficus elastic, Ficus virens*, and *Ficus nerifolia* leaves. Epidermal cells were quite variable in the configuration of the characteristic markers, which have been proven to be systematic value in cuticular characters, epidermis, stomata, subsidiary cells and trichomes [40].

Three different types of epidermal cells were identified (polygonal, hexagonal and pentagonal) for family Moraceae [19]

Polygonal shape of epidermal cells were found on both abaxial and adaxial foliar sides of *Artocarpus integrifolia, Artocarpus lakoocha, Broussonatia papyrifolia, Ficus bengalensis, Ficus benjamina, Ficus lyrata, Ficus infactoria, Ficus macrophylla, Ficus racemosa, Ficus nerifolia,* and *Ficus triangularis* leaves. Polygonal cells were found on adaxial side of *Ficus glomerata*. Hexagonal cells were seen on both abaxial and adaxial foliar sides of *Benjamina comosa, Ficus elastic, Ficus virens, Morus alba* and *Morus nigra* leaves. Hexagonal cells were found on adaxial side of *Ficus glomerata.* Pentagonal cells were present on both abaxial and adaxial sides of *Ficus retusa* and *Ficus pumila* leaves.

Leaf epidermis features like stomata; epidermal cell type; epidermal cell size, shape and size of trichome, and stomata cavity are of taxanomic values in delimiting species of the same genus or genera within a family [41]. On the basis of the cell arrangement 20 species of family Moraceae were separated into different categories during present piece of work. Double and rough layered cells were found on both abaxial and adaxial leaf sides of *Ficus bengalensis Ficus benjamina, Ficus elastic* and *Ficus macrophylla* leaves. Double and wavy layered cells were found on both abaxial and adaxial surface of *Artocarpus integrifolia.* Double and wavy layered cells were found on adaxial surface of *Ficus glomerata* and found on abaxial side of *Ficus lyrata.* Double and smooth layered cells were seen on adaxial side of *Ficus lyrata* and were seen on abaxial side of *Ficus glomerata* leaves. Double and smooth layered cells were found on both abaxial and adaxial surfaces of *Artocarpus lakoocha, Broussonatia papyrifera, Ficus infactoria, Ficus nerifolia, Ficus pumila, Ficus racemosa, Ficus retusa, Ficus religiosa, Ficus triangularis, Ficus virens, Morus alba* and *Morus nigra* leaves.

The size and shape of organelles were different on abaxial and adaxial surface of leaf epidermus because of bifacial leaf, Oil droplets were also recorded in some species of family Moracea. Oil droplets were found in *Artocarpus integrifolia, Ficus bengalensis, Ficus glomerata, Ficus retusa, Ficus pumila,* and *Ficus triangularis*. Whereas oil droplets were not found in *Artocarpus lakoocha, Benjamina comosa, Broussonatia papyrifera, Ficus benjamina, Ficus elastica, Ficus lyrata, Ficus infactoria, Ficus nerifolia, Ficus religiosa, Ficus racemosa, Ficus virens, Ficus macrophylla, Morus alba, Morus nigra.*

Another important anatomical character found in some species of family Moraceae was the presence of silica bodies. On the basis of silica bodies species were separated in 2 categories. Round shape silica bodies were found in *Ficus nerifolia, Ficus racemosa* and *Ficus triangularis.*

Irregular shape of silica bodies were found in *Ficus macrophylla*. Silica bodies were not seen in *Artocarpus integrifolia, Artocarpus lakoocha, Benjamina comosa Broussonatia papyrifera, Ficus bengalensis, Ficus benjamina, Ficus elastic, Ficus glomerata, Ficus infactoria, Ficus lyrata, Ficus pumila, Ficus religiosa, Ficus retusa, Ficus virens, Morus alba* and *Morus nigra* leaves.

Leaf epidermal characters like trichome types, variation of stomata types, shape of cells, arrangement of cells, oil droplets and presence of silica bodies were considered as very important features for the characterization of species of family Moraceae.

Cluster Analysis of Family Moraceae

Dendrogram was constructed on the basis of length/width of epidermal cells, presence or absence of stomata, non glandular and glandular trichomes on adaxial and abaxial surface in order to establish relationship among different taxa of Moraceae (Figure 18, 19).

Cluster analysis based on leaf epidermal anatomical characters revealed that 20 species of family Moraceae were divided into two main clusters 1 and 2 at 100%phylogenetic distance. At 60% genetic linkage group 1 was further divided into two sub clusters 1a and 1b. At 40% genetic linkage group 2 was further divided into 2 sub clusters 2a and 2b. At 14% genetic linkage group 2a was further divided into 2 sub clusters 2ai and 2aii. At 34% genetic linkage was further divided into 2 sub clusters 2bi and 2bii. The species included in 1a group was *Morus nigra* and in 1b group were *Morus alba* and *Ficus triangularis.* The species included in 2ai group was *Ficus racemosa* and 2aii were*Broussonatia papyrifera* and *Ficus bengalensis*. The species included in 2bi was *Benjamina comosa* and *Ficus religiosa.* And the species included in 2bii was *Benjamina comosa* and *Ficus religiosa*. The group 2bii was further divided into more groups. The species included in these groups were *Artocarpus integrifolia, Artocarpus lakoocha, Ficus benjamina, Ficus lyrata, Ficus pumila, Ficus elastic, Ficus nerifolia, Ficus macrophylla, Ficus retusa, Ficus integrifolia Ficus glomerata,* and *Ficus virens*. (Figure 18).

However, when dendrogram was constructed on the basis of leaf epidermal characteristic of abaxial side at 68% genetic linkage group 1 was further divided into two sub clusters 1a and 1b. At 46% genetic linkage group 2 was further divided into 2 sub clusters 2a and 2b. At 42% genetic linkage group 2b was further divided into 2 sub clusters 2bi and 2bii. The species included in 1a group was *Morus alba* and in 1b group was included *Ficus nerifolia, Ficus pumila* and *Ficus lyrata.* The species included in 2a group was *Ficus benjamina.*

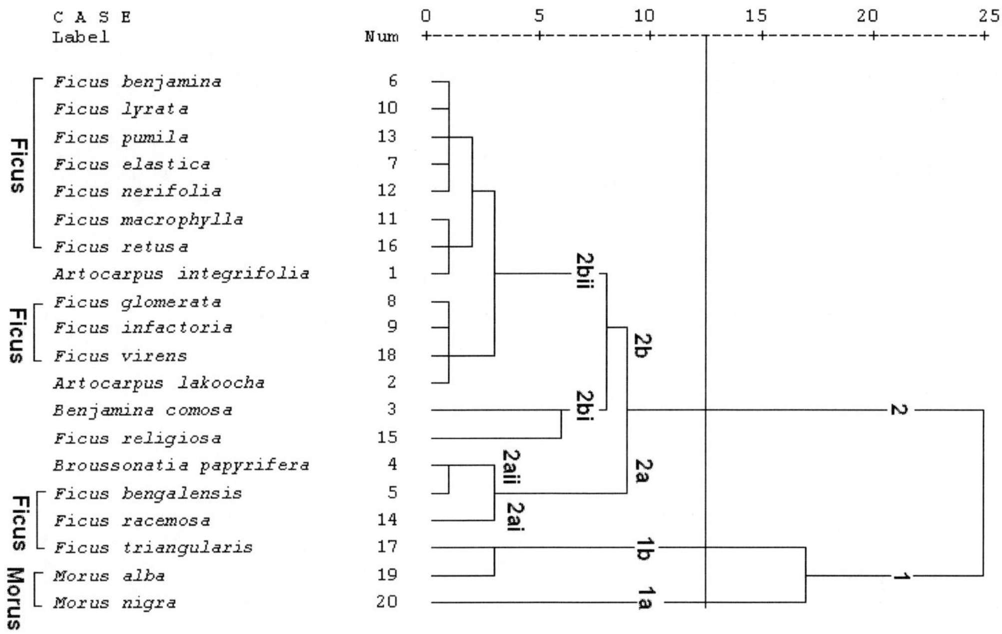

Figure 18. Dendrogram of different species of family Moraceae based on leaf epidermal anatomical characters (Adaxial side).

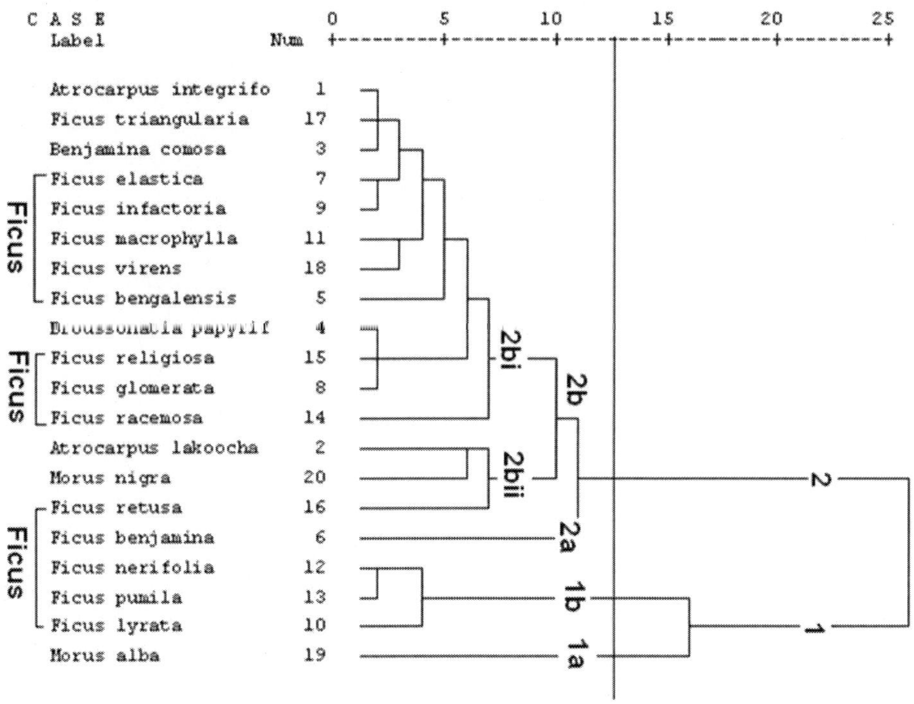

Figure 19. Dendrogram of different species of family Moraceae based on leaf epidermal anatomical characters (Abaxial side).

Table 1. Co-relation among the species of Family Moraceae based on leaf epidermal anatomical characters at adaxial side

Species	1	2	3	4	5	6	7	8	9	10	11	12	13	14	15	16	17	18	19	20
1	1																			
2	0.86	1																		
3	0.57	0.69	1																	
4	0.46	0.39	0.15	1																
5	0.41	0.34	0.12	1.00	1															
6	0.90	0.98	0.68	0.39	0.34	1														
7	0.76	0.88	0.55	0.32	0.27	0.90	1													
8	0.93	0.94	0.62	0.44	0.38	0.94	0.86	1												
9	0.85	0.97	0.67	0.39	0.33	0.96	0.89	0.97	1											
10	0.90	0.98	0.68	0.39	0.34	1.00	0.90	0.94	0.97	1										
11	0.94	0.80	0.47	0.46	0.39	0.82	0.80	0.88	0.78	0.82	1									
12	0.77	0.84	0.53	0.29	0.26	0.86	0.96	0.80	0.83	0.87	0.81	1								
13	0.97	0.92	0.61	0.46	0.40	0.95	0.83	0.93	0.89	0.95	0.93	0.81	1							
14	0.37	0.44	0.21	0.91	0.93	0.44	0.45	0.42	0.45	0.45	0.35	0.44	0.39	1						
15	0.58	0.68	0.72	0.70	0.69	0.69	0.57	0.64	0.69	0.69	0.47	0.54	0.61	0.77	1					
16	0.98	0.85	0.54	0.49	0.42	0.87	0.77	0.95	0.86	0.87	0.96	0.74	0.96	0.36	0.56	1				
17	0.31	0.31	0.13	0.93	0.96	0.32	0.28	0.32	0.32	0.32	0.27	0.28	0.30	0.97	0.71	0.30	1			
18	0.89	0.95	0.66	0.38	0.34	0.96	0.86	0.96	0.98	0.96	0.79	0.84	0.89	0.44	0.70	0.87	0.34	1		
19	0.44	0.51	0.29	0.92	0.93	0.51	0.44	0.49	0.52	0.51	0.37	0.42	0.45	0.98	0.81	0.43	0.97	0.52	1	
20	0.77	0.89	0.58	0.71	0.68	0.88	0.79	0.87	0.90	0.89	0.70	0.75	0.81	0.77	0.86	0.77	0.68	0.89	0.82	1

Key: - 1: Artocarpusintegrifolia 2: Artocarpuslakoocha 3: Benjaminacomosa 4: Broussonatiapapyrifera 5: Ficus bengalensis 6: Ficus benjamina 7: Ficus elastic 8: Ficus glomerata 9: Ficus infactoria 10: Ficus lyrata 11: Ficus macrophylla 12: Ficus nerifolia 13: Ficus pumila 14: Ficus racemose 15: Ficus religiosa 16: Ficus retusa 17: Ficustriangularis 18: Ficus virens 19: Morus alba 20:Morusnigra.

Table 2. Co-relation among the species of Family *Moraceae* based on leaf epidermal anatomical characters at abaxial side

Species	1	2	3	4	5	6	7	8	9	10	11	12	13	14	15	16	17	18	19	20
1	1																			
2	0.93	1																		
3	0.95	0.87	1																	
4	0.78	0.72	0.80	1																
5	0.81	0.65	0.77	0.60	1															
6	0.75	0.67	0.75	0.78	0.63	1														
7	0.97	0.88	0.94	0.75	0.85	0.80	1													
8	0.75	0.73	0.74	0.95	0.56	0.77	0.73	1												
9	0.94	0.83	0.90	0.73	0.78	0.78	0.97	0.72	1											
10	0.43	0.44	0.45	0.84	0.27	0.59	0.39	0.88	0.37	1										
11	0.87	0.82	0.86	0.67	0.69	0.59	0.84	0.64	0.81	0.36	1									
12	0.24	0.22	0.26	0.70	0.18	0.54	0.24	0.77	0.24	0.95	0.18	1								
13	0.22	0.19	0.21	0.66	0.18	0.51	0.22	0.75	0.22	0.92	0.14	0.99	1							
14	0.69	0.74	0.74	0.71	0.54	0.57	0.68	0.74	0.61	0.55	0.58	0.39	0.35	1						
15	0.81	0.75	0.81	0.98	0.62	0.77	0.77	0.96	0.75	0.82	0.70	0.66	0.62	0.70	1					
16	0.96	0.87	0.93	0.73	0.84	0.74	0.96	0.70	0.92	0.39	0.94	0.22	0.19	0.65	0.76	1				
17	0.98	0.92	0.97	0.77	0.84	0.77	0.98	0.75	0.93	0.42	0.86	0.24	0.22	0.74	0.79	0.96	1			
18	0.83	0.86	0.83	0.67	0.65	0.63	0.84	0.65	0.80	0.38	0.82	0.20	0.17	0.76	0.64	0.85	0.84	1		
19	0.38	0.33	0.37	0.74	0.32	0.66	0.40	0.80	0.40	0.90	0.27	0.95	0.95	0.48	0.68	0.36	0.38	0.36	1	
20	0.88	0.88	0.88	0.71	0.68	0.71	0.90	0.71	0.87	0.41	0.76	0.24	0.20	0.82	0.70	0.86	0.90	0.94	0.42	1

Key: - 1: Artocarpusintegrifolia 2: Artocarpuslakoocha 3: Benjaminacomosa 4: Broussonatiapapyrifera 5: Ficus bengalensis 6: Ficus benjamina 7: Ficus elastic 8: Ficus glomerata 9: Ficus infactoria 10: Ficus lyrata 11: Ficus macrophylla 12: Ficus nerifolia 13: Ficus pumila 14: Ficus racemose 15: Ficus religiosa 16: Ficus retusa 17: Ficustriangularis 18: Ficus virens 19: Morus alba 20: Morusnigra.

Table 3. Co-relation among the characters of species of Family Moraceae based on leaf epidermal anatomical characters at adaxial side

Characters	1	2	3	4	5	6	7	8	9	10	11	12	13	14	15	16	17	18
1	1																	
2	0.71	1																
3	0.33	0.30	1															
4	0.48	0.39	0.92	1														
5	0.26	0.31	0.91	0.83	1													
6	0.28	0.36	0.84	0.83	0.92	1												
7	-0.06	0.25	0.13	0.07	0.17	0.07	1											
8	-0.01	0.41	0.28	0.19	0.32	0.22	0.97	1										
9	-0.09	0.16	0.11	0.04	0.17	0.05	0.99	0.91	1									
10	-0.08	0.19	0.12	0.05	0.17	0.05	0.99	0.94	1.00	1								
11	-0.09	0.16	0.11	0.05	0.17	0.05	0.99	0.91	1.00	1.00	1							
12	-0.05	0.27	0.13	0.07	0.17	0.07	1.00	0.98	0.98	0.99	0.98	1						
13	-0.09	0.14	0.11	0.04	0.17	0.04	0.98	0.90	1.00	0.99	1.00	0.97	1					
14	-0.01	0.41	0.14	0.11	0.15	0.10	0.94	1.00	0.87	0.90	0.87	0.96	0.85	1				
15	0.16	0.01	-0.25	-0.20	-0.26	-0.11	-0.19	-0.20	-0.18	-0.19	-0.18	-0.19	-0.18	-0.18	1			
16	0.12	-0.02	-0.26	-0.22	-0.28	-0.12	-0.19	-0.20	-0.18	-0.18	-0.18	-0.19	-0.18	-0.18	1.00	1		
17	0.03	-0.01	-0.31	-0.31	-0.41	-0.47	-0.25	-0.27	-0.24	-0.25	-0.24	-0.25	-0.24	-0.25	-0.05	-0.06	1	
18	0.03	0.01	-0.34	-0.32	-0.40	-0.46	-0.25	-0.27	-0.24	-0.25	-0.24	-0.25	-0.24	-0.24	-0.06	-0.07	0.98	1

Key:- 1: Length of trichomes 2: Width of trichomes 3: Length of long cell 4: Width of long cell 4: Width of long cell 5: Length of short cell 6: Width of short cell 7: Length of stomatal cavity 8: Width of stomatal cavity 9: Length of guard cell 10: Width of guard cell 11: length of long subsidiary cell 12: Width of long subsidiary cell 13: Length of short subsidiary cell 14: Width of short subsidiary cell 15: Length of short subsidiary cell 16: Width of short subsidiary cell 17: length of oil droplets 18: Width of oil droplets

Table 4. Co-relation among the characters of species of Family Moraceae based on leafepidermal anatomical characters at abaxial side

Characters	1	2	3	4	5	6	7	8	9	10	11	12	13	14	15	16	17	18
1	1																	
2	0.62	1																
3	-0.31	-0.22	1															
4	-0.10	-0.07	0.86	1														
5	-0.23	-0.25	0.84	0.80	1													
6	-0.24	-0.24	0.86	0.82	0.93	1												
7	-0.23	-0.45	0.31	0.18	0.32	0.23	1											
8	-0.09	-0.30	-0.08	-0.14	0.03	-0.05	0.79	1										
9	-0.07	-0.28	0.60	0.46	0.51	0.52	0.67	0.28	1									
10	0.06	-0.22	0.41	0.30	0.41	0.41	0.63	0.31	0.94	1								
11	-0.14	-0.25	0.70	0.60	0.52	0.69	0.47	0.02	0.72	0.65	1							
12	-0.23	-0.32	0.73	0.67	0.55	0.73	0.24	-0.08	0.61	0.50	0.63	1						
13	-0.32	-0.03	0.02	-0.27	-0.24	-0.31	0.33	0.13	0.14	0.04	0.07	-0.15	1					
14	-0.28	-0.15	0.33	0.30	0.17	0.08	0.19	-0.10	0.18	0.02	0.19	0.38	0.58	1				
15	-0.07	0.17	0.22	0.34	0.37	0.23	-0.12	-0.09	-0.18	-0.16	-0.19	-0.19	-0.37	-0.29	1			
16	-0.07	0.17	0.22	0.34	0.37	0.23	-0.12	-0.09	-0.18	-0.16	-0.19	-0.19	-0.37	-0.29	1	1		
17	-0.28	-0.36	0.03	-0.19	0.30	-0.03	0.17	0.08	-0.05	-0.04	0.05	0.14	0.09	0.05	-0.10	-0.10	1	
18	-0.27	-0.35	0.06	-0.17	0.03	-0.02	0.23	0.11	-0.01	0.01	0.10	0.11	0.18	0.11	-0.09	-0.09	0.99	1

Key:- 1: Length of trichomes 2: Width of trichomes 3: Length of long cell 4: Width of long cell 4: Width of long cell 5: Length of short cell 6: Width of short cell 7: Length of stomatal cavity 8: Width of stomatal cavity 9: Length of guard cell 10: Width of guard cell 11: length of long subsidiary cell 12: Width of long subsidiary cell 13: Length of short subsidiary cell 14: Width of short subsidiary cell 15: Length of short subsidiary cell 16: Width of short subsidiary cell 17: length of oil droplets 18: Width of oil droplet.

The species included in 2bi were *Artocarpus lakoocha, Morus nigra* and *Ficus retusa.* The group 2bii was further divided into more groups. The species included in these groups were *Artocarpus integrifolia, Ficus triangularis, Benjamina comosa, Ficus elastica, Ficus infactoria, Ficus macrophylla, Ficus virens, Ficus bengalensis, Broussonatia papyrifera, Ficus religiosa, Ficus glomerata,* and *Ficus racemosa.* (Figure 19 (abaxial side)).

Co-Relation among the Species of Family Moraceae

Co-relation was calculated on the basis of length/width of epidermal cells, presence or absence of stomata, non glandular and glandular trichomes on adaxial and abaxial foliar surface in order to establish relationship among different taxa of Moraceae. Table 1 shows similarity index between the species of Family Moraceae. Co-relation of the species ranged from 0.12 – 0.98. Based on leaf epidermal anatomical characters *Artocarpus lakoocha* abaxial side was closely related to *Ficus benjamina,* and *Artocarpus integrifolia* adaxial side was closely related to *Ficus retusa* and *Artocarpus lakoocha* adaxial side was closely related to *Ficus lyrata.* The value of co-relation for *Artocarpus integrifolia, artocarpus lakoocha, Ficus benjamina, Ficus lyrata* and *Ficus retusa* was 0.98 (Table 1). The species that showed minimum co-relation were *Benjamina comosa* with *Ficus bengalensis,* their relationship percentage was 12%. These two species were totaly different from one another on the basis of leaf epidermal anatomical characters. This table also shows that the species belonged to genus *Artocarpus* (*A.integrifolia* and *A.lakoocha*) closely related to each other. Their relationship percentage was 86%. The species of genus *Ficus* showed similarity in the range of 0.27 – 0.98. The species belonging to genus *Morus* were closely related to one another. Their relationship percentage was 77%.

Table 2 shows the relationship between the species of Family Moraceae based on leaf epidermal characters of abaxial side epidermis. Co-relation of the species ranged from 0.14 – 0.98. *Broussonatia papyrifera* was closely related to *Ficus religiosa,* and *Artocarpus integrifolia* to *Ficus triangularis.* The value of co-relation for *Artocarpus integrifolia/ Broussonatia papyrifera and Ficus religiosa /Ficus triangularis* was 0.98 (Table 2). The species that showed minimum co-relation were *Ficus macrophylla* with *Ficus pumila,* their relationship percentage was 14%. These two species were totally different from one another on the basis of leaf epidermal characters. This table shows that the species belonged to genus *Artocarpus* (*A.integrifolia* and *A.lakoocha*) closely related to each other. Their relationship percentage was 93%. The species of genus *Ficus* were intact to one another because the maximum and minimum relationship percentage was belonged to this genus. The species belonged to genus *Morus* were closely related to one another. Their relationship percentage was 88%.

Table 3 shows co-relation among 18 quantitative characters of abaxial epidermal characters of family Moraceae. The similarity index between the characters of family Moraceae ranges from 0.01 – 1.00. Width of trichomes (similarity index 0.01) related with the length of silica body as well as width of oil droplets. Length of silica bodies showed similarity index 1.00 related with the width of silica bodies. The characters showed negative relation ranged from -0.01 – -0.47. Length of trichomes was -0.01 related with width of stomatal cavity as well as width of short subsidiary cell. Width of trichomes showed -0.01 relation with length of oil droplets. Width of short cell -0.47 related with length of oil droplets. Table 4 shows co-relation among 18 quantitative characters of adaxial epidermal anatomy of family Moraceae. The similarity index between the characters of family Moraceae ranges from 0.01 – 1.00. Width of guard cell (similarity index 0.01) related with width of oil droplets. Length of silica bodies showed similarity index of 1.00 with the width of silica bodies. The characters showed negative relation ranged -0.01 – -0.45. Length of guard cell was -0.01 related with width of oil droplet. Width of trichomes was -0.45 related with length of stomata cavity.

REFERENCES

[1] Thurston, E.L. and Lersten. N.R. (1969). The morphology and toxicology of plant stinging trichomes. *Bot. Rew.* 63:393-412.

[2] Pollard, A.J. and Briggs. D.(1984). Genecological studies of Urtica dioica L. III. Stinging troichomes and plant herbivore interaction. *New.Phytol.* 97:507-522.

[3] Lin, C.T., Yao Y.N.and Chung. N.J. (1991). Morphology and habitats of Devils leaf (*Urtica thunbergiana*) in chi-tou forest Recreation Area. *Quart.J. Expt. Forest, NTU.* 5:167-183.

[4] Corsi, G. (1992). The stinging trichome of Urtica membranaceae poiret (Urticaceae) II *histochemistry.Phyton*.32:247-253.

[5] Fodor, F. and Cseh. E. (1993). Effect of different nitrogen forms and iron-chelates on development of stigging nettle. *J.Plant.Nutr.* 16:2239-2253.

[6] Turberville, T.D., Dudly P.G. and Pollard. A.J. (1996). Responses of invertebrates herbivores to stinging trichomes of *Urtica dioica* and *laporiea Canadensis*. *Oikos.* 71:83-88.

[7] Bhatia, R.C. (1984). Foliar epidermal studies of heliotropium suptnum L. Folia Geobot. *Phytotaxon.*, 19:381-385.

[8] Stace, C.A. (1984).The taxonomic importance of leaf surface. In: Heywood , V.H. and D.M. Moore (end.), current concepts in plants Taxonomy. Pp:67-94. London: Academic presss, London.

[9] Jones, J.H. (1986). Evolution of the Fagaceae. The implication of foliar feature *Annl. Missouri Bot. Garden*, 73.228-275.

[10] Baranova, M. (1972). Systematic anatomy of the leaf epidermis in the Magnoliaceae and some related families. *Taxon.*, 21:447-469.

[11] Gangadhara, M. and Inamdar. J.A. (1977). Trichomes and stomata, and their taxonomic significance in the*Urticales. Plant Systematics and Evolution.* 127:121-137.

[12] Lackey, J .A. (1978). Leaflet anatomy of Phaseoleae (Leguminosae; Paplionoidae) and its relation relation to taxonomy. *Bot. Gaz.* 139(4): 436-446.

[13] Ogundipe, O.T. (1992). Leaf epidermal studies of genus *Dautura* (Linn.) Solanaceae. *Phytomorphology*. 42 (34): 207-217.

[14] Yi F.H., Chen S.J. and Kuo-huang. L.L. (2003). Comparative study on the stinging trichomes and some related epidermal structures in the leaves of *Dendrocnide meyeniana,* Girardinia diversifolia, and Urtica thunbergiana. *Taiwania,* 48(4):213-223.

[15] Kanademir, N. (2003). The morphological anatomical and karyological properties of endemic saliva hypargeia Fich and Mey. (Lamiaceae) in Turkey. *Pak.J.Bot.* 35(2):219-236.

[16] Liscovsky, I.J. and Cosa. M.T. (2005). Comparative anatomy of leaf and stem in species of *Cestreae* G.Don (Solanaceae)from *Argentina. Gayana Botanica.*33-43.

[17] Carpenter, R.J., Hill R.S. and Jordant, G.J. (2005). Leaf cuticular morphology links Platanaceae and Proteaceae. *Int.J.Plant Sci.,* 5:846-855.

[18] Ozdemir, C. And Altn. Y. (2006). Morphological and anatomical investigation of some endemic alkana species. *Pak.J.of.Bot.* 38(3):527-537.

[19] Klimko, M. And Truchan. M. (2006). Morphology variability of the leaf epidermis in selected taxa of the genus *Ficus* L.(Moraceae) and its taxonomic implication. acta. *Socities botanicorum poloniae*, 75 (4): 309-324.

[20] Perveen, A., Abid R. and Fatima. R. (2007). Stomatal types of some dicots within flora of Karachi, Pakistan. *Pak.J.Bot.* 39(4):1017-1023.

[21] Munsif, S., Khan, M.A., Ahmed, M., Zafar, M., Shanand G.M and Shaheen, N. (2007). *J.Agri. Soc.* 3(2):43-46.

[22] Rojas, C.B. (2007). Anatomy of the foliar epidermis of the *Solanum nudum* species group of *solanum sect germinate* (Solanaceae). *Acta Hort.* (ISHS) 745: 287-294.

[23] Jiang, Y.L., Li. Q., Li, Y., Deng, W., Li, T., L.M. and Lua, C.J. (2007). Leaf epidermal anatomy of *lonicera japonica* and *Lonicera confuse. J.of Foresty Res.*18(2):103-108.

[24] Yousaf, Z., Shinwari, Z.K., Asghar R. and Perveen. A. (2008). Leaf epidermal anatomy of selected *Allium* species, Family *Alliiaceae* from Pakistan. *Pak.J.Bot.* 40(1):77-90.

[25] Yasmin, G., Khan, M.A., Shaheen N.and Hayat. M.Q. (2010). Taxonomic significance of leaf epidermal anatomy of selected *Persicaria* Mill. Species of family Polygonaceae from Pakistan. African *Journal of Biotechnology* .9(25): 3759-3768.

[26] Azmati, R., Haider. S., Nasreen. H., Aziz, F.and Riaz, M. (2009). A viable alternative mechanism in adapting the plants to heavy metal environment. *Pak. J. Bot.*, 41(6):2729-2738.

[27] Ahmed, K., Khan, M.A., Ahmed, M., Zafar, M., Arshed M. and Ahmed, F. (2009). Taxonomic diversity of stomata in dicot Flora of a district tank (N.W.F.P) in Pakistan. *African.J.Biotec.*, 8(6):1052-1055.

[28] Shaheen, N., M. A. Khan, G. Yasmin and M. Q. Hayat. 2009. Diversity of foliar trichomes and their systematic. Relevance in the Genus *Hibiscus* (Malvaceae). *Int.J.Agr. & Bio.* 11(3):279-284.

[29] Kadiri, A. B. and G. O. Ajayi. 2009. Phyto-anatomical characteristics of the West African (Umbrella tree) *Musanga cercropioides* M. Smithii R. Br. (Moraceae). *Indian Journal of Science and Techonology*. 2.(7.): 0974-6846.

[30] Rahaman, A.A.A., Egbedo E.D. and Oladel. E.A. (2009). Stomatal complex types, stomatal density and the stomatal index in some species of *Discorea. Arch.Biol.Sci. Beelgrade.* 61 (4):847-851.

[31] Shaheed, S.A. and Illoh. H.C. (2010). Taxonomic study of some species in *Cassiae* (Leguminoceae) using leaf epidermal characters. *Not.Bot.Hort.Agro. Bot.Clu.* 38(1):20-27.

[32] Ahmad, K., Khan, M.A., Ahmad, M., Shaheen N. and Nazir. A., (2010). Taxonomic diversity in epidermal cells of some sub-tropical plant species. *Int. J. Agric. Biol.*, 12: 115–118.

[33] Metcalf, C.R. (1960). Anatomy of the monocotyledons I. Gramineae, Oxford University process, London. Pp:1-711.

[34] Kadiri, A.B., Olowokugejo J.D.and Ogundipe. O.T., (2006). Some aspects of Foliar epidermal morphology of cylicodiscus gabunensis (Taub.) Harms (Mimoceae). *J. sci. res.*, 10; 33-38.

[35] Mauseth, J.D. (1988). Plant anatomy. The Benjamin/ Cummings publishing company.

[36] Gillani, S.S., Khan. M.A., Shinwari, Z. K. and Yousaf, Z. (2002). Leaf epidermal anatomy of selected *Digiteria* species, Tribe paniceae, Family Poaceae of Pakistan. *Pak.J.Bot.* 34(3):257-273.

[37] Metcalf, C.R., and Chalk, L .(1979). Anatomy of the dicotyledon, systematic anatomy of the leaf and stem. Vol 1 and 2.nd Ed Clarendon press, oxford.

[38] Stresburger, E. (1886). Ein Beitrag Zur Entwick lungs eschichte der spaltoffungen jahrd. *Wiss Bot.* 5:297-342.

[39] Vesque, M.J. (1889). Del employs descharacto anatomical dams la classification demegataux.

[40] Ellis, R.P. (1976). A procedure for Standardizing comparative leaf anatomy in Poaceae.1.The leaf blade as viewed in transverse section. Bothalia. 12:65-109.

[41] Abdul Rahaman, A. A., and Oladele, F.A. (2005). Stomata, trichomes, and epidermal cells as diagnostic features in six species of genus *Ocimum* L. (Lamiaceae). *Niger. J. Bot.* 18, 214-222.

[42] Webster, G.L., Aguilar, D.A., and Smith, B.A. (1983). Systematic distribution of foliar trichomes types types in Croton (Euphorbiaceae). *Bot. Journal of the Linnean Scoiety,* 121:41-57.

In: Leaf Epidermal Anatomical Characters … ISBN: 978-1-63483-190-1
Editors: Z. Yousaf, A. Younas and A. Aftab

Chapter 6

TAXONOMIC EVALUATION OF FAMILY FABACEAE ON THE BASIS OF LEAF EPIDERMAL ANATOMICAL CHARACTERS

Zahra Fazal [*] ***and Haleema Khalid***
Department of Botany,
Lahore College for Women University,
Lahore, Pakistan

ABSTRACT

Present piece of work was conducted to investigate leaf epidermal anatomy of selected species of Family Fabaceae.As. morphology based taxonomic information of family Fabaceae is contradictory. Morphometry of this family is misleading and it is need of time to investigate new taxonomic tools to resolve existing taxonomic problems. Leaf epidermal anatomy is considered as one of the effective technique used for the generation of more authentic taxonomic information. Important anatomical markers identified for the Family Fabaceae were types and shapes of cells, cell margins variation, stomata cell shape, types and developmental variation, shape and type of trichomes and silica bodies. Cellwere pentagonal/ polygonal in shape with smooth margins in *Dalbergia sissoo* Roxb. The maximum length of long cells ranged from 234.4µm- 295.5µm,were found in *Albizia mimosa* Benthum on abaxial side of mature leaf. In *Albizia procera* (Roxb.) Benthum stellate types of trichomes were present. Silica bodies in the species of Family F were round and irregular shaped. Based on present study observation it is suggested that epidermal features like stomata, trichomes and silica bodies can be used to delimit the specific taxa of Family Fabaceae.

[*] zahrafazal56@yahoo.com.

INTRODUCTION

The family Fabaceae is widely distributed throughout the world and comprised of more than 650 genera and 18000 species, especially in tropical and subtropical regions [1]. According to Joly [2] Fabaceae is among It includes a large number of domesticated species that is harvested as crops for human and animal consumption as well as for oils, fertilizers, medicinal and chemicals [3]. In African traditional medicinal system two species (*Dialium guineense* Wild and *Dialium dinklagei* Harms) from this family are of considerable economic importance because of their culinary and medicinal usefulness. An infusion of the leaves and fruits of *D. guineense*is used for curing fever, and a decoction of the bark is used for stomach-ache in Sierra Leone [4].

Taxonomically, Fabaceae has been divided into three subfamilies, the Caesalpinioideae, Mimosoideae, and Papilionoideae. This family was closely related to the Connaraceae and Sapindaceae on the basis of anatomy, morphology, and bio-geographic distribution [5]. Morphometry provide the basis for taxonomy but it could not provide solution for complex taxonomic problems [6]. Therefore, taxonomist involved other biological techniques like leaf epidermal anatomy, cytology, electrophoresis as taxonomic tools [6]. The genus *Cercis* can be distinguished easily as compared to other genera present in Caesalpinioideae due to its pseudo papilionaceous flowers (Hao*et al.*, 2003) [7]. Although the leaves of *Disanthus* (Hamamelidaceae) more closely resemble to those of *Cercis* than other genera in Caesalpinioideae [8], they can also be distinguished from *Cercis* by irregular epidermal cells with sinuous anticline walls and a paracytic stomata apparatus [9].

Anatomical markers of leaf epidermis were found to be an important tool for taxonomic evaluation at inter and intra generic level [10, 11]. Leaf epidermal anatomical characters involved trichome, stomata, silica bodies and other markers have been proved useful features for potential taxonomical importance [12, 13]. Hence, anatomical features are of particular value to identify small parts of plant material [6].

Trichome features are important character for systematical studies. Variations in types of trichomes can provide insight into evolution and relationship within and among species [14, 15]. The structure and function of glandular trichome occurring in plants of the Lamiaceae is well documented [16, 17, 18 and 19] and they recognized trichome as the site of essential oil biosynthesis, secretion and accumulation [20].

Stace, Metcalfe and Carlquist have stated that the leaf anaytomy is the most varied aspect of angiosperms and it provides a variety of anatomical features that can be employed as useful taxonomic characters [21, 22, and 23]. Especially stomata types have long been regarded of systematic and evolutionary significance [24, 25]. Structural features of individual stoma may be important, but equally valuable in their pattern of distribution, while presence or absence of stomata on the upper leaf surface is often a good diagnostic feature.

On the basis of arrangement of the epidermal cells neighboring to the guard cells, more than 25 main types of stomata have been recognized in dicots [26]. Pettigrew & Watson studied 23 Australian *Acacia* species and reported two types of stomata along with their variants that is paracytic and cyclocytic types [27]. In *Trifolium clypeatum* stomata were surrounded by a 'rosette' of epidermal cells at early maturation of the stomata, an example of heterochrony. In *Scorpiurus muricatus* parallel 'chains' of stomata, each with two subsidiary cells were present.

Stomata development represents a tractable system for analyzing the cell and molecular biology. Stomata development is very critical for plant survival and productivity and is one of the crucial events in land plant evolution and acclimatization [28, 29, 30]. The differentiation of stomata proceeded by an asymmetric cell division of meristemoid mother cells (MMCs; a subset of proto-dermal cells) [29]. Asymmetric cell divisions refer to any cell divisions that give rise to two daughter cells with different developmental potential [31, 32, 33]. The present study was plane to explore leaf epidermal diversity among the members of family Fabaceae to solve existing taxonomic problems.

HISTORICAL BACKGROUND

Adeniji and Ariwaodo [35] studied foliar epidermis of two species of genus *Pericopsis* found in Nigeria by using light microscopy. They identified the species even when only leaf fragments were available by elucidating their taxonomic significance and complementary data. They observed in the leaf section of *P. laxiflora* amphi-stomata and trichome, but these were lacking in *P. elata*. They also observed the shape of cell which was polygonal and irregular on both surfaces and either polygonal or irregular on any of the two surfaces of the leaf and anticline wall was either straight or curved in *P. elata* but only straight in *P. laxiflora*.

Bhagat and Bukhtar [36] described the anatomical standardization of genus *Capparis* L. which was shrub or rarely tree, erect, scandent, or sprawling, shoot generally armed with stipular spines belonging to family Capparadaceae. They explained that anatomical data showed arrangement of tissues in particular pattern.Stomata was anomocytic or ranunculaceous, eglandular tomentose, dorsiventral mesophyll, prevascular fibers, thickened cuticle and also showed the presence of crystals in leaf epidermis near to vascular cylinder of petiole which was the unusual character of *Capparis zeylanica* L.

Vatén and Bergmann [37] explained that Plant development has a significant postembryonic phase that is regulated by interactions between the plant and the outside environment. This is particularly evident in the development, pattern and function of stomata and epidermal pores on the aerial surface of land plants. They observed that Stomata have been found in fossils. They indicated that the morphology of the individual stomata complex is largely unchanged, but their sizes, numbers, arrangements and surrounding cells have diversified tremendously. They studied that in the flowering plant *Arabidopsis thaliana* have established a basic molecular framework for the acquisition of cell fate and generation of cell polarity in these compartments, as well as describing some of the key signals and receptors required to produce stomata in organized patterns and in environmentally optimized numbers. They also presented parallel analyses of stomata developmental pathways at morphological and molecular levels.

Khan et al., [38] identified the medicinally important arboreal flora which was important for sustainable and effective utilization. Their study was conducted for taxonomic evaluation of the selected sixteen tropical and subtropical arboreal species, belonging to four orders and six angiosperm families based on leaf epidermal anatomical markers. Their results showed substantial variation in size, shape and arrangement of epidermal cells, stomata, trichome number and shape of subsidiary cells and silica bodies. They observed eight different shapes of stomata viz anomcytic, anomotetracytic, actinocytic, amphianisocytic, brachyparacytic, anisocytic, amphicyclocytic and stauricytic. Anomotetracytic was the most dominant type of stomata. Staurocytic stomata were found in *Erythrina subrosa* L. They concluded that leaf epidermal anatomical markers could be utilized significantly in delimiting the closely related taxa of medicinally important arboreal flora.

Shaheen et al., [39] studied the micro morphological diversity linked to different elements silica bodies, epidermal leaf anatomy by using elemental dispersive spectrophotometer (EDS) analysis and light & scanning electron microscopy of five species of genus *Brachiaria.* They first time used EDS analysis technique in Pakistan for grass taxonomy. This examination showed variations in shapes of silica bodies, prickles, micro hair and macro hair form. They determined new structure of phytolith used as an aid in plant systematic characterization by EDS analysis. In their work quantitative analysis provided silicon as a new taxonomic character to distinguish different species.

Saheed and Illoh [40] studied foliar epidermal characters of ten species from the genus Cassia (Leguminosae). Emend. Gaertner, *Senna* Miller and *Chamaecrista* Moench found in South Western Nigeria were studied. Their aim was to establish some useful diagnostic features for their delimitation. Their results showed exciting features that was helpful in the identification of each species. These include guard cell area, stomata index and frequency, presence or absence, types and length of trichome on epidermal surfaces and wall types. Their results suggested that diagnostic features were found on the epidermal surface for the separation of the new genera *Senna* and Chamaecrista from their initial genus *Cassia*.

According to Yasmin et al., [41] worked out the micro morphological investigation of foliar anatomy of the genera *Acononon* and *Bistorta* family Polygonaceae. They collected the material from the herbarium of Quaid-e-Azam University Islamabad. They studied anatomical features such as shape of epidermal stomata distribution on the abaxial and adaxial surface of leaves and trichome types under the light microscope. Among the species they found six types of stomata (paracytic, cyclocytic, staurocytic, anisocytic, anomocytic, and dicytic) and variation among glandular and non-glandular trichome. They just emphasized on the anatomical features of these species. They ignored the arrangement of epidermal cell, stoma variety and variation in trichome which could provide very important taxonomic tool for identification of species.

Ahmed et al., [42] worked out taxonomic diversity in epidermal cell of some sub-tropical plant species. For this purpose they selected 40 species belonging to 38 genera of 22 families. The shape and type of large epidermal cells among these varies from straight and polygonal. They concluded that large variation in epidermal cells configuration at different taxonomic level. But they ignored other anatomical characters that could provide an inter generic and intra generic variation in many species. There is a need of more comprehensive work on the shape and type of leaf epidermal cells for their further elaboration on taxonomic characters.

Shaheen et al., [43] worked on diversity of foliar trichome and their systematic relevance in the genus *Hibiscus* (Malvaceae). They collected the material from the different localities of Pakistan and dried leaves from the herbarium of Quaid-e-Azam University Islamabad. They examined qualitative and quantitative morphological characters. They distributed the systematic relevance of both glandular and non-glandular foliar trichome in species of genus *Hibiscus*. Glandular capitate trichome was the most dominant on leaf surface. Among of these species they just emphasized only on trichome anatomical marker.

Ahmad et al., [44] investigated 36 dicot species for stomata diversity from 34 genera and 20 families. They found seven types of stomata in which amphi anisocytic was the dominant one found in 12 species while staurocytic and diacytic were found in 7 and 6 species respectively. In 7 species two and in one species three different types of stomata were found.

Zou et al., [45] observed the micro morphology of leaf epidermal cells and stomata of all eight species and one form (11 samples) of *Cercis* by scanning electron microscopy and light microscopy (SEM & LM). They also observed that epidermal cells of both sides adaxial and abaxial were polygonal or irregular in shape; the anticline walls were straight and arched or undulate. They found two types of stomata, which occur only on the abaxial surface of the leaves. Atypical paracytic type is present in only one species, *Cercis chingii*, and the anomocytic type is present in all other species. Interspecific differences of leaf epidermal characters are minor in the genus, except for *C. chingii*, characterized by typical paracytic-type stomata.

Abid et al., [46] were examined the 3 types of Stomata (tetracytic, paracytic and anomocytic) of 54 monocot species having 42 genera in 10 families by using light and scanning electron microscopy. Their results showed that only tetracytic type was present in the family Commelinaceae and in the family Liliaceae both tetracytic and anomocytic types were found. While paracytic type was found in Typhaceae, Gramineae, Cyperaceae, Palmae and Juncaceae. Stomata was absent in the remaining three families i.e., Potamogetonaceae, Najadaceae and Hydrocharitaceae. Their findings clearly indicate that among the monocots of Karachi the paracytic type is most common irrespective of advance and primitive taxa.

Perveen et al., [47] examined Stomata morphology of 69 dicot species by using light and scanning electron microscopy distributed in 64 genera and 28 families. In all they recognized 6 types of stomata i.e., anomocytic, paracytic, diacytic, parallelocytic, cyclocytic and anisocytic. Most dominant was anomocytic found in 54 taxa out of 69 species, followed by paracytic and diacytic type, represented by 9 and 3 taxa respectively, but they recognized parallelocytic, cyclocytic and anisocytic types only in one species.

Wang et al., [48] explained that in *Arabidopsis thaliana*, stomata development is preceded by asymmetric cell divisions, and stomata distribution follows the one-cell spacing rule that reflecting the coordination of cell fate specification. They observed that Stomata development and patterning are regulated by both genetic and environmental signals. They reported that *Arabidopsis* MITOGEN-ACTIVATED PROTEIN KINASE 3 (MPK3) and MPK6, two environmentally responsive mitogen-activated protein kinases (MAPKs), and their upstream MAPK kinases, MKK4 and MKK5, are key regulators of stomata development and patterning. They explained that activation of MKK4/MKK5-MPK3/MPK6 causes the suppression of asymmetric cell divisions and stomata cell fate specification, resulting in a lack of stomata differentiation.

Agbagwa and Okoli [49] compared the leaf epidermis of three species such as *A. canescens* Welw. Ex Bak. *A. precatorius* L. and *A. pulchellus* Wall. Ex Thw. and also a new collection of *Abrus adanson* (Simly referred to as Abrus sp.) which is found in tropical West Africa. Comparison was undertaken by simple microscopy. They determined that epidermal cells were irregular with wavy, sinuous or arcuate anticline walls while the coastal cells were parallel, elongated and rectangular, which were terminat obliquely. They found six types of stomata such as paracytic, diacytic, anomocytic, anisocytic, staurocytic and contiguous in the species. *A. canescens* lacks anomocytic stomata. The new collection *Abrus* sp. was distinguished from the other three species because of its amphi stomatic nature and possession of stomata in-groups of two or three bordered by six continuous subsidiary cells on its abaxial surface. It also contains stomatal Index (SI) of 18.19, which is highest in the genus. The species contain simple unbranched trichomes occur on both surfaces of the leaf. Trichome index (TI) of 3.79 on the abaxial surface of *A. canescens*was observed to be the highest in the genus *Abrus*. The study was aimed at improve systematic information on the tropical West African species of *Abrus*.

Ozdemir and Altn, [50] investigated morphology and anatomy of some endemic *alkana* species. Two *alkana* species were examined for morphological and anatomical differences. Aim of study was to provide information of morphological and anatomical properties of *alkana friedinii*, Rech, Fie and *A. cardifalia* L. Koch. It has been observed that *A. froedinii* has some different characteristics from *A. cardiofolia* such as numerous setiform and glandular hairs.

According to Klimko et al., [51] worked on the morphological variability of the leaf epidemis in selected taxa of the genus *Ficus* L. (Moraceae) and its taxonomic implication. They collected the material from the botanical garden, Adam Mickiewicz University, Pozan. They selected the leaf epidermal of 36 taxa (Species, subspecies, varieties and cultivars) of the genus *Ficus* (subgenera *Ficus, Synoecia, Sycamores, osyces and Urostigmapharmac*) that was observed under light microscope and simple electron microscope. They revealed the number of micro morphological characters at taxonomic level. They optimized the plant growth under cultivation condition to minimize the modifying effect of environmental condition on the epidermal leaf traits. They emphasized on leaf epidermal characters and ignored other anatomical characters. If they observed these anatomical characters that could be helpful in producing better result.

Bergmann and Sack [52] studied the pathway of stomata formation in Arabidopsis thaliana where stomata are produced through a series of divisions in a dispersed stem cell compartment. The stomata pathway is considered as an accessible system for analyzing developmental processes such as position-dependent patterning via intercellular signaling. They contrast the function of genes that act earlier in the pathway, including receptors, kinases, and proteases, with those that act later in the cell lineage. They also explained the relationships between environmental signals, stomatal development genes, and the capacity for controlling shoot gas exchange.

Liscovsky and Cosa [53] worked on comparative anatomy of leaf and stem in species of *Cestreae* G. Don (Solanaceae) from Argentina. Leaf and stem anatomy of thirteen native and cultivated species of *lestrum* L. and *sessa vestioides* was studied. Leaf characters that best distinguished among species were secondary vein Architecture, marginal ultimate venation and stomata in hairs. Characteristic cuticle and presence of fibers adjacent to external and internal phloem in the middle vein, scleried in the middle vein, scleried presence and abundance in mesophyll and middle vein.

Carpenter et al., [54] worked out leaf cuticlar morphology link planteaceae and proteaceae. They collected the specimen from the foliar museum of natural history, university of Floride. For this purpose, they used the light and scanning electron microscopy for the investigation of leaf cuticular morphology of extent species of plantantaceae, they observed the anatomical characters types of stomata, epidermal cells on abaxial and adaxial surface. They studied the morphology linking plantanaceae and proteaceae. Although it is good piece of work yet they just emphasized on the particular characters among these species. But they ignored all other anatomical characters like shape and size of epidermal cells and arrangement of cells, variation in types of stomata cells and silica bodies etc. This could be proven very important taxonomic tool for identification of species.

Kanademir [55] studied the morphological, anatomical and karyological properties of endemic *Salvia hypergia Fich* and *Mey* (Lamiaceae). It was an endemic plant that was investigated. In these studies author observed that these species have perennial root system and herbaceous stem is rectangular, leaves are simple type, glandular and non-glandular hairs were observed on both sides of leaves. Anatomically, the internal morphological properties of root, stem, leaf, and petiole were observed.

According to Snir and Sachs [56] stomata are homologous structures that develop differently even in related species. They analyzed the developmental processes that generate cellular configurations in the pea epidermis. They explained that their variability is expressed by quantitative changes in the number of cell divisions and by local cell growth. They found the larger quantitative changes in the timing and degree of cell growth in stressed plants. They also explained that these processes were a basis for a survey of the epidermal development of 70 Fabaceae species such as In *Trifolium clypeatum* stomata were surrounded by a 'rosette' of epidermal cells result from an early maturation of the stomata, an example of heterochrony. In *Scorpiurus muricatus* there were parallel 'chains' of stomata, each with two subsidiary cells. So they resulted that the early developmental stages and unusual mature forms in the same *Scorpiurus* leaves show that this pattern could arise by small modifications of the orientation of divisions and the location of cell growth.

Kong [57] investigated leaf epidermis of 23 samples representing 16 species of all the four extant genera of the Chloranthaceae, i.e., *Sarcandra, Chloranthus, Ascarina* and *Hedyosm*um, by using both light microscope and scanning electron microscope. They observed some characters of leaf epidermis, such as pattern of epidermal cells, type of stomata, shape of guard cell pairs and cuticular ornamentation, were constant and thus of great significance in understanding the relationships between and within genera. They also explained that evidence from leaf epidermis, just as that from stem anatomy and cytology as well as sequence analysis of ITS region, suggested the separation of the genus on the basis of characteristics of androecial organs into two groups.

Idu et al., [58] described epidermal morphology, structure and development of stomata in 10 hardwood species of Fabaceae in *Nigerian*. In their result they observed that epidermal cells varied from irregular to straight-walled and in some taxa sinuous patterns were also observed. Leaves of family Fabaceae were hypo stomatic. They observed 2 types of stomata such as anomocytic and paracytic stomata. Anomocytic stomata followed an agenous ontogenetic pathway, while paracytic stomata were eumesogenously derived. These features of stomata were considered to be of systematic value in delimiting hardwood species in family Fabaceae. They analyzed the relative frequencies of stomata types bring out the taxonomic as well as phylogenetic significancies of stomata characteristics.

Sheteolu and ayodele [59] studied the epidermal morphology of adaxial and abaxial leaflet surfaces of five species of *Dialiirrn* (Fabaceae: Caesalpinoideae) by using light microscopy. They observed isodiametric, irregular or both types of epidermal cells. In epidermal cells anticline walls were straight, curved or undulate. They observed paracytic and staurocytic stomata. Paracytic stomata occur in all species except *D.guineense* that possess both types of stomata. In their result they observed variation in some features of the epidermis include stomatal size and density, size and density of epidermal cells, and type, size, shape, frequency and basal cells of the trichomes.

Ogundipe [60] studied leaf epidermal study of genus *Datura* L. of family Solanaceae. Five West African species of *Dautura* with narcotic properties were studied. The most useful anatomical characters for identification were palisade ratio, stomata density and types of crystals. *D. candida, D. suaveolens, D. dinnoxia, D. metal and D. stramonium* was studied and differentiated on the basis of these anatomical characters.

Kagan and Sachs [61] explained the development of immature stomata in *Sansevieria trifasciata.* They observed that the development of all stomata structures started at the same time and the early stages of the development of immature stomata had no special characteristics. They used statistical analysis that showed that the mature stomata were more evenly spaced than all potential stomata, both mature and immature. Furthermore, the distribution of mature stomata per unit area was more predictable or orderly than comparable structures of a random model that developed in the same way. These facts indicate that a nonrandom loss of many stomata by "immaturity" is a major determinant, acting during rather than preceding development, of the distribution of the mature, functional stomata. Thus they resulted that in *Sansevieria* there is a selection of an epidermal pattern from an excess of cells that undergo the early stages of stomata development.

Lackey [62] worked on leaflet anatomy of Phaseoleae (Leguminous: sub family papilionoideae) and its relation to taxonomy. In department of Botany, Smithsonian Institution, Washington, Author studied the leaflet anatomy of the tribe phaseolese by light microscopy of cleaned material and by scanning electron microscopy of leaflet surfaces. Anatomical results compare for the species belonging to recently revised sub tribal classification of the tribe. Taxonomically important characters that were found Bulbous based and vesicular glands are restricted to the sub tribe Canjaninae hooked hairs are present in three generic groups. Glandular hairs were particularly present, branched hairs were also present.

RESULTS

1. *Acacia catechu (Hook.)* Raf

a. Adaxial Side

i. Fully mature leaf

Trichomes, Silica bodies and oil droplets were absent. Cells were polygonal with smooth margins. Length of long cells ranged from 124.5μm-128.4μm (Av: 126.4μm). Width of long cells ranged from 27μm-28μm (Av: 27.5μm). Similarly length of short cells ranged from 40.7μm-60.2μm (Av: 50.4μm). Width of short cells ranged from 28.8μm-29.2μm (Av: 29μm). Stomata were present. Number of stomata per unit area was 5. Type of stomata was diacytic. Length of stomatal cavity ranged from 11.1μm-35.2μm (Av: 23.1μm). Width of stomatal cavity ranged from 4.1μm- 5.1μm (Av: 4.6μm). Length of guard cells were 35.4μm-71.5μm (Av: 53.4μm). Width of guard cells were 23.5μm-44.9μm (Av: 34.2μm). Length of long subsidiary cells ranged from 79.2μm-85.6μm (Av: 82.4μm). similarly length of short subsidiary cells ranged from 39.1μm-47μm (Av: 43μm). Width of long cells ranged from

17.1μm-22.5μm (Av: 19.8μm). Width of short cells ranged from 7.6μm-9.1μm (Av: 8.35μm). (Figure 1)

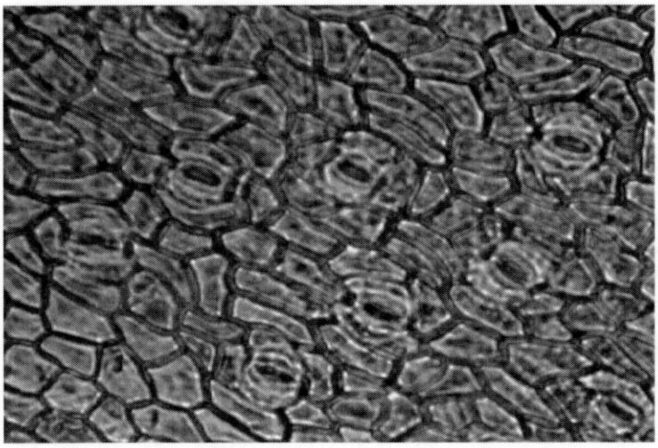

Figure 1. Leaf Epidermal Anatomy of *Acacia catechu.* Adaxial side.

ii. Mature leaf

Trichomes, Silica bodies and oil droplets were absent. Cells were polygonal with smooth margins. Length of long cells ranged from 66μm-90.2μm (Av: 78.1μm). Width of long cells ranged from 35μm-45.3μm (Av: 40.1μm). Length of short cells ranged from 27.8μm-33μm (Av: 30.4μm). Width of short cells ranged from 29.4μm- 34.2μm (Av: 31.8μm). Stomata were present. Type of stomata was diacytic. Length of stomatal cavity ranged from 25.2μm-34.7μm (Av: 29.9μm). Width of stomatal cavity ranged from 7.1μm-8.1μm (Av: 7.6μm). Length of guard cells were 27.5μm- 52.4μm (Av: 16.4μm). Width of guard cells were 24μm-37.7μm (Av: 30.8μm). Length of long subsidiary cells 58.5μm-69.4μm (Av: 63.9μm) Length of short subsidiary cells ranged from 36.6μm-42.2μm (Av: 39.4μm). Width of long cells ranged from 21.3μm-24.4μm (Av: 22.8μm). Width of short cells ranged from 17μm- 21.3μm (Av: 19.1μm).

iii. Immature leaf

Trichomes, Silica bodies and oil droplets were absent. Polygonal cells had smooth margins. Length of long cells ranged from 103μm-109.8μm (Av: 106.4μm). Width of long cells ranged from 32.8μm-33.6μm (Av: 33.2μm). Length of short cells ranged from 23.3μm-26.4μm (Av: 24.8μm). Width of short cells ranged from 42.8μm-46.4μm (Av: 44.6μm). Stomata were present. Type of stomata was diacytic. Length of stomatal cavity were 25.1μm-36.8μm (Av: 30.9μm). Width of stomatal cavity were 13μm-15.6μm (Av: 28.6μm). Length of guard cells were 42.9μm-52.3μm (Av: 46.7μm). Width of guard cells were 34μm-42.4μm (Av: 38.2μm). Length of long subsidiary cells ranged from 69.6μm-81.9μm (Av: 75.7μm). Length of short subsidiary cells ranged from 44.7μm- 45.3μm (Av: 45μm).Width of long cells ranged from 23.3μm-31.4μm (Av: 27.3μm). Width of short cells ranged from 8.5μm – 17.8μm (Av: 13.1μm).

b. Abaxial Side

i. Fully mature leaf

Trichomes, Silica bodies and oil droplets were absent. Polygonal cells had smooth margins. Length of long cells ranged from 83.9µm-87.4µm (Av: 85.6µm). Width of long cells ranged from 38.1µm-41.4µm (Av: 39.7µm). similarly length of short cells ranged from 25µm –33.5µm (Av: 29.2µm). Width of short cells ranged from 26µm-42µm (Av: 34µm). Stomata were present. Number of stomata per unit area was 10. Type of stomata was diacytic. Length of stomatal cavity ranged from 25.1µm-36.8µm (Av: 30.9µm). Width of stomatal cavity ranged from 436.8µm-41.5µm (Av: 39.1µm). Length of guard cells ranged from 19.9µm-22.8µm (Av: 21.3µm). Width of guard cells ranged from 54.2µm-69.6µm (Av: 61.9µm). Length of long subsidiary cells ranged from 78.8µm-83.9µm (Av: 81.35µm) Length of short subsidiary cells ranged from 29.8µm-32.3µm (Av: 31µm). Width of long cells ranged from 16.5µm- 38.3µm (Av: 27.6µm). Width of short cells ranged from 30.9µm-39.1µm (Av: 35µm).

ii. Mature leaf

Trichomes, Silica bodies and oil droplets were absent. Polygonal cells had smooth margins. Length of long cells ranged from 90.2µm-124.8µm (Av: 107.5µm). Width of long cells ranged from 31.9µm-49.8µm (Av: 40.8µm). Length of short cells ranged from 36.1µm-43.4µm (Av: 39.7µm). Width of short cells ranged from 35.6µm-37.6µm (Av: 36.6µm). Stomata were present. Number of stomata per unit area was 6. Type of stomata was diacytic. Length of stomatal cavity ranged from 44.4µm-52.8µm (Av: 48.6µm). Width of stomatal cavity ranged from 33.8µm-35.4µm (Av: 34.6µm). Length of guard cells ranged from 56.6µm-79.2µm (Av: 67.9µm). Width of guard cells ranged from 67.9µm-78.8µm (Av: 73.3µm). Length of long subsidiary cells ranged from 94.2µm-99.1µm (Av: 96.6µm) Length of short subsidiary cells ranged from 44.3µm-63.0µm (Av: 53.6µm). Width of long cells ranged from 19.2µm-19.8µm (Av: 19.5µm). Width of short cells ranged from 13.6µm – 21.4µm (Av: 17.5µm).

iii. Immature leaf

Trichomes, Silica bodies and oil droplets were absent. Polygonal cells had double layer and smooth margins. Length of long cells ranged from 90µm-93.6µm (Av: 91.8µm). Width of long cells ranged from 16µm-28.2µm (Av: 22.1µm). Similarly length of short cells ranged from 32.6µm-33µm (Av: 32.8µm). Width of short cells ranged from 21µm-36.3µm (Av: 28.6µm). Stomata were present. Number of stomata per unit area was 8. Type of stomata was diacytic. Length of stomatal cavity ranged from 42µm-120.2µm (Av: 81.1µm). Width of stomatal cavity ranged from 7.6µm-17µm (Av: 12.3µm). Length of guard cells ranged from 44.9µm-76.2µm (Av: 60.5µm). Width of guard cells ranged from 35.4µm-45.3µm (Av: 40.3µm). Length of long subsidiary cells ranged from 87.3µm-94.1µm (Av: 87.7µm) Length of short subsidiary cells ranged from 18.4µm-30.5µm (Av: 24.4µm). Width of long cells ranged from 8.1µm-11µm (Av: 9.5µm). Width of short cells ranged from 8.1µm-11.7µm (Av: 9.9µm).

2. *Acacia modesta* Wall

a. Adaxial Side

i. Fully mature leaf

Silica bodies and oil droplets were absent. Glandular trichomes were present. Length of glandular trichomes ranged from 32.7μm-77.2μm (Av: 54.7μm). Width of glandular trichomes ranged from 30μm-37.2μm (Av: 33.6μm). Polygonal cells had double layer and smooth margins. Length of long cells ranged from 91.8μm-97.0μm (Av: 94.4μm). Width of long cells ranged from 45.2μm-49.8μm (Av: 47.5μm). Length of short cells ranged from 30.6μm-36.9μm (Av: 33.7μm). Width of short cells ranged from 30.5μm-44.2μm (Av: 37.35μm).

ii. Mature leaf

Oil droplets were absent. Irregular silica bodies were present. Number of silica bodies per unit area was 2. Length of silica bodies ranged from 24.0μm- 13.0μm (Av: 18.5μm). Width of silica bodies ranged from 14.0μm-19.0μ (Av: 16.5μm). Polygonal cells had double layer and smooth margins. Length of long cells ranged from 97.1μm-100.4μm (Av: 98.7μm). Width of long cells ranged from 53.3μm-55.4μm (Av: 54.3μm). Length of short cells ranged from 45.0μm-46.0μm (Av: 45.5μm). Width of short cells ranged from 59.2μm- 68.1μm (Av: 63.6μm). Non-glandular trichomes with blunt tip were present. Length of non-glandular trichomes ranged from 37.4μm-202.1μm (Av: 119.7μm). Width of non-glandular trichomes ranged from 4.0μm-7.1μm (Av: 5.5μm).

iii. Immature leaf

Silica bodies and oil droplets were absent. Glandular trichomes were present. Length of glandular trichomes ranged from 26.9μm-57.3μm (Av: 42.1μm). Width of glandular trichomes ranged from 26.2μm-34.4μm (Av: 30.3μm). Polygonal cells had double layer and smooth margins. Length of long cells ranged from 70.2μm- 73.2μm (Av: 71.7μm). Width of long cells ranged from 19.4μm-33.0μm (Av: 26.2μm). Length of short cells ranged from 21.0μm-23.0μm (Av: 22μm). Width of short cells ranged from 26.0μm-29.3μm (Av: 27.6μm).

b. Abaxial Side

i. Fully mature leaf

Silica bodies, trichomes and oil droplets were absent. Polygonal cells had double layer and smooth margins. Length of long cells ranged from 87.2μm-96.4μm (Av: 91.8μm). Width of long cells ranged from 30.5μm-59.8μm (Av: 45.1μm). Length of short cells ranged from 25.0μm-34.1μm (Av: 29.5μm). Width of short cells ranged from 22.0μm-38.0μm (Av: 30.0μm). Stomata were present. Number of stomata per unit area was 10-11. Type of stomata was cyclocytic. Length of stomatal cavity ranged from 24.2μm-32.4μm (Av: 28.3μm). Width of stomatal cavity ranged from 21.4μm-26.4μm (Av: 23.9μm). Length of guard cells ranged from 57.2μm- 71.6μm (Av: 64.4μm). Width of guard cells ranged from 58.5μm-64.4μm (Av: 61.4μm). Length of long subsidiary cells ranged from 67.2μm- 82.3μm (Av: 74.7μm) Length of short subsidiary cells ranged from 25.0μm-34.1μm (Av: 29.5μm). Width of long cells ranged from 30.5μm-59.8μm (Av: 45.1μm). Width of short cells ranged from 22.0μm-38.0μm (Av: 30.0μm).

ii. Mature leaf

Oil droplets were absent. Rounded silica bodies were present. Number of silica bodies per unit area was 2. Length of silica bodies ranged from 32.0μm-77.2μm (Av: 54.6μm). Width of silica bodies ranged from 30.3μm-76.1μm (Av: 53.2μm). Polygonal cells had double layer and smooth margins. Length of long cells ranged from 109.2μm-116.0μm (Av: 112.6μm). Width of long cells ranged from 46.3μm- 52.0μm (Av: 49.1μm). Length of short cells ranged from 21.0μm-30.0μm (Av: 25.5μm). Width of short cells ranged from 28.0μm-34.1μm (Av: 31.0μm). Non-glandular trichomes were present. Length of non-glandular trichomes ranged from 90.4μm-317.2μm (Av: 203.8μm). Width of non-glandular trichomes ranged from 21.2μm-24.1μm (Av: 22.6μm). Stomata were present. Number of stomata per unit area were10-11. Type of stomata was cyclocytic. Length of stomatal cavity ranged from 21.4μm-34.1μm (Av: 27.7μm). Width of stomatal cavity ranged from 24.1μm- 32μm (Av: 28.0μm). Length of guard cells ranged from 53.9μm-70.7μm (Av: 62.3μm). Width of guard cells ranged from 45.2μm-64.0μm (Av: 54.6μm). Length of long subsidiary cells ranged from 58.8μm-64.5μm (Av: 61.6μm) Length of short subsidiary cells ranged from 23.3μm -26.1μm (Av: 24.7μm). Width of long cell ranged from 13.4μm- 17.1μm (Av: 15.2μm). Width of short cells ranged from 9.4μm -9.8μm (Av: 9.6μm)

iii. Immature leaf

Silica bodies, trichomes and oil droplets were absent. Polygonal cells had double layer and smooth margins. Length of long cells ranged from 97.5μm- 105.5μm (Av: 101.5μm). Width of long cells ranged from 26.2μm-27.3μm (Av: 26.7μm). similarly length of short cells ranged from 20.0μm-25.0μm (Av: 22.5μm). Width of short cells ranged from 30.3μm-31.1μm (Av:30.7μm). Stomata were present. Number of stomata per unit area was 5-6. Type of stomata was cyclocytic. Length of stomatal cavity ranged from 22.2μm-27.2μm (Av: 24.7μm). Width of stomatal cavity ranged from 19.1μm-30.0μm (Av: 24.5μm). Length of guard cells ranged from 51.2μm-.6μm). Length of long subsidiary cells ranged from 41.2μm-49.3μm (Av: 45.2μm). Length of short subsidiary cells ranged from 17.0μm-24.1μm (Av: 20.5μm). Width of long cell ranged from 10.2μm-11.7μm (Av: 10.9μm). Width of short cells ranged from 17.1μm-27.0μm (Av: 22.0μm). (Figure 2)

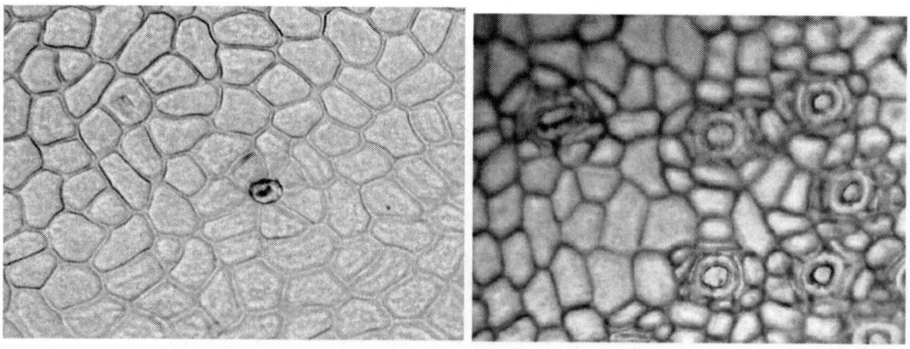

Figure 2. Leaf Epidermal Anatomy of *Acacia modesta.* Adaxial side (Right) Abaxial side (Left).

3. *Albizia lebbek* (L.) Benth

a. Adaxial Side

i. Fully mature leaf

Trichomes, Silica bodies and oil droplets were absent. Tetragonal cells had smooth margins. Length of long cells ranged from 137.8μm-139.6μm (Av: 138.7μm). Width of long cells ranged from 52.2μm-72.8μm (Av: 62.5μm). Length of short cells ranged from 49.5μm-30μm (Av: 39.7μm). Width of short cells ranged from 3.8μm-30.1μm (Av: 16.9μm). Stomata were present. Type of stomata was diacytic. Length of stomatal cavity ranged from 55.3μm-20.6μm (Av: 37.9μm). Width of stomatal cavity ranged from 4.1μm-4.5μm (Av: 4.3μm). Length of guard cells ranged from 26.2μm-70.9μm (Av: 48.5μm). Width of guard cells ranged from 40.8μm-42μm (Av: 41.4μm). Length of long subsidiary cells ranged from 113.6μm-123.4μm (Av: 118.5μm). Length of short subsidiary cells ranged from 64.3μm-68.3μm (Av: 66.3μm). Width of long cells ranged from 39.3μm-61.1μm (Av: 50.2μm). Width of short cells ranged from 36.7μm-45.2μm (Av: 4.9μm).

ii. Mature leaf

Irregular silica bodies were present. Number of silica bodies per unit area was 2. Length of silica bodies ranged from 42μm- 53.1μm (Av: 47.5μm). Width of silica bodies ranged from 27.1μm-30.1μm (Av: 28.6μm). Polygonal cells had wavy margins. Non-glandular trichomes were present. Length of non-glandular trichomes ranged from 322μm-22.8μm (Av: 275.3μm). Width of trichomes ranged from 6.3μm-12.6μm (Av: 9.4μm).Length of long cells ranged from 200.6μm-292.5μm (Av: 246.5μm). Width of long cells ranged from 121.1μm-177.1μm (Av: 149.1μm). Length of short cells ranged from 68.1μm-85.6μm (Av: 76.8μm). Width of short cells ranged from 91μm-92.8μm (Av: 91.9μm).Stomata were present. Type of stomata was diacytic. Length of stomatal cavity ranged from 140μm-143.1μm (Av: 141.5μm). Width of stomatal cavity ranged from 30.4μm - 36μm (Av: 33.2μm). Length of guard cells ranged from 143.1μm-140μm (Av: 141.5μm). Width of guard cells ranged from 111.0μm- 198.5μm (Av: 154.7μm). Length of long subsidiary cells ranged from 195.0μm-198.1μm (Av: 196.5μm). Length of short subsidiary cells ranged from 97μm-100.1μm (Av: 98.5μm). Width of long cells ranged from 41.2μm -52.2μm (Av: 46.7μm). Width of short cells ranged from 45.2μm-51.1μm (Av: 48.1μm).

iii. Immature leaf

Trichomes, silica bodies and oil droplets were absent. Polygonal cells had wavy margins. Length of long cells ranged from 193.4μm-207.9μm (Av: 200.6μm). Width of long cells ranged from 62.4μm-72μm (Av: 67.2μm). Length of short cells ranged from 113μm-115μm (Av: 114μm). Width of short cells ranged from 76.4μm-76.5μm (Av: 76.4μm). Stomata were present. Type of stomata was diacytic. Length of stomatal cavity ranged from 10.3μm-10.4μm (Av: 10.3μm). Width of stomatal cavity ranged from 6.3μm –6.9μm (Av: 6.6μm). Length of guard cells ranged from 77.5μm - 123μm (Av: 100.2μm). Width of guard cells ranged from 52.2μm-55.5μm (Av: 107.7μm). Length of long subsidiary cells ranged from 154μm-193.4μm (Av: 200.6μm). Length of short subsidiary cells ranged from 100.6μm-

102.9μm (Av: 101.7μm). Width of long cells ranged from 40μm-57.6μm (Av: 48.8μm). Width of short cells ranged from 26.9μm-42μm (Av: 34.4μm).

b. Abaxial Side

i. Fully mature leaf

Trichomes and oil droplets were absent. Rounded silica bodies were present. Number of silica bodies per unit area was 2. Length of silica bodies ranged from 103μm-104μm (Av: 103.5μm). Width of silica bodies ranged from 109.1μm-110μm (Av: 109.5μm). Tetragonal cells had wavy margins. Length of long cells ranged from 140.5μm-188.3μm (Av: 164.4μm). Width of long cells ranged from 69.6μm-87.3μm (Av: 87.4μm). Length of short cells ranged from 80.5μm-85μm (Av: 182.7μm). Width of short cells ranged from 85.4μm-107.1μm (Av: 96.2 μm).

ii. Mature leaf

Trichomes, silica bodies and oil droplets were absent. Polygonal cells had double layer and smooth margins. Length of long cells ranged from 126.8μm-132.1μm (Av: 129.4μm). Width of long cells ranged from 2.2μm-70.1μm (Av: 66.1μm). Length of short cells ranged from 59.1μm-70.6μm (Av:64.8μm). Width of short cells ranged from 58.1μm-46.5μm (Av: 52.8μm). Stomata were present. Type of stomata was diacytic. Length of stomatal cavity ranged from 20.6μm-34.5μm (Av: 27.5μm). Width of stomatal cavity ranged from 5.1μm-5.8μm (Av: 5.4μm). Length of guard cells ranged from 53.6μm-86.4μm (Av: 70μm). Width of guard cells ranged from 32.0μm-47.6μm (Av: 39.8μm). Length of long subsidiary cells ranged from 117μm-144.9μm (Av: 130.9μm). Length of short subsidiary cells ranged from 71.2μm-73.1μm (Av: 72.1μm). Width of long cells ranged from 35μm-48.0μm (Av: 41.5μm). Width of short cells ranged from 32.6μm-32.7μm (Av: 32.6μm). (Figure 3)

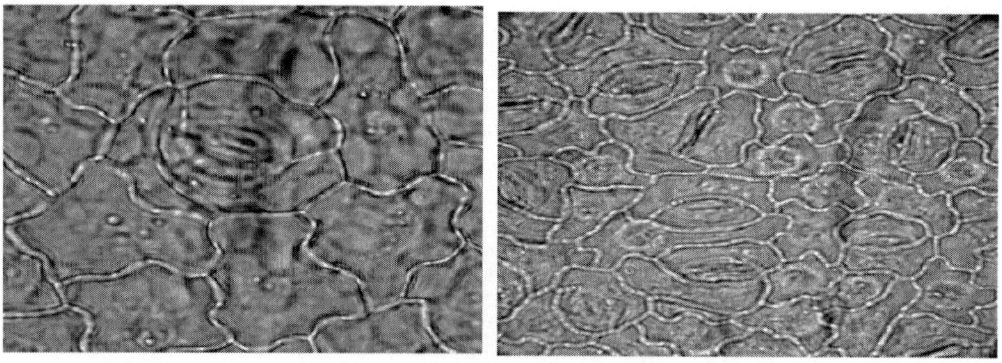

Figure 3. Leaf Epidermal Anatomy of *Albizia lebbek.* Adaxial side (Right) Abaxial side (Left).

iii. Immature leaf

Trichomes, silica bodies and oil droplets were absent. Polygonal cells had smooth margins. Length of long cells ranged from 161μm-104.8μm (Av: 132.9μm). Width of long cells ranged from 47.2μm – 58.7μm (Av: 52.9μm). Length of short cells ranged from 55μm – 81.1μm (Av: 68μm). Width of short cells ranged from 88.0μm-99.0μm (Av: 93.5 μm).Stomata were present. Type of stomata was diacytic. Length of stomatal cavity ranged from 22.4µm-82.9µm (Av: 52.6µm). Width of stomatal cavity ranged from 9.0µm-10.0µm (Av: 9.5µm). Length of guard cells ranged from 80.1µm-109.8µm (Av: 94.9µm). Width of guard cells ranged from 37µm -59µm (Av: 48µm). Length of long subsidiary cells ranged from 130.3µm-237.6µm (Av: 183.9µm) Length of short subsidiary cells ranged from 49µm-59µm (Av: 183.9µm). Width of long cells ranged from 30µm-59.7µm (Av: 44.8µm). Width of short cells ranged from 21µm-29µm (Av: 44.8µm). (Figure 3)

4. *Albizia mimosa* (Leaf), Benth

a. Adaxial Side

i. Fully mature leaf

Trichomes, stomata and oil droplets were absent. Rounded silica bodies were present. Number of silica bodies per unit area was 2. Length of silica bodies ranged from 50µm-88µm (Av: 69µm). Width of silica bodies ranged from 47µm-97 µm (Av: 72µm). Polygonal cells had double layer and wavy margins. Length of long cells ranged from 219.7µm-230.1 µm (Av: 224.9µm).Width of long cells ranged from 96.5µm-12.4µm (Av: 54.4µm). Length of short cells ranged from 89.4-13.9 (Av: 51.6µm). Width of short cells ranged from 103.1-126.1 (Av: 114.6µm).

ii. Mature leaf

Trichomes, Silica bodies and oil droplets were absent. Polygonal cells had double layer and wavy margins. Length of long cells ranged from 234.5µm-225.4µm (Av: 229.9µm). Width of long cells ranged from 45.2 µm-66.9 µm (Av: 56.0µm). Length of short cells ranged from 84.1µm-86.0µm (Av: 85.0 µm). Width of short cells ranged from 92.3µm-132.5µm (Av: 112.4µm). Stomata were present. Type of stomata was anomocytic. Length of stomatal cavity ranged from 38.9µm-47.5µm (Av: 43.2µm). Width of stomatal cavity ranged from 17.0µm - 19.8µm (Av: 18.4µm). Length of guard cells ranged from 65.0µm -80.1µm (Av: 72.5µm). Width of guard cells ranged from 43µm - 50.6µm (Av: 46.8µm). Length of long subsidiary cells ranged from 156.1µm-170.6µm (Av: 163.3µm). Length of short subsidiary cells ranged from 55.3µm -83.2µm (Av. 69.2µm). Width of long cells ranged from 64.0µm -76.1µm (Av. 70.0µm). Width of short cells ranged from 23.0µm-24.4µm (Av: 23.7µm).

iii. Immature leaf

Trichomes, stomata, Silica bodies and oil droplets were absent. Polygonal cells had double layer and wavy margins. Length of long cells ranged from 154.9µm-162.8µm (Av: 158.8 µm). Width of long cells ranged from 84µm-106.3µm (Av: 95.1µm). Length of short cells ranged from 88.2µm-105.6µm (Av: 96.9µm). Width of short cells ranged from 100.0µm-105.7µm (Av: 102.8µm).

b. Abaxial Side

i. Fully mature leaf

Trichomes, Silica bodies and oil droplets were absent. Polygonal cells had double layer and wavy margins. Length of long cells ranged from 106.1μm-120.1μm (Av: 113.1μm). Width of long cells ranged from 49.0μm-52.2μm (Av: 50.6μm). Length of short cells ranged from 53.0μm-58.0μm (Av: 55.5μm). Width of short cells ranged from 61.0μm-75.1μm (Av:68.0μm). Stomata were present. Type of stomata was anomocytic. Length of stomatal cavity ranged from 22μm-26.3μm (Av: 24.1μm). Width of stomatal cavity ranged from 17.0μm-14.1μm (Av: 15.5μm). Length of guard cells ranged from 62.8μm-66.6μm (Av: 64.7μm). Width of guard cells ranged from 49.2μm-54.7μm (Av: 51.9μm). Length of long subsidiary cells ranged from 112.0μm -122.3μm (Av: 117.5μm) Length of short subsidiary cells ranged from 54.3μm-62.2μm (Av: 58.2μm). Width of long cells ranged from 38.1μm-59.2μm (Av: 48.6μm). Width of short cells ranged from 9.4μm-15.1μm (Av: 12.2μm). (Figure 4)

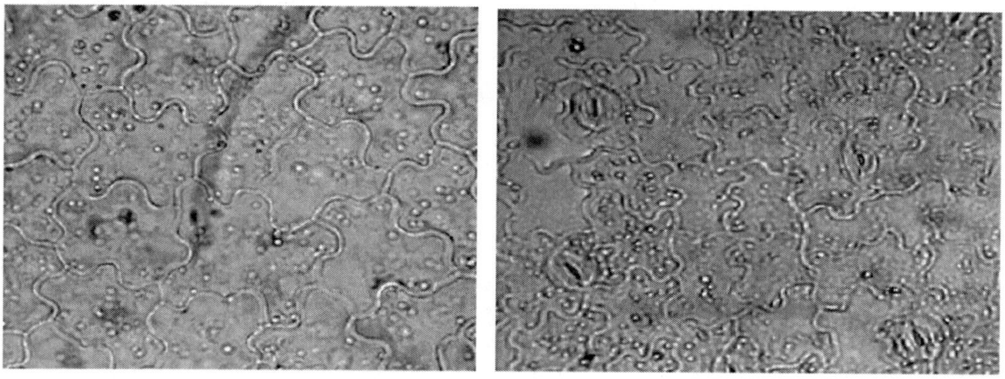

Figure 4. Leaf Epidermal Anatomy of *Albizia mimosa*. Adaxial side (Right) Abaxial side (Left).

ii. Mature leaf

Trichomes and oil droplets were absent. Rounded silica bodies were present. Length of silica bodies ranged from 30.0μm-30.9μm (Av: 30.4μm). Width of silica bodies ranged from 30.1μm-31.0μm (Av: 30.5μm). Polygonal cells had double layer and wavy margins. Length of long cells were 234.4μm-295.5μm (Av: 264.9μm). Width of long cells were 82.8μm-236.3μm (Av: 159.5μm). Length of short cells ranged from 160.8μm-121.8μm (Av: 141.3μm). Width of short cells ranged from 192.8μm-214.2μm (Av:203.5μm). Stomata were present. Type of stomata was anomocytic. Length of stomatal cavity ranged from 18.0μm-79.7μm (Av: 48.8μm). Width of stomatal cavity ranged from 8.5μm -17.3μm (Av:12.9μm). Length of guard cells ranged from 37.1μm-99.3μm (Av: 68.2μm). Width of guard cells ranged from 27.5μm-41.8μm (Av: 34.6μm). Length of long subsidiary cells ranged from 145.0μm -185.8μm (Av: 165.4μm) Length of short subsidiary cells ranged from 36.2μm-47.5μm (Av: 41.8μm). Width of long cells ranged from 39μm-66.3μm (Av: 52.6μm). Width of short cells ranged from 14.9μm- 26.6μm (Av: 20.7μm). (Figure 4)

iii. Immature leaf

Trichomes, stomata, Silica bodies and oil droplets were absent. Polygonal cells had double layer and wavy margins. Length of long cells ranged from 154.7µm-173.4µm (Av: 164.0µm). Width of long cells ranged from 68.4µm-84.0µm (Av: 76.2µm). Length of short cells ranged from 68.1µm-74.1µm (Av: 71.1µm). Width of short cells ranged from 92.3µm-98.0µm (Av: 95.1µm).

5. *Albizia procera*, (Roxb.) Benth

a. Adaxial Side

i. Fully mature leaf

Silica bodies and oil droplets were absent. Glandular and stellate trichomes with blunt end present. Length of trichomes ranged from 278.8µm-309.1µm (Av: 290.6µm). Width of trichomes ranged from 24.1µm-32.7µm (Av: 26.2µm). Polygonal cells had double layer and smooth margins. Length of long cells ranged from 142.1µm- 143.7µm (Av: 142.9µm). Width of long cells ranged from 62.2µm-72.2µm (Av: 67.2µm). Length of short cells ranged from 63.5µm- 68.0µm (Av: 65.7µm). Width of short cells ranged from 39.1µm- 75.0µm (Av: 57.0µm).

ii. Mature leaf

Silica bodies and oil droplets were absent. Glandular and non-glandular trichomes were present. Number of trichomes per unit area was 2. Length of trichomes ranged from 310.0µm-314.7µm (Av: 312.3µm). Width of trichomes ranged from 77.8µm- 78.1µm (Av: 77.9µm). Polygonal cells had double layer and smooth margins. Length of long cells ranged from 112.3µm-137.1µm (Av: 124.7µm). Width of long cells ranged from 57.2µm-118.1µm (Av: 87.6µm). Length of short cells ranged from 48.5µm-50.5µm (Av: 99.0µm). Width of short cells ranged from 56.2µm- 99.3µm (Av: 77.7µm).

iii. Immature leaf

Silica bodies and oil droplets were absent. Glandular and stellate trichomes were present with blunt end. Length of trichomes ranged from 181.2µm-390.7µm (Av: 280.2µm). Width of trichomes ranged from 17.7µm-23.0µm (Av: 20.7µm). Polygonal cells had double layer and smooth margins. Length of long cells ranged from 145.5µm-151.8µm (Av: 148.6µm). Width of long cells ranged from 75.2µm- 97.7µm (Av: 86.4µm). Length of short cells ranged from 49.4µm- 54.4µm (Av: 51.9µm). Width of short cells ranged from 58.7µm- 84.6µm (Av: 71.6µm)

b. Abaxial Side

i. Fully mature leaf

Silica bodies and oil droplets were absent. Glandular and stellate trichomes were present with blunt end. Length of trichomes ranged from 376.4µm- 420.0µm (Av: 391.2µm). Width of trichomes ranged from 25.0µm-32.0µm (Av: 43.9µm). Polygonal cells had double layer and wavy margins. Length of long cells ranged from 112.3μm-119.5μm (Av: 115.9μm). Width of long cells ranged from 56.6μm-66.9μm (Av: 61.7μm). Length of short cells ranged from 46.3μm- 51.0μm (Av: 97.3μm). Width of short cells were 56.0μm- 70.0μm (Av:63μm). Stomata were present. Type of stomata was diacytic. Length of stomatal cavity ranged from 33.3μm-52.2μm (Av: 42.7μm). Width of stomatal cavity ranged from 13.6μm-23.3μm (Av: 18.4μm). Length of guard cells were 81.0μm- 84.7μm (Av: 82.8μm). Width of guard cells were 56.3μm- 63.1μm (Av: 59.7μm). Length of long subsidiary cells ranged from 124.0μm-136.3μm (Av: 130.1μm) Length of short subsidiary cells ranged from 68.9μm-76.2μm (Av: 72.5μm). Width of long cell ranged from 18.8μm- 63.3μm (Av: 41.0μm). Width of short cells ranged from 16.6μm-30.0μm (Av: 23.3μm) (Figure 5)

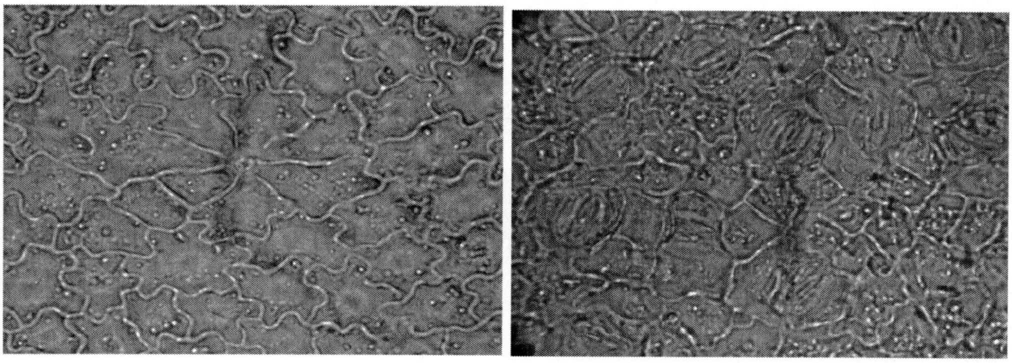

Figure 5. Leaf Epidermal Anatomy of *Albizia procera*. Adaxial side (Right) Abaxial side (Left).

ii. Mature leaf

Silica bodies and oil droplets were absent. Glandular and non-glandular trichomes were present. Length of non-glandular trichomes were 246.0μm – 677.0μm (Av: 461.5μm). Width of trichomes were 23.1μm – 34.9μm (Av: 29.0μm). Polygonal cells had double layer and wavy margins. Length of long cells ranged from 101.1μm-20.1μm (Av: 110.5μm). Width of long cells were 85.6μm-69.3μm (Av: 77.4μm). Length of short cells were 42.0μm – 58.1μm (Av: 50.0μm). Width of short cells were 69.1μm-85.2μm (Av: 77.1μm). Stomata were present. Type of stomata was diacytic. Length of stomatal cavity ranged from 27.1μm-53.2μm (Av: 40.1μm). Width of stomatal cavity ranged from 13.1μm – 23.3μm (Av: 18.2μm). Length of guard cells ranged from 61.3μm- 89.9μm (Av: 75.6μm). Width of guard cells ranged from 48.5μm-57.4μm (Av: 52.9μm). Length of long subsidiary cells 111.1μm-113.5μm (Av: 112.3μm). Length of short subsidiary cells ranged from 56.1μm-70.4μm (Av: 63.2μm). Width of long cells ranged from 31.4μm- 39.1μm (Av: 35.2μm). Width of short cells ranged from 21.1μm- 24.2μm (Av: 22.6μm).

iii. Immature leaf

Silica bodies and oil droplets were absent. Glandular and non-glandular trichomes were present. Length of non-glandular trichomes were 381.1μm-389.5μm (Av: 385.3μm). Width of trichomes were 42.1μm-46.0μm (Av: 44.0μm). Polygonal cells had double layer and wavy margins. Length of long cells were 137.4μm-208.7μm (Av: 173.0μm). Width of long cells were 39.4μm-66.6μm (Av: 53.0μm). Length of short cells were 75.3μm-78.0μm (Av: 76.6μm). Width of short cells were 48.8μm-116.0μm (Av:82.4μm). Stomata were present.

Type of stomata was diacytic. Length of stomatal cavity ranged from 28.9μm-54.2μm (Av: 41.5μm). Width of stomatal cavity ranged from 12.0μm-22.0μm (Av: 17.0μm). Length of guard cells ranged from 70.0μm-98.0μm (Av: 84.0μm). Width of guard cells ranged from 52.2μm-53.3μm (Av: 105.5μm). Length of long subsidiary cells ranged from 145.0μm-165.6μm (Av: 155.3μm) Length of short subsidiary cells ranged from 81.4μm-101.0μm (Av: 182.4μm). Width of long cells ranged from 39.1μm-40.2μm (Av: 39.6μm). Width of short cells ranged from 28.1μm-40.0μm (Av: 34.0μm) (Figure 5)

6. *Cassia fistula* L.

a. Adaxial Side

i. Fully mature leaf

Silica bodies and stomata were absent. Pentagonal cells had double layer and smooth margins. Length of long cells 104μm-111μm (Av: 107.5μm). Width of long cells ranged from 80.6μm – 116.4μm (Av: 98.5μm). Length of short cells ranged from 57.8μm-77.9μm (Av: 67.8μm). Width of short cells ranged from 75.9μm-8.4μm (Av: 87.1μm). Length of non-glandular trichomes ranged from 208.0μm-5.8μm (Av: 131.9μm). Width of non-glandular trichomes ranged from 11.5μm- 36.4μm (Av: 23.9μm).

ii. Mature leaf

Silica bodies and stomata were absent. Polygonal cells had double layer and wavy margins. Length of long cells ranged from 118.2μm-149.6μm (Av: 133.9μm). Width of long cells ranged from 56.3μm-83.5μm (Av: 69.9μm). Length of short cells ranged from 76.8μm-86μm (Av: 81.4μm). Width of short cells ranged from 87.1μm-102.5μm (Av: 94.8μm). Non-glandular trichomes were present. Length of non-glandular trichomes were 56.7μm-223.2μm (Av: 279.9μm). Width of non-glandular trichomes were 12.2μm-28.1μm (Av: 120.1μm).

iii. Immature leaf

Silica bodies and stomata were absent. Pentagonal cells had double layer and smooth margins. Length of long cells ranged from 116.2μm-135.7μm (Av: 125.9μm). Width of long cells ranged from 37.2μm-45.3μm (Av: 41.2μm). Length of short cells ranged from 35.7μm-49μm (Av: 42.3μm). Width of short cells ranged from 58.3μm-59.0μm (Av:58.6μm). Non-glandular trichomes were present. Length of non-glandular trichomes were 67.3μm-218.5μm (Av: 142.9μm). Width of non-glandular trichomes were 18.4μm- 30μm (Av: 24.2μm).

b. Abaxial Side

i. Fully mature leaf

Silica bodies and stomata were absent. Polygonal cells had double layer and wavy margins. Length of long cells ranged from 44.9μm-53.5μm (Av: 49.2μm). Width of long cells ranged from 18.1μm-20.2μm (Av: 19.1μm). Length of short cells ranged from 22μm-22.6μm (Av: 22.3μm). Width of short cells ranged from 8.2μm- 19.0μm (Av:13.6μm).

Non-glandular trichomes were present. Length of non-glandular trichomes were 62.0µm-44.9µm (Av: 53.4µm). Width of non-glandular trichomes were 13µm-15.4µm (Av: 14.2µm). (Figure 6)

ii. Mature leaf

Silica bodies and stomata were absent. Polygonal cells had double layer and wavy margins. Length of long cells ranged from 92.4µm-94.6µm (Av: 93.5µm). Width of long cells ranged from 37.7µm-37.9µm (Av: 37.8µm). Length of short cells ranged from 39.5µm-41.9µm (Av: 40.7µm). Width of short cells ranged from 46.1µm-3.1µm (Av: 59.6µm). Non-glandular trichomes were present with blunt end. Length of non-glandular trichomes were 100.2µm-107.6µm (Av: 103.9µm). Width of non-glandular trichomes were 16.6µm-19.9µm (Av: 18.2µm). Stomata were present. Type of stomata was diacytic. Length of stomatal cavity ranged from 21.9µm-42.5µm (Av: 32.2µm). Width of stomatal cavity ranged from 5µm-12µm (Av: 8.5µm). Length of guard cells were 61µm-78.5µm (Av: 69.7µm). Width of guard cells were 50.8µm-58.1µm (Av: 54.4µm). Length of long subsidiary cells ranged from 92.8µm-108.7µm (Av: 100.7 µm) Length of short subsidiary cells ranged from 31.3µm-60.1µm (Av: 15.8µm). Width of long cell ranged from 13.6µm-18µm (Av: 45.7µm). Width of short cells ranged from 10.3µm-13µm (Av: 11.6µm) (Figure 6)

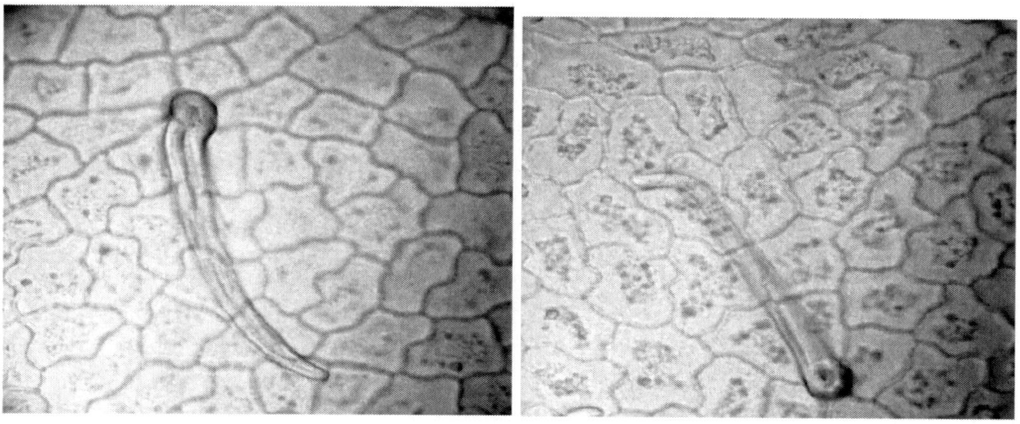

Figure 6. Leaf Epidermal Anatomy of *Cassia fistula.* Adaxial side (Right) Abaxial side (Left).

iii. Immature leaf

Silica bodies and stomata were absent. Polygonal cells had double layer and wavy margins. Length of long cells ranged from 92.4µm-94.6µm (Av: 93.5µm). Width of long cells ranged from 37.7µm-37.9µm (Av: 37.8µm). Length of short cells ranged from 39.5µm-41.9µm (Av: 40.7µm). Width of short cells 46.1µm-73.1µm (Av:59.6µm). Non-glandular trichomes with blunt ends were present. Length of non-glandular trichomes ranged from 62.3µm-320.1µm (Av: 19.1µm). Width of non-glandular trichomes ranged from 40.8µm-51µm (Av: 45.9µm).

7. *Cassia occidentalis* L.

a. Adaxial Side

i. Fully mature leaf

Stomata were absent. Irregular silica bodies were present. Number of silica bodies per unit area was 1. Length of silica bodies were 26.1µm-28µm (Av: 27µm). Width of silica bodies ranged from 34µm-23.1µm (Av: 28.5µm). Cells were polygonal and smooth margins. Length of long cells ranged from 161.3µm-168µm (Av: 164.6µm). Width of long cells ranged from 87.9µm-104µm (Av: 95.9µm). Length of short cells were 91.3µm-98.4µm (Av: 94.8µm). Width of short cells were 83.7µm-106.2µm (Av:94.9µm). Non-glandular trichomes were present. Number of trichomes per unit area was 27-28. Length of non-glandular trichomes ranged from 79.2µm- 153.6µm (Av: 116.4µm). Width of non-glandular trichomes ranged from 8µm- 8.1µm (Av: 8.1µm).

ii. Mature leaf

Silica bodies and stomata were absent. Polygonal cells had double layer and wavy margins. Length of long cells ranged from 162.9µm-176.5µm (Av: 169.7µm). Width of long cells ranged from 114.5µm-172.4µm (Av: 143.4µm). Length of short cells were 135µm-137.3µm (Av: 136.1µm). Width of short cells 7were 3µm-166.1µm (Av:119.5µm). Non-glandular trichomes were present. Length of non-glandular trichomes were 85.9µm-161µm (Av: 123.4µm). Width of non-glandular trichomes were 9.2µm- 14µm (Av: 11.6µm).

iii. Immature leaf

Silica bodies and stomata were absent. Polygonal cells had double layer and wavy margins. Length of long cells ranged from 124.9µm-135.6µm (Av: 130.2µm). Width of long cells ranged from 99.2µm-70µm (Av: 84.6µm). Length of short cells ranged from 74.7µm-91.9µm (Av: 83.3µm). Width of short cells ranged from 54µm-62.1µm (Av:58µm). Non-glandular trichomes were present. Length of non-glandular trichomes were 93.8µm- 497µm (Av: 370.2µm). Width of non-glandular trichomes were 9.1µm-46.2µm (Av: 32.9µm). (Figure 7)

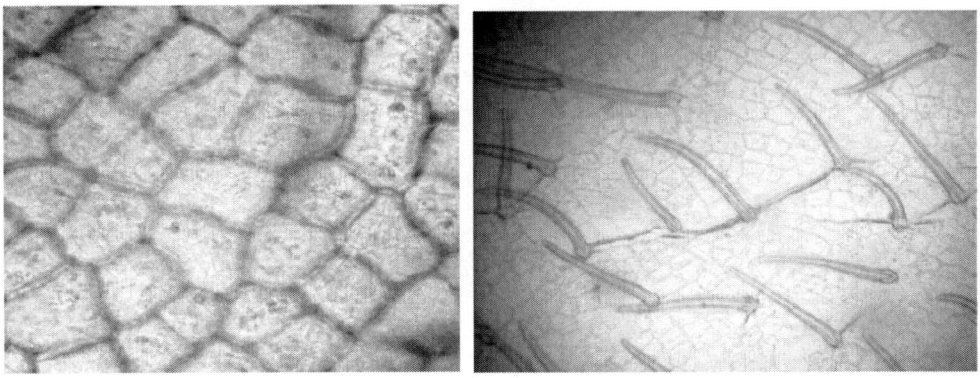

Figure 7. Leaf Epidermal Anatomy of *Cassia occidentalis.* Adaxial side (Right) Abaxial side (Left).

b. Abaxial Side

i. Fully mature leaf

Silica bodies and stomata were absent. Polygonal cells had double layer and smooth margins. Length of long cells were 89.7µm-91.8µm (Av: 90.7µm). Width of long cells were 48.5µm-57.6µm (Av: 53µm). Length of short cells were 48.4µm- 66.1µm (Av: 57.2µm). Width of short cells were 26µm-40µm (Av:33µm). Non-glandular trichomes were present. Number of trichomes per unit area was 27. Length of non-glandular trichomes ranged from 13.5µm- 399.7µm (Av: 256.6µm). Width of non-glandular trichomes ranged from 11µm-22.5µm (Av: 16.7µm).

ii. Mature leaf

Silica bodies and stomata were absent. Polygonal cells had double layer and smooth margins. Length of long cells were 125µm-130.6µm (Av: 127.8µm). Width of long cells were 69.3µm-76.6µm (Av: 72.9µm). Length of short cells were 62.2µm-72.2µm (Av: 67.2µm). Width of short cells were 48.3µm-59.3µm (Av:53.8µm). Non-glandular trichomes were present. Number of trichomes per unit area was 30. Length of non-glandular trichomes ranged from 111.7µm-325.5µm (Av: 218.6µm). Width of non-glandular trichomes ranged from 12.8µm-15µm (Av: 13.9µm).

iii. Immature leaf

Silica bodies and stomata were absent. Polygonal cells had double layer and smooth margins. Length of long cells were 70.6µm-78.3µm (Av: 74.4µm). Width of long cells were 42µm-42.8µm (Av: 42.4µm). Length of short cells were 31.4µm- 31.6µm (Av: 31.5µm). Width of short cells were 28.3µm-30.1µm (Av: 29.2µm). Non-glandular trichomes were present. Number of trichomes per unit area was 17. Length of non-glandular trichomes ranged from 101µm-245.1µm (Av: 173µm). Width of non-glandular trichomes ranged from 9.8µm-16.1µm (Av: 12.9µm). (Figure 7)

8. *Dalbergia sisssoo* Roxb

a. Adaxial Side

i. Fully mature leaf

Silica bodies and oil droplets were absent. Glandular type of trichomes was present. Pentagonal cells had double layer and smooth margins. Length of long cells ranged from 192.4µm-217.1µm (Av: 204.7µm). Width of long cells ranged from 96.5µm- 116.7µm (Av: 106.6µm). Length of short cells ranged from 91.2µm-145.3µm (Av: 118.2µm). Width of short cells ranged from 68.9µm-86.3µm (Av: 77.6µm). Stomata were present. Type of stomata was diacytic. Length of stomatal cavity ranged from 38.3 µm-86.1µm (Av: 62.2µm). Width of stomatal cavity ranged from 13.6µm-55.0µm (Av: 34.3µm). Length of guard cells were 51.1µm-126.7µm (Av: 88.9µm). Width of guard cells were 46.4µm-106.3µm (Av: 76.0µm). Length of long subsidiary cells ranged from 145.4µm-191.3µm (Av: 168.3µm). Length of short subsidiary cells ranged from 55.9µm-67.2µm (Av: 61.5µm). Width of long cells ranged from 99.6µm-103.7µm (Av: 101.6µm). Width of short cells ranged from 41.8µm-48.1µm (Av: 44.9µm).

ii. Mature leaf

Oil droplets were absent. Irregular silica bodies were present. Number of silica bodies per unit area was 2. Length of silica bodies were 28.0µm-39.8µm (Av: 33.9µm). Width of silica bodies were 24.0µm-66.0µm (Av: 45.0µm). Pentagonal cells had double layer and smooth margins. Length of long cells ranged from 216.0µm-255.3µm (Av: 235.6µm). Width of long cells ranged from 82.8µm-102.2µm (Av: 92.5µm). Length of short cells ranged from 137.1µm-143.1µm (Av: 140.1µm). Width of short cells ranged from 130.1µm-168.0µm (Av: 149.0µm). Non-glandular trichomes were present. Length of non-glandular trichomes were 219.7µm-486.0µm (Av: 352.8µm). Width of non-glandular trichomes were 19.3µm-25.0µm (Av: 22.1µm). Stomata were present. Type of stomata was diacytic. Length of stomatal cavity ranged from 35.8µm-79.6µm (Av: 57.7µm). Width of stomatal cavity ranged from 12µm-21.4µm (Av: 16.7µm). Length of guard cells ranged from 58.6µm- 83.4µm (Av: 71.0µm). Width of guard cells ranged from 40.7µm-68.1µm (Av: 54.4µm). Length of long subsidiary cells were 109.0µm-136.2µm (Av: 122.6µm) Length of short subsidiary cells were 91.1µm-103.2µm (Av: 97.1µm). Width of long cells were 75.5µm- 78.8µm (Av: 77.1µm). Width of short cells were 60.3µm- 85.1µm (Av: 72.7µm).

iii. Immature leaf

Stomata and oil droplets were absent. Irregular silica bodies were present. Number of silica bodies per unit area was 2. Length of silica bodies were 64.0µm-69.0µm (Av: 66.5µm). Width of silica bodies were 63.0µm-74.0µm (Av:68.5µm). Pentagonal cells had double layer and smooth margins. Length of long cells ranged from 134.4µm-137.2µm (Av: 135.8µm). Width of long cells ranged from 57.9µm-81.2µm (Av: 69.5µm). Length of short cells ranged from 43.6µm-49.0µm (Av: 46.3µm). Width of short cells ranged from 68.0µm-67.0µm (Av:67.5µm). (Figure 8)

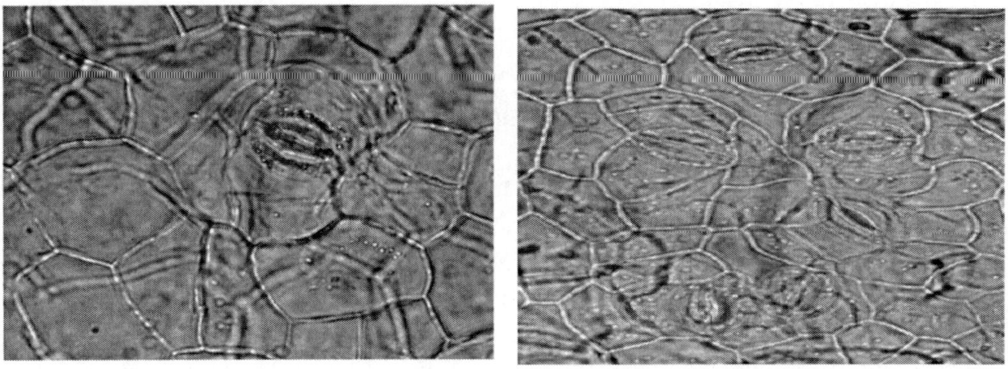

Figure 8. Leaf Epidermal Anatomy of *Dalbergia sissoo.* Adaxial side (Right) Abaxial side (Left).

b. Abaxial Side

i. Fully mature leaf

Oil droplets were absent. Irregular silica bodies were present. Number of silica bodies per unit area was 2. Length of silica bodies were 73.4µm-80.1µm (Av: 76.7µm). Width of silica bodies were 80.1µm- 89.0µm (Av: 84.5µm). Pentagonal cells had double layer and smooth margins. Length of long cells ranged from 181.2µm-201.1µm (Av: 191.1µm). Width of long cells ranged from 82.1µm-90.7µm (Av: 86.4µm). Length of short cells ranged from 93.9µm-109.8µm (Av: 101.8µm). Width of short cells ranged from 42.5µm-99.5µm (Av:71.0µm). Stomata were present. Glandular and non-glandular trichomes were present. Length of non-glandular trichomes were 169.2µm-170.1µm (Av: 169.6µm). Width of non-glandular trichomes were 14.2µm-14.3µm (Av: 14.2µm). Type of stomata was diacytic. Length of stomatal cavity ranged from 61.5µm-77.9µm (Av: 69.7µm). Width of stomatal cavity ranged from 25.1µm- 38.6µm (Av: 31.8µm). Length of guard cells ranged from 111.0µm-132.2µm (Av: 121.6µm). Width of guard cells ranged from 86.1µm- 74.9µm (Av: 80.5µm). Length of long subsidiary cells were 149.0µm-157.3µm (Av: 153.1µm). Length of short subsidiary cells ranged from 105.9µm-110.9µm (Av: 108.4µm). Width of long cell ranged from 32.3µm-38.6µm (Av: 35.4µm). Width of short cells ranged from 43.0µm – 100.7µm (Av: 71.8µm).

ii. Mature leaf

Silica bodies and oil droplets were absent. Pentagonal cells had double layer and smooth margins. Length of long cells ranged from 143.9µm-147.1µm (Av: 145.5µm). Width of long cells ranged from 78.1µm – 105.1µm (Av: 91.6µm). Length of short cells ranged from 100.2µm-103.2µm (Av: 101.7µm). Width of short cells ranged from 58.2µm-88.1µm (Av:73.1µm).Stomata were present. Glandular and non-glandular trichomes were present. Length of non-glandular trichomes were 25.5µm-254.4µm (Av: 139.9µm). Width of non-glandular trichomes were 13µm-13.2µm (Av: 13.1µm). Type of stomata was diaytic. Length of stomatal cavity ranged from 40.9µm-57.7µm (Av: 49.3µm). Width of stomatal cavity ranged from 20.2µm- 22.0µm (Av: 21.1µm). Length of guard cells ranged from 99.3µm-108.8µm (Av: 104.0µm). Width of guard cells ranged from 70.9µm-90.8µm (Av: 80.8µm). Length of long subsidiary cells were 180.2µm-189.3µm (Av: 184.7µm) Length of short subsidiary cells were 101.8µm-124.2µm (Av: 113.0µm). Width of long cell ranged from 39.3µm-49.4µm (Av: 44.3µm). Width of short cells ranged from 47.9µm-70.1µm (Av: 59.0µm).

iii. Immature leaf

Silica bodies and oil droplets were absent. Non-glandular type of trichomes was present. Length of non-glandular trichomes were 347.8µm-374.4µm (Av: 361.1µm). Width of non-glandular trichomes were 27.2µm-42.1µm (Av: 57.15µm). Pentagonal cells had double layer and smooth margins. Length of long cells ranged from 176.0µm-183.9µm (Av: 179.9µm). Width of long cells ranged from 95µm-100.3µm (Av: 97.6µm). Length of short cells were 99.7µm-137.7µm (Av:118.7µm). Width of short cells wereb73.8µm-95.3µm (Av:84.5µm). Stomata were present. Type of stomata was diacytic. Length of stomatal cavity ranged from 31.3µm-99.0µm (Av: 65.1µm). Width of stomatal cavity ranged from 31.4µm-67.1µm (Av: 49.2µm). Length of guard cells ranged from 99.6µm-141.7µm (Av: 120.6µm). Width of guard cells ranged from 87.3μm-105.0μm (Av: 96.1μm). Length of long subsidiary cells were 203.4μm-209.3μm (Av: 206.3μm). Length of short subsidiary cells were 118.6μm-155.9μm (Av: 137.2μm). Width of long cell were 52.5μm-67.2μm (Av: 59.8μm). Width of short cells were 50.2μm-54.1μm (Av: 52.15μm).(Figure 8)

9. *Dalbergia lanceolaria* L.

a. Adaxial Side

i. Fully mature leaf

Trichomes, stomata, Silica bodies and oil droplets were absent. Hexagonal cells had double layer and smooth margins. Length of long cells were 157.2μm-174.2μm (Av: 165.7μm). Width of long cells were 77.8μm-158.5μm (Av: 118.1μm). Length of short cells were 70.6μm-82.0μm (Av: 76.3μm). Width of short cells were 59.5μm-83.8μm (Av: 71.6μm).

ii. Mature leaf

Silica bodies, trichomes and oil droplets were absent. Hexagonal cells had double layer and smooth margins. Length of long cells were 104.4μm-122.8μm (Av: 113.6μm). Width of long cells were 52.2μm-84.1μm (Av: 68.1μm). Length of short cells were 22.5μm-26.9μm (Av: 24.7μm). Width of short cells were 51.9μm-67.4μm (Av:59.6μm). Stomata were present. Type of stomata was actinocytic. Length of stomatal cavity ranged from 18.4μm-44.1μm (Av: 31.2μm). Width of stomatal cavity ranged from 4.0μm-5.7μm (Av: 4.8μm). Length of guard cells ranged from 71.2μm-88.6μm (Av: 79.9μm). Width of guard cells ranged from 56.6μm-68.0μm (Av: 66.9μm). Length of long subsidiary cells ranged from 57.7μm-76.7μm (Av: 67.2μm) Length of short subsidiary cells 29.5μm-43.9μm (Av: 36.7μm). Width of long cell ranged from 53.0μm-81.0μm (Av: 67μm). Width of short cells ranged from 25.6μm- 45.6μm (Av: 35.2μm). (Figure 9)

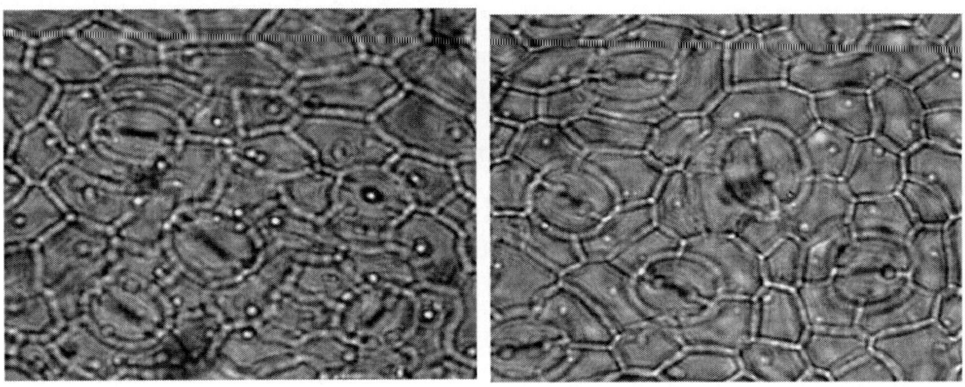

Figure 9. Leaf Epidermal Anatomy of *Dalbergia lanceolaria.* Adaxial side (Right) Abaxial side (Left).

iii. Immature leaf

Trichomes, stomata, Silica bodies and oil droplets were absent. Hexagonal cells had double layer and smooth margins. Length of long cells ranged from 204.8μm-211.5μm (Av: 208.1μm).Width of long ranged from 107.8μm-124.3μm (Av: 116μm). Length of short cells ranged from 156.5μm-162.6μm (Av: 159.5μm). Width of short cells ranged from 137.0μm-239.3μm (Av: 188.1μm).

b. Abaxial Side

i. Fully mature leaf

Silica bodies and oil droplets were absent. Glandular trichomes were present. Hexagonal cells had double layer and smooth margins. Length of long cells ranged from 54.6μm-59.2μm (Av: 56.9μm). Width of long cells ranged from 61.3μm-22.1μm (Av: 32.3μm). Length of short cells ranged from 24.6μm-32.3μm (Av: 28.4μm). Width of short cells ranged from 19.1μm-25.5μm (Av: 22.9μm). Stomata were present. Type of stomata was actinocytic. Length of stomatal cavity ranged from 13.4μm-31.9μm (Av: 22.6μm). Width of stomatal cavity ranged from 4.5μm- 6.4μm (Av: 5.4μm). Length of guard cells were 45.2μm-66.4μm (Av: 55.8μm). Width of guard cells were 40.7μm-47.8μm (Av: 44.2μm). Length of long subsidiary cells were 37.7μm-45.2μm (Av: 41.4μm). Length of short subsidiary cells were 16.1μm- 20.0μm (Av: 35.4μm). Width of long cells were 28.8μm-42.0μm (Av: 18.0μm). Width of short cells were 29.3μm-32.1μm (Av: 30.7μm). (Figure 9)

ii. Mature leaf

Oil droplets were absent. Silica bodies were present. Length of silica bodies were 14.0μm-67.1μm (Av: 40.5μm). Width of silica bodies were 12.0μm-39.0μm (Av: 33μm). Hexagonal cells had double layer and smooth margins. Length of long cells ranged from 54.6μm-66.3μm (Av: 60.4μm). Width of long cells ranged from 19.8μm-21.3μm (Av: 20.5μm). Length of short cells ranged from 15.6μm – 25.1μm (Av: 20.3μm). Width of short cells ranged from 23.1μm – 31.1μm (Av: 27.1μm). Non-glandular trichomes were present. Length of non-glandular trichomes ranged from 79.3μm-267.7μm (Av: 73.5μm). Width of non-glandular trichomes ranged from 14.0μm-19.4μm (Av: 66.7μm). Stomata were present. Type of stomata was actinocytic. Length of stomatal cavity ranged from 6.7μm-26.9μm (Av: 26.9μm). Width of stomatal cavity ranged from 5.4μm-9.1μm (Av: 7.2μm). Length of guard cells ranged from 33.5μm-47.8μm (Av: 40.6μm). Width of guard cells 32.8μm- 37.4μm (Av: 35.1μm). Length of long subsidiary cells were 26.6μm-30.4μm (Av: 28.5μm). Length of short subsidiary cells were 14.9μm-15.1μm (Av: 15μm). Width of long cell were 20.9μm-30.8μm (Av: 25.8μm). Width of short cells were 27.0μm-28.5μm (Av: 27.7μm).

iii. Immature leaf

Silica bodies, trichomes and oil droplets were absent. Hexagonal cells had double layer and smooth margins. Length of long cells ranged from 93.3μm- 99.1μm (Av: 96.2μm). Width of long cells ranged from 42.0μm-63.0μm (Av: 52.5μm). Length of short cells ranged from 39.0μm-40.0μm (Av: 39.5μm). Width of short cells ranged from 60.0μm-66.8μm (Av: 63.4μm). Stomata were present. Type of stomata was actinocytic. Length of stomatal cavity ranged from 25.1μm-48.2μm (Av: 36.6μm). Width of stomatal cavity ranged from 6.4μm-

7.2μm (Av: 6.8μm). Length of guard cells ranged from 65.0μm-89.9μm (Av: 77.4μm). Width of guard cells ranged from 60.2μm-63.5μm (Av: 61.8μm). Length of long subsidiary cells ranged from 68.9μm – 83.2μm (Av: 76.0μm) Length of short subsidiary cells ranged from 23.0μm-27.0μm (Av: 52.7μm). Width of long cell were 48.0μm-57.4μm (Av: 25μm). Width of short cells were 43.0μm-84.8μm (Av: 63.9μm). (Figure 9)

10. *Delonix regia* (Hook.) Raf.

a. Adaxial Side

i. Fully mature leaf

Silica bodies, trichomes and oil droplets were absent. Polygonal cells with wavy margins were present. Length of long cells ranged from 85.1μm-183.1μm (Av: 84.1μm). Width of long cells ranged from 49.8μm-45.2μm (Av: 47.5μm). Length of short cells ranged from 46μm-51.6μm (Av: 48.8μm). Width of short cells ranged from 35μm-41.1μm (Av: 38μm). Type of stomata was anomocytic. Number of stomata per unit area was 7. Length of stomatal cavity ranged from 8.5μm -18.7μm (Av: 13.6μm). Width of stomatal cavity ranged from 6.4μm-6.1μm (Av:6.2μm). Length of guard cells were 54μm-73.8μm (Av: 63.9μm). Width of guard cells were 40.5μm – 54.1μm (Av: 47.3μm). Length of long subsidiary cells were 60.2μm-74.6μm (Av: 67.4μm) Length of short subsidiary cells were 27.9μm-32.6μm (Av: 30.2μm). Width of long cell ranged from 39.7μm- 56.3μm (Av: 52.2μm). Width of short cells ranged from 47.4μm-77.1μm (Av: 62.2μm).

ii. Mature leaf

Stomata, trichomes and oil droplets were absent. Irregular silica body was present. Number of silica bodies per unit area was 2. Length of silica bodies ranged from 111μm-114.1μm (Av: 112.5μm). Width of silica bodies ranged from 147.1μm- 158μm (Av: 152.5μm). Polygonal cells with wavy margins were present. Length of long cells ranged from 134.1μm-143.1μm (Av: 138.6μm). Width of long cells ranged from 78.1μm-78.7μm (Av: 78.4μm). Length of short cells ranged from 64.9μm-80.4μm (Av: 72.6μm). Width of short cells ranged from 60.2μm-67.7μm (Av: 63.9μm). (Figure 10)

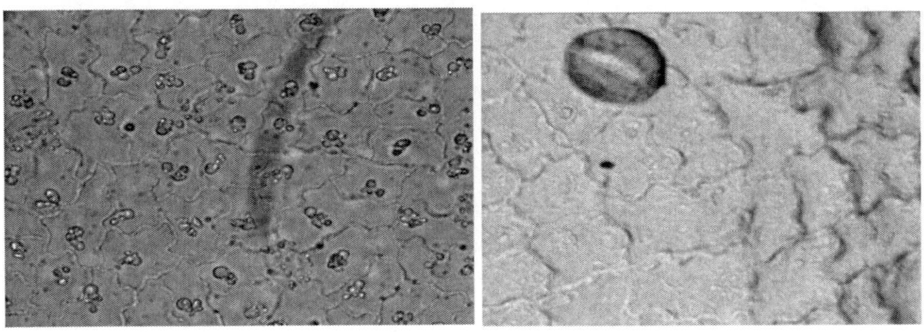

Figure 10. Leaf Epidermal Anatomy of *Delonix regia.* Adaxial side (Right) Abaxial side (Left).

iii. Immature leaf

Stomata, silica bodies and trichomes were absent. Cells were polygonal with wavy margins. Length of long cells ranged from 71.8µm-98.6µm (Av: 85.2µm). Width of long cells ranged from 51.9µm-54.4µm (Av: 53.1µm). Length of short cells ranged from 40.3µm-44.6µm (Av: 42.4µm). Width of short cells ranged from 34.4µm-53.6µm (Av: 44µm)

b. Abaxial Side

i. Fully mature leaf

Silica bodies, trichomes and oil droplets were absent. Cells were polygonal with wavy margins. Length of long cells were 41.1µm-42.8µm (Av: 41.9µm). Width of long cells were 52µm-57µm (Av: 54.5µm). Length of short cells were 87.8µm-89.4µm (Av: 88.6µm). Width of short cells were 41.8µm-44.4µm (Av: 43.1µm). Number of stomata per unit area was 7. Type of stomata was anomocytic. Length of stomatal cavity ranged from 36.1µm-49.9µm (Av: 43µm). Width of stomatal cavity ranged from 9.8µm-49.2µm (Av: 29.5µm). Length of guard cells were 53.2µm-90µm (Av: 71.6µm). Width of guard cells were 55.7µm-72.2µm (Av: 63.9µm). Length of long subsidiary cells ranged from 78.6µm-83µm (Av: 80.8µm). Length of short subsidiary cells ranged from 31.4µm-47.8µm (Av: 39.6µm). Width of long cell ranged from 51.6µm-65.7µm (Av: 58.6µm). Width of short cells ranged from 50.4µm-76.2µm (Av: 63.3µm).

ii. Mature leaf

Silica bodies, trichomes and oil droplets were absent. Polygonal cells with wavy margins. Length of long cells were 119.1µm-128.3µm (Av: 123.7µm). Width of long cells were 59.2µm-71.6µm (Av: 65.4µm). Length of short cells were 59µm-67.2µm (Av: 63.1µm). Width of short cells were 70.6µm-72.8µm (Av: 71.7µm). Stomata were present. Number of stomata per unit area was 7. Type of stomata was anomocytic. Length of stomatal cavity ranged from 33µm-57.2µm (Av: 45.1µm). Width of stomatal cavity ranged from 9µm-17.5µm (Av: 13.2µm). Length of guard cells were 82.2µm-106µm (Av: 94.1µm). Width of guard cells were 76.2µm-82.4µm (Av: 79.3µm). Length of long subsidiary cells ranged from 114.8µm- 126.1µm (Av: 120.4µm) Length of short subsidiary cells ranged from 48µm-57µm (Av: 52.5µm). Width of long cell were 90.2µm-41.6µm (Av: 40.9µm). Width of short cells were 47µm- 69.3µm (Av: 58.1µm).

iii. Immature leaf

Silica bodies, trichomes and oil droplets were absent. Cells were polygonal with wavy margins. Length of long cells ranged from 95.3µm-95.9µm (Av: 95.6µm). Width of long cells ranged from 56.1µm-83µm (Av: 69.5µm). Length of short cells ranged from 44µm-65.1µm (Av: 54.5µm). Width of short cells ranged from 78.2µm-93.6µm (Av: 85.9µm). Number of stomata per unit area was 6. Type of stomata was anomocytic. Length of stomatal cavity ranged from 47.5µm-55.6µm (Av: 51.5µm). Width of stomatal cavity ranged from 9.9µm-11.2µm (Av: 10.5µm). Length of guard cells were 79.2µm-102.8µm (Av: 71µm). Width of guard cells were 77.9µm- 81.1µm (Av: 79.5µm). Length of long subsidiary cells were 66.7µm-104µm (Av: 85.3µm) Length of short subsidiary cells were 42µm-49.8µm (Av:

45.4μm). Width of long cell ranged from 63.9μm-83.3μm (Av: 73.6μm). Width of short cells ranged from 68.1μm-76μm (Av: 72μm).(Figure 10)

11. *Erythrina suberosa* Roxb.

a. Adaxial Side

i. Fully mature leaf

Trichomes, silica bodies and oil droplets were absent. Polygonal cells had double layer and smooth margins. Length of long cells ranged from176.5μm-196.4μm (Av: 186.4μm). Width of long cells ranged from 67.5μm-72.2μm (Av: 69.8μm). Length of short cells ranged from 88.6μm-99.2μm (Av: 93.9μm). Width of short cells ranged from 64.0μm-73.0μm (Av: 68.5μm). Stomata were present. Type of stomata was staurocytic. Length of stomatal cavity ranged from 20.1μm-32.8μm (Av: 26.4μm). Width of stomatal cavity ranged from 6.3μm-130.0μm (Av: 68.1μm). Length of guard cells were 43.8μm-69.6μm (Av: 56.8μm). Width of guard cells were 36.9μm-58.1μm (Av: 47.5μm). Length of long subsidiary cells were 69.9μm-78.9μm (Av: 74.4μm). Length of short subsidiary cells were 25.0μm-30.5μm (Av: 27.7μm). Width of long cell ranged from 39-38.4μm (Av: 39μm). Width of short cells ranged from 53.8μm-55.8μm (Av: 54.8μm). (Figure 11)

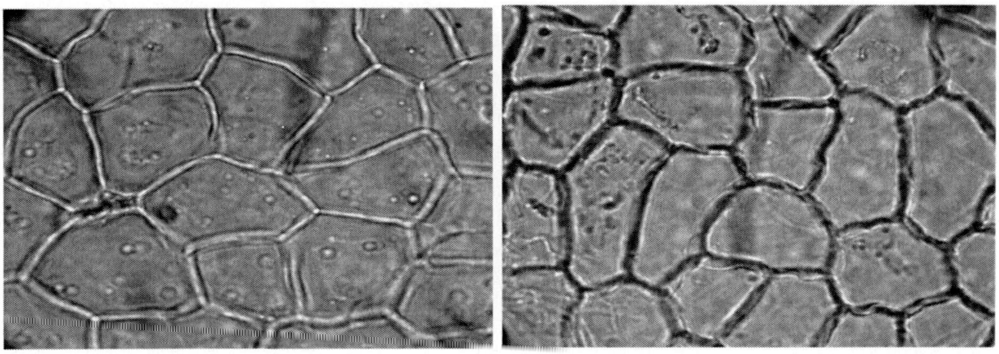

Figure 11. Leaf Epidermal Anatomy of *Erythrina suberosa.* Adaxial side (Right) Abaxial side (Left).

ii. Mature leaf

Trichomes, Silica bodies and oil droplets were absent. Polygonal cells had double layer and smooth margins. Length of long cells ranged from 165.7μm-185.6μm (Av: 175.6μm). Width of long cells ranged from 155.5μm-181.0μm (Av: 168.2μm). Length of short cells ranged from 112.7μm-130.0 μm (Av: 121.3μm). Width of short cells ranged from 109.6μm-168.1μm (Av: 138.8μm). Stomata were present. Type of stomata was staurocytic. Length of stomatal cavity ranged from 47.6μm-58.7μm (Av: 53.1μm). Width of stomatal cavity ranged from 9.2μm-14.9μm (Av: 12.0μm). Length of guard cells were 80.1μm-90.6μm (Av: 85.3μm). Width of guard cells were 63.3μm-72.7μm (Av: 68μm). Length of long subsidiary cells ranged from 110.7μm-130.7μm (Av: 120.7μm) Length of short subsidiary cells ranged from 78.2μm- 94.8μm (Av: 86.5μm). Width of long cell ranged from 41.0μm-43.7μm (Av: 42.3μm). Width of short cells ranged from 29.7μm-55.2μm (Av: 42.4μm).

iii. Immature leaf

Stomata and oil droplets were absent. Glandular trichomes were present. Silica bodies were present. Length of silica bodies were 53.0μm-30.1μm (Av: 41.5μm). Width of silica bodies were 55.0μm- 26.0μm (Av: 40.5μm). Polygonal cells had double layer and smooth margins.

Length of long cells were 144.0μm-192.1μm (Av: 168μm). Width of long cells were 80.2μm-91.9μm (Av: 86μm). Length of short cells were 66.0μm-85.3μm (Av: 75.6μm). Width of short cells were 46.7μm-91.0μm (Av: 68.8μm).

b. Abaxial Side

i. Fully mature leaf

Stomata and oil droplets were absent. Irregular silica bodies were present. Number of silica bodies per unit area were 2. Length of silica bodies were 40.3μm-41.3μm (Av: 40.8μm). Width of silica bodies were 17.8μm-18.1μm (Av: 98μm). Non-glandular trichomes were present.

Length of non-glandular trichomes were 84.9μm-138.6μm (Av: 111.7μm). Width of non-glandular trichomes were 13.3μm-30.6μm (Av: 21.9μm). Polygonal cells had double layer and smooth margins. Length of long cells ranged from 100.0μm-103.8μm (Av: 101.9μm).Width of long cells ranged from 48.8μm-59.9μm (Av: 38.5μm). Length of short cells ranged from 32.0μm-45.0μm (Av: 54.3μm). Width of short cells ranged from 23.8μm-37.2μm (Av: 30.5μm).

ii. Mature leaf

Stomata and oil droplets were absent. Glandular trichomes were present. Silica bodies were present. Number of silica body per unit area was 2. Length of silica body were 27.0μm-28μm (Av: 27.5μm). Width of silica body were 31μm-32.0μm (Av: 31.5μm). Polygonal cells had double layer and smooth margins. Length of long cells ranged from 88.6μm-94.8μm (Av: 91.7μm).

Width of long cells ranged from 39.0μm-42.0μm (Av: 33μm). Length of short cells ranged from 30.0μm-36.0μm (Av: 67.5μm). Width of short cells ranged from 54.0μm-65.0μm (Av: 59.5μm).

iii. Immature leaf

Stomata and oil droplets were absent. Glandular trichomes were present. Silica bodies were present. Number of silica bodies per unit area was 2. Length of silica bodies were 44.1μm-55.4μm (Av: 34.8μm). Width of silica bodies were 33.9-35.8μm(Av: 49.7μm). Polygonal cells had double layer and smooth margins. Length of long cells ranged from 199.5μm-246.1μm (Av: 222.8μm). Width of long cells ranged from 89.0μm-98.3μm (Av: 104.8μm).

Length of short cells ranged from 98.0μm-111.6μm (Av: 93.6μm). Width of short cells ranged from 100.0μm-122.0μm (Av: 111μm). (Figure 11)

12. *Gleditsia triacanthos* L.

a. Adaxial Side

i. Fully mature leaf

Silica bodies, trichomes and oil droplets were absent Polygonal cells had double layer and wavy margins. Length of long cells ranged from 141.4μm-14.5μm (Av: 143.4μm). Width of long cells ranged from 55.5μm-59.6μm (Av: 57.5μm). Length of short cells ranged from 54.6μm-78.9μm (Av: 66.7μm). Width of short cells ranged from 35μm-44μm (Av: 39.5μm). Stomata were present. Type of stomata was cyclocytic. Length of stomatal cavity ranged from 67.7 μm-80.1μm (Av: 73.9μm). Width of stomatal cavity ranged from 17.1μm-18.4μm (Av: 17.7μm). Length of guard cells ranged from 100.6μm-109.6μm (Av: 105.1μm). Width of guard cells ranged from 58.3μm-62.8μm (Av: 60.5μm). Length of long subsidiary cells were 90.5μm- 94.6μm (Av: 92.5μm). Length of short subsidiary cells were 31.3μm-48.2μm (Av: 39.7μm). Width of long cells were 42.4μm-52μm (Av: 47.2μm). Width of short cells were 74.7μm-75.6μm (Av: 75.1μm).

ii. Mature leaf

Silica bodies, trichomes and stomata were absent. Polygonal cells had double layer and wavy margins. Length of long cells were 123.6μm-134.2μm (Av: 128.9μm). Width of long cells were 77.4μm-95μm (Av: 86.2μm). Length of short cells ranged from 20.9μm- 33.4μm (Av: 27.1μm). Width of short cells ranged from 38.9μm-55.5μm (Av: 47.2μm). Length of non-glandular trichomes were 113.5μm- 358.4μm (Av: 235.9μm). Width of non-glandular trichomes were 61.9μm-28.2μm (Av: 45μm).

iii. Immature leaf

Silica bodies, trichomes and oil droplets were absent. Polygonal cells had double layer and wavy margins. Length of long cells ranged from 163.8μm-178.1μm (Av: 170.9μm). Width of long cells ranged from 50.4μm-70μm (Av: 60.2μm). Length of short cells ranged from 76.2μm-77.1μm (Av: 76.6μm). Width of short cells ranged from 84μm-90.4μm (Av: 69.2μm). Glandular trichomes were present. Type of stomata was cyclocytic. Length of stomatal cavity ranged from 21.9μm-80μm (Av: 50.9μm). Width of stomatal cavity ranged from 8.1μm- 24.2μm (Av: 16.1μm). Length of guard cells were 28.4μm-126.3μm (Av: 77.3μm). Width of guard cells were 33.3μm-73.4μm (Av: 53.3μm). Length of long subsidiary cells ranged from 104μm-177.3μm (Av: 140.6μm). Length of short subsidiary cells ranged from 58.2μm-59.1μm (Av: 58.6μm). Width of long cells ranged from 47.8μm-65.5μm (Av: 56.6μm). Width of short cells ranged from 50μm-63.1μm (Av: 56.5μm). (Figure 12)

b. Abaxial Side

i. Fully mature leaf

Silica bodies, trichomes and oil droplets were absent. Polygonal cells had double layer and smooth margins. Length of long cells were 138.1μm-162.1μm (Av: 150.1μm). Width of long cells were 40.5μm-72.2μm (Av: 56.3μm). Length of short cells were 55μm-73.3μm (Av: 64.1μm). Width of short cells were 38μm-55.5μm (Av: 46.7μm). Stomata were present. Type of stomata was cyclocytic. Length of stomatal cavity ranged from 72.4µm-76.1µm (Av: 74.2µm). Width of stomatal cavity ranged from 18.6µm-21.5µm (Av: 20µm). Length of guard cells were 107.5µm-121.1µm (Av: 114.3µm). Width of guard cells were 64.1µm-69.1µm (Av: 80.1µm). Length of long subsidiary cells were 113.5µm-144.1µm (Av: 128.8µm). Length of short subsidiary cells were 36.2µm-42.1µm (Av: 39.1µm). Width of long cell ranged from 36µm-50.3µm (Av: 43.1µm). Width of short cells ranged from 64.1µm-69.1µm (Av: 80.1µm).

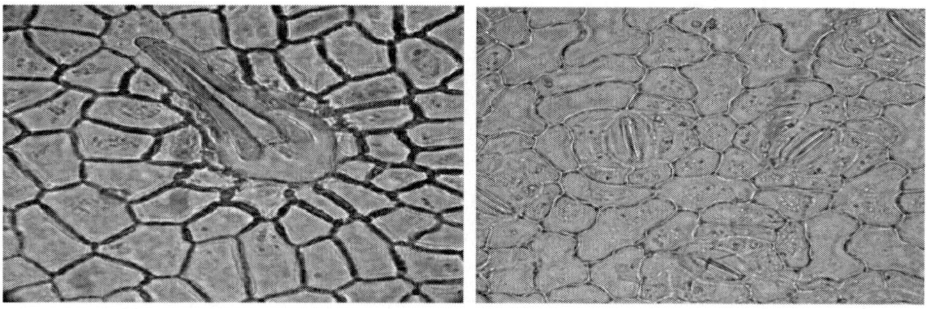

Figure 12. Leaf Epidermal Anatomy of *Gleditsia triacanthos*. Adaxial side (Right) Abaxial side (Left).

ii. Mature leaf

Silica bodies, trichomes and oil droplets were absent. Polygonal cells had double layer and wavy margins. Length of long cells ranged from 155.3µm-226.7µm (Av: 191µm). Width of long cells ranged from 62.1µm-63.8µm (Av: 62.9µm). Length of short cells ranged from 70.7µm-96µm (Av: 83.3µm). Width of short cells ranged from 52µm-63.6µm (Av: 57.8µm). Stomata were present. Type of stomata was cyclocytic. Length of stomatal cavity ranged from 77.5µm-104.6µm (Av: 91.0µm). Width of stomatal cavity ranged from 20.1µm-18.4µm (Av: 19.2µm). Length of guard cells were 146.5µm-155µm (Av: 150.7µm). Width of guard cells were 94.9µm-110µm (Av: 102.4µm). Length of long subsidiary cells were 150.8µm-153.7µm (Av: 152.2µm). Length of short subsidiary cells were 70.7µm-96µm (Av: 83.3µm). Width of long cells ranged from 62.1µm-63.8µm (Av: 62.9µm). Width of short cells ranged from 52µm-63.6µm (Av: 57.8µm).

iii. Immature leaf

Silica bodies, trichomes and oil droplets were absent. Polygonal cells had double layer and wavy margins. Length of long cells ranged from 177.2µm-221.6µm (Av: 199.3µm). Width of long cells ranged from 37.5µm-66.6µm (Av: 52µm). Length of short cells ranged from 36.8µm-61.3µm (Av: 49µm). Width of short cells ranged from 42.4µm-87.3µm (Av: 64.8µm). Stomata were present. Type of stomata was cyclocytic. Length of stomatal cavity ranged from 56.1µm-66.5µm (Av: 61.3µm). Width of stomatal cavity ranged from 26.6µm-29µm (Av: 27.8µm). Length of guard cells were 101.3µm-107.5µm (Av: 104.4µm). Width of guard cells were 73.4µm-75.3µm (Av: 74.3µm). Length of long subsidiary cells were 75.2µm-79.3µm (Av: 77.2µm). Length of short subsidiary cells were 26.7µm-49.2µm (Av: 37.9µm). Width of long cells ranged from 25.2µm-31.8µm (Av: 28.5µm). Width of short cells ranged from 55.3µm-100.7µm (Av: 78µm).(Figure 12)

13. *Leucaena glauca* (L.) Bent

a. Adaxial Side

i. Fully mature leaf

Trichomes and oil droplets were absent. Silica bodies were present. Number of silica bodies per unit area was 2. Length of silica bodies were 74.1μm-79μm (Av: 76.5μm). Width of silica bodies were 86.3μm-89.1μm (Av: 87.7μm). Polygonal cells had double layer and smooth margins. Length of long cells ranged from 27.1μm-39μm (Av: 33μm). Width of long cells ranged from 38.6μm-41μm (Av: 39.8μm). Length of short cells ranged from 89.3μm-110.2μm (Av: 99.7μm). Width of short cells ranged from 32.1μm-66μm (Av: 49μm. Stomata were present. Number of stomata per unit area was 14. Type of stomata was diacytic. Length of stomatal cavity ranged from 13.9μm-21.4μm (Av: 17.6μm). Width of stomatal cavity ranged from 5.1μm-8μm (Av: 6.5μm). Length of guard cells were 23.1μm-47.1μm (Av: 35.1μm). Width of guard cells were 13.6μm-23.4μm (Av: 18.5μm). Length of long subsidiary cells were 57.1μm- 72.2μm (Av: 64.6μm). Length of short subsidiary cells were 36μm-40.2μm (Av: 38.1μm). Width of long cell were 12μm-14.1μm (Av: 13.0μm). Width of short cells were 16.3μm- 17.1μm (Av: 16.7μm).

ii. Mature leaf

Silica bodies and oil droplets were absent. Polygonal cells had double layer and smooth margins. Length of long cells ranged from 154.2μm-164.2μm (Av: 157.6μm). Width of long cells ranged from 55.7μm-64.2μm (Av: 59.9μm). Length of short cells ranged from 55.2μm-70μm (Av: 62.6μm). Width of short cells ranged from 58.3μm-94.9μm (Av: 76.6μm). Length of non-glandular trichomes were 267μm- 306.1μm (Av: 268.5μm). Width of non-glandular trichomes were 5μm- 239.5μm (Av: 19.4μm). Stomata were present. Type of stomata was diacytic. Length of stomatal cavity ranged from 11μm-20.6μm (Av: 15.8μm). Width of stomatal cavity ranged from 4.1μm-4.5μm (Av: 4.3μm). Length of guard cells were 64.6μm-75.4μm (Av: 70μm). Width of guard cells were 37.5μm-43.3μm (Av: 40.4μm). Length of long subsidiary cells were 102μm-120.8μm (Av: 111.4μm). Length of short subsidiary cells were 62.4μm-98μm (Av: 80.2μm). Width of long cells ranged from 39.6μm- 49.3μm (Av: 44.4μm). Width of short cells ranged from 26.2μm-29.4μm (Av: 28μm). (Figure 13)

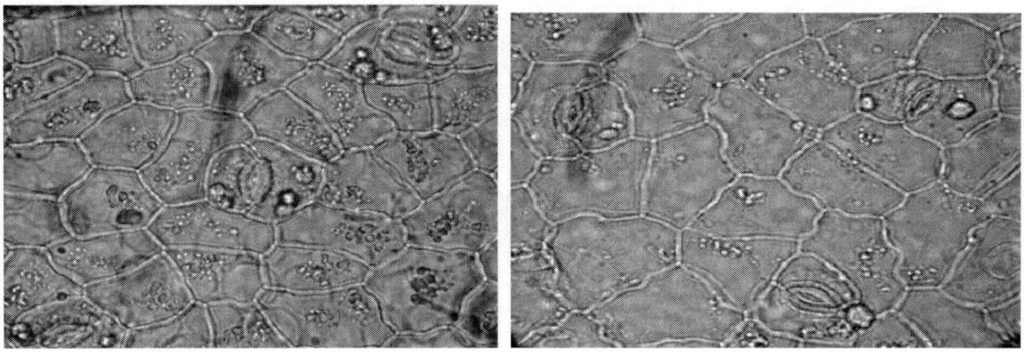

Figure 13. Leaf Epidermal Anatomy of *Leucaena glauca.* Adaxial side (Right) Abaxial side (Left).

iii. Immature leaf

Trichomes and oil droplets were absent. Silica bodies were present. Number of silica bodies per unit area was 2. Length of silica bodies were 105μm-124μm (Av: 114.5μm). Width of silica bodies were 98μm- 120μm (Av: 109μm). Polygonal cells had double layer and smooth margins. Length of long cells ranged from 135.4μm-150.7μm (Av: 143μm). Width of long cells ranged from 58.2μm-67.1μm (Av: 62.6μm). Length of short cells ranged from 84.5μm-90μm (Av: 87.2μm). Width of short cells ranged from 76.8μm – 84.5μm (Av: 80.6μm). Stomata were present. Type of stomata was diacytic. Length of stomatal cavity ranged from 15.1μm-43.9μm (Av: 29.5μm). Width of stomatal cavity ranged from 3μm-16.3μm (Av: 9.6μm). Length of guard cells were 67.7μm-78.9μm (Av: 73.3μm). Width of guard cells were54.9μm-48.2μm (Av: 51.5μm). Length of long subsidiary cells ranged from 112.4μm-137.2μm (Av: 124.8μm) Length of short subsidiary cells ranged from 67.6μm-88.1μm (Av: 382μm). Width of long cell ranged from 27.2μm-36.1μm (Av: 31.6μm). Width of short cells ranged from 42.7μm-48.2μm (Av: 45.4μm).

b. Abaxial Side

i. Fully mature leaf

Trichomes and oil droplets were absent. Silica bodies were present. Number of silica bodies per unit area was 2. Length of silica bodies were 120μm-135μm (Av: 127.5μm). Width of silica bodies were 119.2μm-154μm (Av: 136.6μm). Polygonal cells had double layer and smooth margins. Length of long cells ranged from 206.8μm-213.6μm (Av: 210.2μm). Width of long cells ranged from 107μm-137.1μm (Av: 122μm). Length of short cells ranged from 134.3μm-153.9μm (Av: 144.1μm). Width of short cells ranged from 89.5μm-68.2μm (Av:78.8μm). Type of stomata was diacytic. Number of stomata per unit area was 8. Length of stomatal cavity ranged from 63.3μm-76.9μm (Av: 70.1μm). Width of stomatal cavity ranged from 21.8μm- 26.4μm (Av: 24.1μm). Length of guard cells were 165.4μm-174.9μm (Av: 170.1μm). Width of guard cells were 62.3μm-53.3μm (Av: 57.8μm). Length of long subsidiary cells were 165.4μm-174.9μm (Av: 170.1μm). Length of short subsidiary cells were 109.1μm-113μm (Av: 111μm). Width of long cells were 36.8μm-84.6μm (Av: 60.7μm). Width of short cells 37.6μm – 80.4μm (Av: 59μm).

ii. Mature leaf

Silica bodies and oil droplets were absent. Polygonal cells had double layer and smooth margins. Length of long cells ranged from 171.7μm-177.4μm (Av: 174.5μm). Width of long cells ranged from 74.4μm-101.2μm (Av: 87.8μm). Length of short cells ranged from 96.3μm-99.2μm (Av: 97.7μm). Width of short cells ranged from 95.5μm-148.1μm (Av:95.9μm). Non-glandular trichomes with blunt ends were present. Length of non-glandular trichomes were 132.1μm-156.1μm (Av: 144.1μm). Width of non-glandular trichomes were 17μm-19.8μm (Av: 18.4μm). Stomata were present. Type of stomata was diacytic. Number of stomata per unit area was 3. Length of stomatal cavity ranged from 22.2μm-32.8μm (Av: 27.5μm). Width of stomatal cavity ranged from 8.5μm-9μm (Av: 8.7μm). Length of guard cells were 76.1μm-86.1μm (Av: 81.1μm). Width of guard cells were 43.3μm-55.2μm (Av: 49.2μm). Length of long subsidiary cells ranged from 83.4μm- 79.9μm (Av: 90.6μm) Length of short subsidiary cells ranged from 121.1μm150.6μm (Av: 135.8μm). Width of long cells ranged from 23.3µm-34.1µm (Av: 28.7µm). Width of short cells 46.2µm-73.5µm (Av: 59.8µm).

iii. Immature leaf

Silica bodies, trichomes and oil droplets were absent. Polygonal cells had double layer and smooth margins. Length of long cells ranged from 150.2µm-156.3µm (Av: 153.2µm). Width of long cells ranged from 86.9µm-88.4µm (Av: 87.6µm). Length of short cells ranged from 75.5µm-78.1µm (Av: 79.6µm). Width of short cells ranged from 101µm-122.1µm (Av: 111.5µm). Stomata were present. Type of stomata was diacytic. Length of stomatal cavity ranged from 15.5µm- 43.9µm (Av: 29.7µm). Width of stomatal cavity ranged from 3.6µm-10.8µm (Av: 7.2µm). Length of guard cells were 47.3µm-71.7µm (Av: 59.5µm). Width of guard cells were 45.2µm-52.8µm (Av: 49µm). Length of long subsidiary cells ranged from 127.8µm-129.5 µm (Av: 128.6µm). Length of short subsidiary cells ranged from 91.9µm-93.5µm (Av: 92.7µm). Width of long cells were 33.2µm-36.8µm (Av: 35µm). Width of short cells were 20.6µm-30.1µm (Av: 25.3µm).(Figure 13)

14. *Millettia ovalifolia* (Wight & Arn.) Kurz

a. Adaxial Side

i. Fully mature leaf

Stomata, trichomes and oil droplets were absent. Silica bodies were present. Length of silica bodies were 12.2µm-38µm (Av: 25.1µm). Width of silica bodies were 10µm-37µm (Av: 23.5µm). Polygonal cells had smooth margins. Length of long cells ranged from 156µm-165.6µm (Av: 160.8µm). Width of long cells ranged from 48.5µm-56.1µm (Av: 52.3µm). Length of short cells ranged from 58.7µm-70µm (Av: 64.3µm). Width of short cells ranged from 82.9µm- 91.8µm (Av: 87.3µm).

ii. Mature leaf

Silica bodies, stomata, trichomes and oil droplets were absent. Polygonal cells had double layer and smooth margins. Length of long cells were 208.8µm-242.7µm (Av: 225.7µm). Width of long cells were 105.5µm-78.1µm (Av: 91.8µm). Length of short cells were 115µm-150µm (Av: 132.5µm). Width of short cells were 113.6µm- 122.3µm (Av: 117.9µm). (Figure 14)

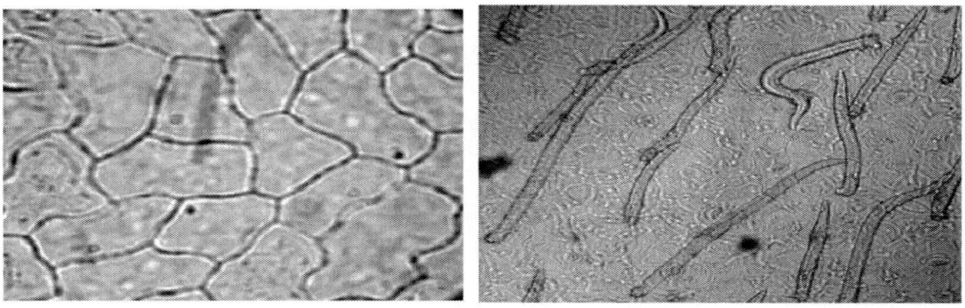

Figure 14. Leaf Epidermal Anatomy of *Millettia ovalifolia.* Adaxial side (Right) Abaxial side (Left).

iii. Immature leaf

Silica bodies, stomata, trichomes and oil droplets were absent. Polygonal cells had double layer and smooth margins. Length of long cells were 118.1μm-135.6μm (Av: 126.8μm). Width of long cells were 30.4μm-37.5μm (Av: 33.9μm). Length of short cells were 37μm-42μm (Av: 39.5μm). Width of short cells were 50.8μm- 96μm (Av: 59.9μm).

b. Abaxial Side

i. Fully mature leaf

Silica bodies and oil droplets were absent. Polygonal cells had double layer and smooth margins. Length of long cells were 120μm-128μm (Av: 124μm). Width of long cells were 99.9μm-105.1μm (Av: 102.5μm). Length of short cells were 34.5μm-37μm (Av: 35.7μm). Width of short cells were 39μm-124.5μm (Av: 31.7μm). Trichomes were present. Length of non-glandular trichomes were 312.2μm- 299.2μm (Av: 305.7μm). Width of non-glandular trichomes were 13.6μm-26.3μm (Av: 19.9μm). Stomata were present. Type of stomata was diacytic. Length of stomatal cavity ranged from 10μm-20.2μm (Av: 15.1μm). Width of stomatal cavity ranged from 4μm-6.1μm (Av: 5.6μm). Length of guard cells were 31.1μm-41.1μm (Av: 36.1μm). Width of guard cells were 21.1μm-28.3μm (Av: 21.9μm). Length of long subsidiary cells were 88μm-90μm (Av: 89μm). Length of short subsidiary cells were 35.1μm-40.7μm (Av: 37.9μm). Width of long cells were 19.2μm-26μm (Av: 22.6μm). Width of short cells were 8.2μm-11.7μm (Av: μm).

ii. Mature leaf

Silica bodies, trichomes and oil droplets were absent. Polygonal cells had smooth margins. Length of long cells ranged from 151.4μm-247.4μm (Av: 199.4μm). Width of long cells ranged from 71.5μm-83.7μm (Av: 77.6μm). Length of short cells ranged from 88.8μm-91.8μm (Av: 90.3μm). Width of short cells ranged from 88.9μm-129.3μm (Av: 218.2μm). Stomata were present. Type of stomata was diacytic. Length of stomatal cavity ranged from 44.3μm-38.9μm (Av: 39.6μm). Width of stomatal cavity ranged from 11.7μm-13μm (Av: 12.5μm). Length of guard cells were 85.1μm-85.1μm (Av: 39.6μm). Width of guard cells were 58μm-65.9μm (Av: 61.9μm). Length of long subsidiary cells ranged from 150.9μm-156.9μm (Av: 153.9μm). Length of short subsidiary cells ranged from 101μm-106μm (Av: 103.5μm). Width of long cell ranged from 33.2μm-43.9μm (Av: 38.5μm). Width of short cells ranged from 18.2μm-27.7μm (Av: 22.9μm). (Figure 14)

iii. Immature leaf

Silica bodies and oil droplets were absent. Polygonal cells had double layer and smooth margins. Length of long cells ranged from 11.2μm-122.1μm (Av: 116.6μm). Width of long cells ranged from 41.1μm-53.2μm (Av: 47.1μm). Length of short cells were 33.5μm-47.5μm (Av: 40.5μm). Width of short cells were 26.3μm- 39.4μm (Av: 32.8μm). Length of non-glandular trichomes were 14.6μm-159.9μm (Av: 87.2μm). Width of non-glandular trichomes were 77.7μm-14μm (Av: 45.8μm). Stomata were present. Type of stomata was diacytic. Length of stomatal cavity ranged from 27.5μm-53.4μm (Av: 40.4μm). Width of stomatal cavity ranged from 7.2μm-12.5μm (Av: 59.8μm). Length of guard cells were 49μm-73.8μm (Av: 70.1μm). Width of guard cells were 23.6μm-49.5μm (Av: 46.5μm). Length of long subsidiary cells were 61.6μm-70μm (Av: 65.8μm). Length of short subsidiary cells were

41.5µm -46.7µm (Av: 44.1µm). Width of long cell were 20.5µm-20.8µm (Av: 20.6µm). Width of short cells were 10.2µm – 14.3µm (Av: 12.2µm).

15. *Vicia faba* L.

a. Adaxial Side

i. Fully mature leaf

Trichomes and oil droplets were absent. Silica bodies were present. Length of silica bodies were 26.5µm-60.0µm (Av: 43.2µm). Width of silica bodies were 13.3µm- 62.0µm (Av: 7.6µm). Hexagonal cells had double layer and smooth margins. Length of long cells ranged from 124.5µm-140.8µm (Av: 132.6µm). Width of long cells ranged from 44.8µm-55.1µm (Av: 49.9µm). Length of short cells ranged from 79.1µm-108.8µm (Av: 93.9µm). Width of short cells ranged from 36.9µm-93.1µm (Av:65µm). Type of stomata was anomocytic. Length of stomatal cavity ranged from 26.4µm- 48.1µm (Av: 37.2µm). Width of stomatal cavity ranged from 8.9µm-12.2µm (Av: 10.5µm). Length of guard cells were 60.1µm-77.0µm (Av: 68.5µm). Width of guard cells were 53.6µm- 53.8µm (Av: 53.7µm). Length of long subsidiary cells were 162.1µm-164.6µm (Av: 163.3µm). Length of short subsidiary cells were 61.0µm- 77.6µm (Av: 69.3µm). Width of long cells were 8.3µm-93.1µm (Av: 50.7µm). Width of short cells were 59.0µm-33.5µm (Av: 46.2µm).

ii. Mature leaf

Trichomes and oil droplets were absent. Silica bodies were present. Length of silica bodies were 30.4µm-51.0µm (Av: 40.7µm). Width of silica bodies were 32.4µm-33.1µm (Av: 32.7µm). Hexagonal cells had double layer and smooth margins. Length of long cells ranged from 92.7µm8.2µm (Av: 95.4µm). Width of long cells ranged from 63.5µm-66.2µm (Av: 64.8µm). Length of short cells ranged from 40.9µm-71.9µm (Av: 56.4µm). Width of short cells ranged from 33.4µm-33.5µm (Av: 33.4µm). Stomata were present. Type of stomata was anomocytic. Length of stomatal cavity ranged from 16.3µm-25.0µm (Av: 20.6µm). Width of stomatal cavity ranged from 7.1µm-10.3µm (Av: 8.7µm). Length of guard cells were 77.8µm-91.9µm (Av: 84.8µm). Width of guard cells were 58.8µm-66.2µm (Av: 62.5µm). Length of long subsidiary cells ranged from 99.4µm-234.0µm (Av: 166.5µm). Length of short subsidiary cells ranged from 61.0µm-99.9µm (Av: 80.4µm). Width of long cells ranged from 75.2µm-80.6µm (Av: 77.9µm). Width of short cells ranged from 41.0µm-150.6µm (Av: 95.8µm). (Figure 15)

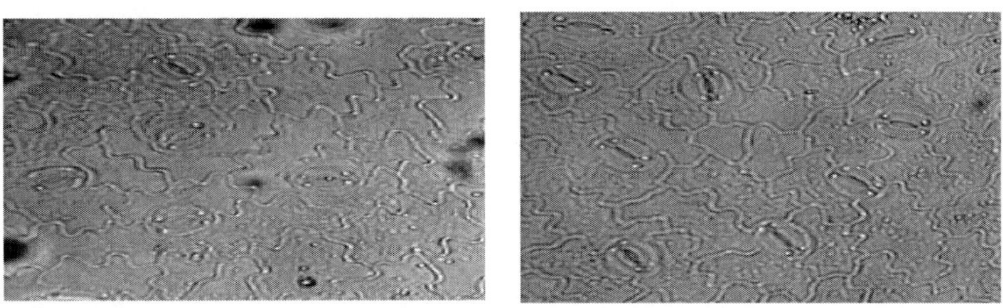

Figure 15. Leaf Epidermal Anatomy of *Vicia faba.* Adaxial side (Right), Abaxial side (Left).

iii. Immature leaf

Silica bodies, oil droplets were absent. Length of silica bodies were 15.1μm- 28.1μm (Av: 21.6μm). Width of silica bodies were 13.2μm-31.1μm (Av: 22.1μm). Hexagonal cells had double layer and smooth margins. Length of long cells ranged from 67.0μm-81.2μm (Av: 74.1μm). Width of long cells ranged from 22.1μm-24.2μm (Av: 23.1μm). Length of short cells ranged from 61.7μm-62.8μm (Av: 62.2μm). Width of short cells ranged from 32.2μm-38.3μm (Av:35.2μm). Stomata were present. Type of stomata was anomocytic. Length of stomatal cavity ranged from 25.5μm- 35.5μm (Av: 30.5μm). Width of stomatal cavity ranged from 15.8μm- 19.2μm (Av: 17.5μm). Length of guard cells were 36.2μm-56.9μm (Av: 46.5μm). Width of guard cells were 35.0μm-49.0μm (Av: 42μm). Length of long subsidiary cells were 93.4μm-132.5μm (Av: 112.9μm). Length of short subsidiary cells ranged from 31.9μm-35.8μm (Av: 33.8μm). Width of long cell ranged from 23.0μm-29.1μm (Av: 26μm). Width of short cells ranged from 12.6μm-20.2μm (Av: 16.4μm).

b. Abaxial Side

i. Fully mature leaf

Silica bodies, trichomes and oil droplets were absent. Polygonal cells had double layer and wavy margins. Length of long cells ranged from 56.5μm-87.3μm (Av:71.9μm). Width of long cells ranged from 58.2μm-65.1μm (Av: 61.6μm). Length of short cells were 79.1μm-119.1μm (Av: 75.6μm). Width of short cells were 67.1μm-84.2μm (Av:99.1). Stomata were present. Type of stomata was anomocytic. Length of stomatal cavity ranged from 26.9μm-36.9μm (Av: 31.9μm). Width of stomatal cavity ranged from 18.4μm-20.1μm (Av: 9.2μm). Length of guard cells were 60.4μm-81.1μm (Av: 70.7μm). Width of guard cells were 55.9μm-58.3μm (Av: 57.1μm). Length of long subsidiary cells ranged from 165.2μm-176.2μm (Av: 170.7μm). Length of short subsidiary cells ranged from 78.8μm-95.3μm (Av: 78μm). Width of long cells ranged from 51.9μm-109.7μm (Av: 80.8μm). Width of short cells ranged from 73.8μm- 113.1μm (Av: 93.4μm).

ii. Mature leaf

Oil droplets were absent. Silica bodies were present. Length of silica bodies were 23.0μm-24.0μm (Av: 23.5μm). Width of silica bodies were 20.0μm- 24.0μm (Av: 22μm). Polygonal cells had double layer and wavy margins. Length of long cells ranged from 117.7μm-122.7μm (Av: 120.2μm). Width of long clls ranged from 34.1μm-37.0μm (Av: 35.5μm). Length of short cells ranged from 63.0μm-69.9μm (Av: 66.4μm). Width of short cells ranged from 38.6μm-42.5μm (Av: 40.5μm). Non-glandular trichomes were present. Length of non-glandular trichomes were 110.3μm- 295.9μm (Av: 165.9μm). Width of non-glandular trichomes were 10.8μm- 22.8μm (Av: 17.0μm). Stomata were present. Type of stomata was anomocytic. Length of stomatal cavity ranged from 21.0μm-34.1μm (Av: 27.5μm). Width of stomatal cavity ranged from 7.1μm-11.2μm (Av: 91.5μm). Length of guard cells were 45.1μm- 59.4μm (Av: 52.2μm). Width of guard cells were 41.0μm-42.6μm (Av: 41.8μm). Length of long subsidiary cells were 131.5μm-153.7μm (Av: 142.6μm). Length of short subsidiary cells ranged from 59.0μm-66.7μm (Av: 62.8μm). Width of long cell ranged from 27.5μm-41.0μm (Av: 34.2μm). Width of short cells ranged from 35.1μm - 46.4μm (Av: 40.7μm). (Figure 15)

iii. Immature leaf

Trichomes and oil droplets were absent. Silica bodies were present. Length of silica bodies were 62.0µm-76.0µm (Av: 69µm). Width of silica bodies were 64.0µm-65.0µm (Av: 64.5µm). Hexagonal cells had double layer and smooth margins. Length of long cells ranged from 66.1µm-68.1µm (Av: 67.1µm). Width of long cells ranged from 37.7µm-40.7µm (Av: 39.2µm). Length of short cells ranged from 65.0µm-62.2µm (Av: 63.6µm). Width of short cells ranged from 29.0µm-34.4µm (Av: 34.4µm). Stomata were present. Type of stomata was anomocytic. Length of stomatal cavity ranged from 11.2µm-21.1µm (Av: 16.1µm). Width of stomatal cavity ranged from 8.2µm-13.0µm (Av: 10.6µm). Length of guard cells ranged from 29.4µm-45.5µm (Av: 37.4µm). Width of guard cells were 28.4µm-30.5µm (Av: 29.4µm). Length of long subsidiary cells ranged from 64.2µm-74.3µm (Av: 69.2µm) Length of short subsidiary cells ranged from 38.9µm-44.3µm (Av: 41.6µm). Width of long cell 33.4µm-34.8µm (Av: 34.1µm). Width of short cells ranged from 12.1µm-16.3µm (Av: 14.2µm). (Figure 15)

CO-RELATION AMONG THE SPECIES OF FAMILY FABACEAE

Table 1 shows relationship between the species of Family Fabaceae. Co-relation of this family ranged from 0.01- 0.92. Based on leaf epidermal anatomical characters *Acacia modesta* abaxial side was closely related to *Dalbergia lanceolaria* and *Dalbergia lanceolaria* abaxial side was closely related to *Delonix regia.* The value of co-relation for *Acacia modesta, Dalbergia lanceolaria* and *Delonix regia* was 0.92 Table (1). Minimum co-relation was observed for *Dalbergia lanceolaria* with *Erythrina suberosa,* relationship was1%. The species that showed negative co-relation ranged from -0.01 to -0.12, the value of co-relation for *Albizia mimosa and Cassia occidentalis* was -0.01. Table 2 shows the relationship between the species of Family Fabaceae. Co- ranged from 0.01- 0.93. Based on leaf epidermal anatomical characters *Albizia procera* adaxial side was closely related to *Delonix regia.* The value of co-relation for *Albizia procera* and *Delonix regia* was 0.93 (Table 2). The species that showed minimum co-relation were *Cassia fistula* with *Cassia occidentalis,* their relationship was1%. These two species were totally different from one another on the basis of leaf epidermal anatomical characters.

Table 3 shows the relationship between the characters of Family Fabaceae. Co-relation ranged from 0.01- 0.98. The value of co-relation for the length of stomatal cell and width of stomatal cell was 0.98 (Table 3). The characters that showed minimum co-relation were width/length of stomata cell, length/width of silica body, and width of trichome. Negative co-relation of various characters ranged from -0.01 -0.15, the value of co-relation for length of trichome and width of stomatal cavity was -0.01. Minimum negative co-relation was observed between length of trichome and length of silica body(-0.15).

Table 4 shows the relationship between the characters of Family Fabaceae. Co-relation on adaxial side of leaf.Values for co-relation ranged from 0.03- 0.97. The value of co-relation for length of stomata cell and length of long subsidiary cell was 0.97 (Table 3). The characters that showed minimum co-relation are length of long subsidiary cell and length of short subsidiary cell. The Negative co-relation ranged from -0.01 to -0.51. the value of co-relation for length of short subsidiary cell, length of silica body, width of trichome and length of stomatal cell was -0.01. The characters that showed minimum negative co-relation were width of short cell and length of short subsidiary cell (– 0.51).

CLUSTER ANALYSIS OF FAMILY FABACEAE

Cluster analysis based on leaf epidermal anatomical characters on abaxial side revealed that 15 species of family Fabaceae were divided in to two main clusters 1 and 2 with polygenetic distance 100% (Figure 16). At 60% genetic linkage distance group 1 is further divided in to two subclusters 1a and 1b. The species include in 1a was *Cassia occidentalis* and in 1b comprised of *Albizia procera* and *Vicia faba.* At 67% genetic linkage group 2 is further divided in to 2a and 2b. The species included in 2a *Dalbergia sisssoo* and *Leucaena glauca.* At 40% genetic linkage group 2b is further divided in to 2bi and 2bii. The species included in 2bi *Albizia lebbek.* The species included in 2bii were *Erythrina suberosa, Cassia fistula, Acacia catechu, Gleditsia triacanthos, Millettia ovalifolia, Albizia mimosa, Delonix regia, Dalbergia lanceolaria and Acacia modesta.* Cluster analysis based on leaf epidermal anatomical characters on adaxial side revealed that 15 species of family Fabaceae were divided into two main clusters 1 and 2. At polygenetic distance 100% (Figure 17). The species include in cluster 1 was *Albizia procera.* At 51% genetic linkage group 2 is further divided in to 2a and 2b.At 31% genetic linkage group 2a is further divided in to 2ai and 2aii. The species included in 2ai was *Albizia mimosa.* The species included in 2aii were *Acacia modesta, Millettia ovalifolia, Dalbergia lanceolaria, Cassia occidentalis and Cassia fistula.* At 43% genetic linkage group 2b is further divided in to 2bi and 2bii. The species included in 2bi was *Dalbergia sisssoo.* The species included in 2bii were *Acacia catechu, Albizia lebbek, Delonix regia, Gleditsia triacanthos, Vicia faba, Erythrina suberosa* and *Leucaena glauca.*

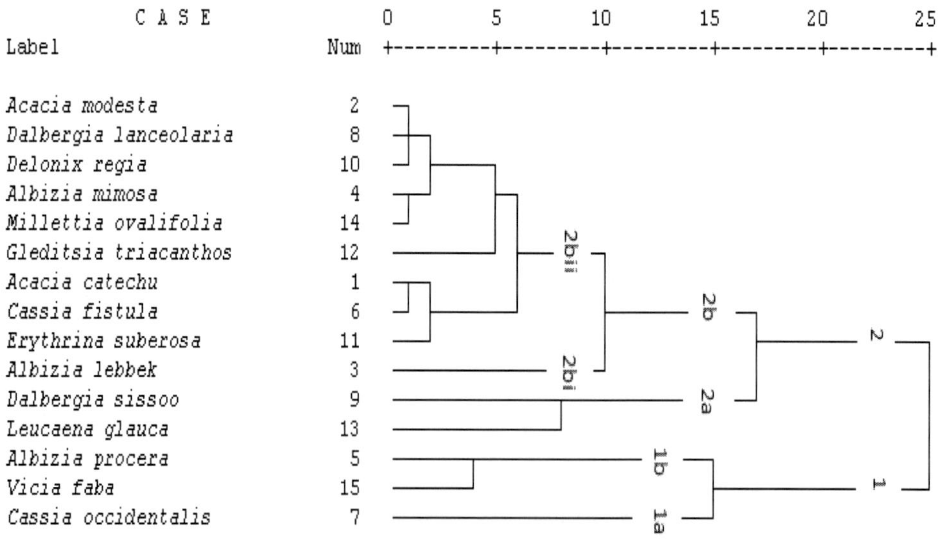

Figure 16. cluster analysis of species of Family Fabaceae on the basis of leaf epidermal anatomical charcters on abaxial side.

Table 1. Co-relation between the species of Family Fabaceae based on leaf epidermal anatomical characters on Abaxial surface

Species	1	2	3	4	5	6	7	8	9	10	11	12	13	14	15
1	1														
2	0.54	1													
3	0.75	0.20	1												
4	0.56	0.91	0.27	1											
5	0.11	0.14	-0.06	0.18	1										
6	0.65	0.17	0.43	0.21	0.74	1									
7	0.28	-0.05	0.15	-0.01	0.84	0.85	1								
8	0.47	0.92	0.15	0.81	0.09	0.12	-0.07	1							
9	0.47	0.61	0.37	0.64	0.68	0.60	0.50	0.51	1						
10	0.43	0.91	0.10	0.85	0.08	0.05	-0.12	0.92	0.52	1					
11	0.57	0.08	0.57	0.08	0.62	0.88	0.79	0.01	0.60	-0.08	1				
12	0.51	0.77	0.19	0.74	0.07	0.18	-0.05	0.78	0.53	0.76	0.07	1			
13	0.59	0.63	0.68	0.72	-0.02	0.17	-0.06	0.56	0.67	0.56	0.28	0.55	1		
14	0.71	0.76	0.53	0.87	0.17	0.40	0.11	0.66	0.65	0.68	0.27	0.65	0.79	1	
15	0.05	0.21	-0.14	0.30	0.93	0.58	0.70	0.19	0.67	0.25	0.43	0.10	0.05	0.23	1

Key: 1: Acacia catechu; 2: Acacia modesta; 3:Albizia lebbek; 4:Albizia mimosa; 5: Albizia procera; 6: Cassia fistula; 7: Cassia occidentalis; 8: Dalbergia lanceolaria; 9: Dalbergia sisssoo; 10: Delonix regia; 11: Erythrina suberosa; 12: Gleditsia triacanthos; 13: Leucaena glauca; 14: Millettia ovalifolia; 15: Vicia faba.

Table 2. Co-relation between the species of Family Fabaceae based on leaf epidermal anatomical characters on Adaxial surface

	1	2	3	4	5	6	7	8	9	10	11	12	13	14	15
1	1														
2	0.47	1													
3	0.92	0.37	1												
4	0.56	0.73	0.43	1											
5	0.16	0.77	0.08	0.34	1										
6	0.28	0.91	0.20	0.58	0.85	1									
7	0.43	0.93	0.31	0.80	0.79	0.93	1								
8	0.61	0.81	0.55	0.88	0.40	0.69	0.83	1							
9	0.88	0.59	0.87	0.47	0.37	0.49	0.52	0.62	1						
10	0.81	0.28	0.86	0.37	0.01	0.18	0.26	0.50	0.81	1					
11	0.83	0.64	0.75	0.66	0.29	0.47	0.60	0.76	0.84	0.75	1				
12	0.87	0.32	0.87	0.40	0.03	0.16	0.27	0.52	0.81	0.91	0.79	1			
13	0.31	0.03	0.22	0.46	-0.11	0.06	0.24	0.27	0.21	0.23	0.19	0.19	1		
14	0.63	0.79	0.49	0.97	0.38	0.62	0.83	0.92	0.55	0.44	0.74	0.48	0.36	1	
15	0.87	0.25	0.87	0.46	0.00	0.14	0.28	0.46	0.79	0.81	0.67	0.78	0.54	0.49	1

Key: 1: Acacia catechu; 2: Acacia modesta; 3:Albizia lebbek; 4:Albizia mimosa; 5: Albizia procera; 6: Cassia fistula; 7: Cassia occidentalis; 8: Dalbergia lanceolaria; 9: Dalbergia sisssoo; 10: Delonix regia; 11: Erythrina suberosa; 12: Gleditsia triacanthos; 13: Leucaena glauca; 14: Millettia ovalifolia; 15: Vicia faba.

Table 3. Co-relation between the Characters of Family Fabaceae based on leaf epidermal anatomical characters on Abaxial surface

	1	2	3	4	5	6	7	8	9	10	11	12	13	14	15	16
1	1															
2	0.87	1														
3	0.78	0.75	1													
4	0.56	0.74	0.77	1												
5	-0.11	0.13	0.08	0.32	1											
6	-0.14	0.04	0.00	0.14	0.91	1										
7	0.61	0.57	0.26	0.40	0.04	-0.04	1									
8	0.47	0.44	0.11	0.30	-0.01	-0.07	0.87	1								
9	0.47	0.43	0.16	0.36	0.11	0.01	0.96	0.88	1							
10	0.35	0.34	0.07	0.32	0.08	-0.03	0.91	0.90	0.98	1						
11	0.41	0.54	0.34	0.55	0.31	0.17	0.70	0.68	0.76	0.74	1					
12	0.18	0.31	0.07	0.43	0.09	-0.07	0.77	0.71	0.83	0.88	0.79	1				
13	0.60	0.58	0.33	0.49	0.17	0.05	0.94	0.73	0.92	0.85	0.77	0.79	1			
14	0.25	0.43	0.14	0.59	0.22	0.02	0.75	0.70	0.78	0.79	0.77	0.85	0.75	1		
15	0.80	0.85	0.77	0.58	-0.15	-0.13	0.27	0.18	0.10	0.01	0.22	-0.06	0.23	0.15	1	
16	0.73	0.79	0.66	0.49	-0.12	-0.04	0.18	0.10	0.01	-0.08	0.12	-0.14	0.14	0.08	0.96	1

Key: 1: length of long cell; 2: width of long cell; 3: length of short cell; 4: width of short cell; 5: length of trichome; 6: width of trichome; 7: length of stomatal cavity; 8: width of of stomatal cavity; 9: length of stomatal cell; 10:width of stomatal cell; 11: Length of long subsidiary cell; 12: width of long subsidiary cell; 13: length of short subsidiary cell; 14: width of short subsidiary cell; 15: length of silica body; 16: width of silica body.

Table 4. Co-relation between the Characters of Family Fabaceae based on leaf epidermal anatomical characters on Adaxial surface

	1	2	3	4	5	6	7	8	9	10	11	12	13	14	15	16
1	1															
2	0.40	1														
3	0.16	0.45	1													
4	0.60	0.47	0.40	1												
5	0.03	0.33	0.10	0.18	1											
6	0.15	0.42	0.27	0.13	0.57	1										
7	0.09	-0.05	0.31	-0.38	-0.29	0.09	1									
8	0.30	0.12	0.49	0.00	-0.14	0.33	0.50	1								
9	-0.03	-0.19	0.28	-0.45	-0.37	-0.01	0.94	0.55	1							
10	0.05	-0.10	0.33	-0.40	-0.34	0.10	0.91	0.62	0.97	1						
11	0.00	-0.14	0.38	-0.39	-0.34	0.07	0.85	0.47	0.88	0.93	1					
12	0.13	0.06	0.37	-0.30	-0.24	0.26	0.84	0.56	0.88	0.95	0.91	1				
13	-0.12	-0.24	0.26	-0.51	-0.39	-0.07	0.81	0.36	0.84	0.87	0.97	0.83	1			
14	-0.04	-0.15	0.18	-0.42	-0.37	-0.07	0.82	0.60	0.91	0.90	0.75	0.81	0.71	1		
15	-0.03	-0.31	0.25	0.41	-0.26	-0.39	-0.19	-0.22	-0.19	-0.23	-0.06	-0.25	-0.01	-0.23	1	
16	-0.06	-0.27	0.19	0.38	-0.21	-0.34	-0.26	-0.22	-0.27	-0.33	-0.22	-0.33	-0.15	-0.30	0.94	1

Key: 1:length of long cell; 2: width of long cell; 3: length of short cell; 4: width of short cell; 5: length of trichome; 6: width of trichome; 7: length of stomatal cavity; 8: width of of stomatal cavity; 9: length of stomatal cell; 10:width of stomatal cell; 11: length of long subsidiary cell; 12: width of long subsidiary cell; 13: length of short subsidiary cell; 14: width of short subsidiary cell; 15: length of silica body;16: width of silica body.

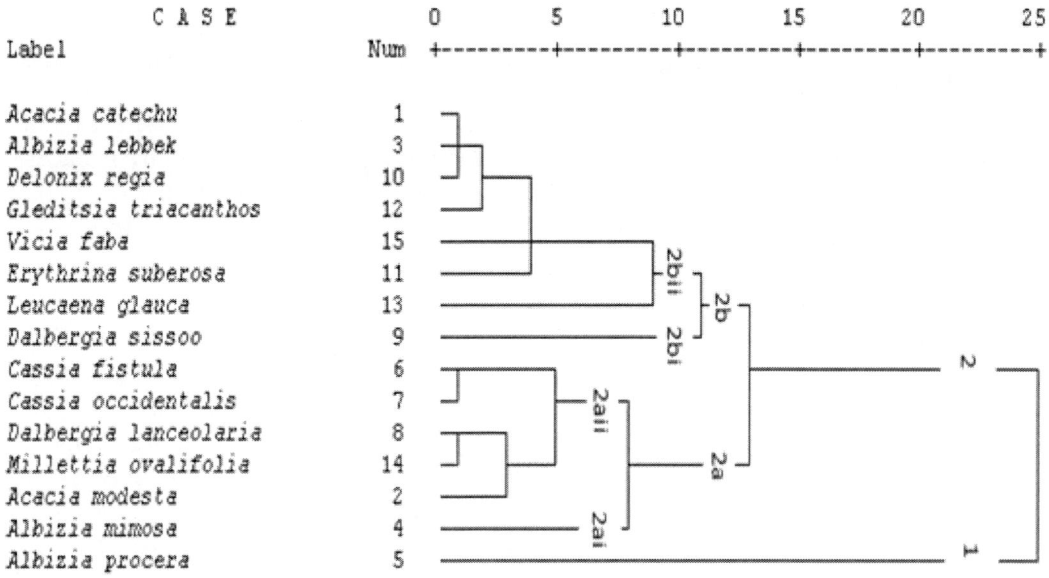

Figure 17. cluster analysis of species of Family Fabaceae on the basis of leaf epidermal anatomical charcters on adaxial side.

Discussion

Epidermal Characters of Family Fabaceae

Many species of family Fabaceae are medicinally important. The data generated from leaf epidermal surface is use in resolving the taxonomic problems of taxa has gained much recognition for a long time [63]. Sufficient research has not been conducted on the use of anatomical features for the delineation and classification of various taxa [64]. Leaf epidermis features like stomatal types, density, size, epidermal cell types, epidermal cell size and trichomes are considered as taxonomically important in delimiting species of the same genus or genera within family [65, 66, 67]. It has been observed that anatomical studies showed variation in size, shape and arrangement of epidermal cells, stomata, trichomes in number and shape of subsidiary cells and silica bodies [68]. Same observation of anatomical features was made during the present research as in *Dalbergia sissoo* most of the cells were pentagonal or polygonal in shape with smooth. During the present work the family Fabaceae was found to be characterized by different types of cells that is polygonal, tetragonal, pentagonal and hexagonalwith smooth and wavy margins. Single and double layered cells were also observed. The size and shape were different on adaxial and abaxial surface of leaf epidermis because of bifacial nature of leaf, which was also confirmed by the findings of Webster [69]. The maximum length of long cells ranged from 234.4μm-295.5μm which was found in *Albizia mimosa* on abaxial side of mature leaf epidermis. Similarly maximum length of short cells ranged from 82.8μm-236.3 μm, exhibited by the same species which is a unique leaf character and is importrant character to differentiate it from other species of family Fabaceae.

Stomata are important taxonomic character of family Fabaceae. In many cases, stomata were found to be more abundant on abaxial than adaxial side of leaf, sometimes totally absent from both sides. The development and pattern of stomata are regulated by both genetic and environmental signals [70]. and Snir and Sachs, [71] observed that stomata are homologous structures that develop differently even in related species. They analyze the developmental processes and variability expressed by quantitative changes in the number of cell divisions and by local cell growth. The ridge of stomata is present in some species of family Fabaceae. This structure is important for opening and closing of stomata, and outer shape of stomata [72]. Later [73] observed that the development of all stomatal structures initiated at the same time and the early stages of the development of immature stomata had no particular differentiating characteristics.

The present study shows that stomata were kidney shaped. Stomatal types vary from species to species. In family Fabaceae major stomatal types include diacytic, actinocytic, staurocytic, anomocytic and cyclocytic. Stomata were totally absent on both adaxial and abaxial surface of mature, fully mature and immature leaf of *Cassia occidentalis,* While in *Gleditsia triacantho s*cyclocytic stomata were not found on adaxial surface of mature leaf and in *Erythrina suberosa* staurocytic stomata were absent on abaxial surface of leaf. In *Acacia catechu* and *Vicia faba* stomata were not fully developed on both adaxial and abaxial surface of immature leaf, while fully developed in mature as well as in fully mature leaf. In *Acacia catechu,* diacytic type of stomata were present, while in *Vicia faba* anomocytic type of stomata were present. The maximum length of stomatal cells ranged from 140μm-143.1μm which was found in *Albizia lebbek* on adaxial side of mature leaf epidermis.

Trichomes were also considered as important taxonomic and identifiable character of family Fabaceae. The length and types of trichomes varies from species to species such as in *Albizia procera* stellate type of trichomes were present and length of trichomes ranged from 246μm-677μm which differentiate it from other species of family Fabaceae. Findings of the present studies show that trichomes could be used for identification of the species of family Fabaceae. Trichomes not only showed variation in length and types but also in origin, it might be intracellular, from the surface of outer/ inner membrane. Trichomes with acute or blunt ends were also found. Same observation was made by [74] while they were working for the leaf epidermal anatomy of *Corton.* Blunt ends were mostly found in *Cassia fistula* and *Millettia ovalifoli.* Trichomes were altogether absent on adaxial and abaxial side of mature, fully mature and immature leaf in *Acacia catechu* and *Albizia mimosa.* Trichomes were glandular, non-glandular, stellate type in Family Fabaceae. Presence of silica bodies also considered as one of taxonomically important characters of family Fabaceae. In *Acacia catechu, Albizia procera, Cassia fistula* and *Gleditsia triacanthos* silica bodies were absent on both adaxial and abaxial surface of mature, fully mature and immature leaf. In family Fabaceae silica bodies were round and irregular in shape. From the result of present studies it is concluded that leaf epidermal anatomical features are important taxonomic markers and they can effectively utilized for resolving the persistent taxonomic problem of family Fabaceae.

CONCLUSION

During the present work the family Fabaceae was found to be characterized by different types of cells that is polygonal, tetragonal, pentagonal and hexagonal and their margins were may be smooth and wavy. The maximum length of long cells ranged from 234.4μm-295.5μm which was found in *Albizia mimosa* on abaxial side of mature leaf epidermis. Similarly maximum length of short cells ranged from82.8μm-236.3μm, exhibited by the same species which is a unique leaf character and is sufficient to differentiate it from other species of family Fabaceae. Stomata were totally absent on both adaxial and abaxial surface of mature, fully mature and immature leaf in *Cassia occidentalis,* While in *Gleditsia triacanthos* stomata were absent in mature leaf and in *Erythrina suberosa* stomata were absent in immature leafThe maximum length of stomatal cells ranged from 140μm-143.1μm which was found in *Albizia lebbek* on adaxial side of mature leaf epidermis. Trichomes were altogether absent on adaxial and abaxial side of mature, fully mature and immature leaf in *Acacia catechu* and *Albizia mimosa.* In *Albizia procera* stellate type of trichomes was present. Presence of silica bodies also considered as one of taxonomically important characters of family Fabaceae. In *Acacia catechu, Albizia procera, Cassia fistula* and *Gleditsia triacanthos* silica bodies were absent on both adaxial and abaxial sides of mature, fully mature and immature leaf. In family Fabaceae silica bodies were round and irregular shaped.

REFERENCES

[1] Willis, J.C. and Airy-Shaw, H.K. (1973). *A Dictionary of the Flowering Plants and Ferns.* ISBN 10: 052108699X / ISBN 13: 9780521086998.

[2] Joly, A.B. (1998). Botânica-Introduçao à Taxonomia vegetal. Companhia Editora Nacional. 12 edição. *São Paulo.* P 420-421.

[3] Lewis, G., Schrire, B., MacKinder B. and Lock, M. (2005). *Legumes of the world.* Royal Botanical Gardens, Kew, UK.

[4] Dalziel, J.M. (1956). *The useful Plants of West Tropical Africa' Crown Agens,* London. pp. 612.

[5] Polhill, R.M. Raven, P.H. and stirton C.H. (1981). Evolution and systematic of leguminosae. In: Polhill, R. M. and Raven P. H, Eds *Advances in legume systematics,* part 1. Kew: Royal Botanical Gardens, 1- 26.

[6] Stace, C. (1980). *Plant taxonomy and Dicotyledonous,* vol.I.Clarendon press, Oxford. Pp 23-25.

[7] Hao, G., Zhang D.X., Zhang M.Y., Guo, L.X. and Li, S. J. (2003). Phylogenetics of *Bauhinia* subgenus *Phanera.* (Leguminosae: Caesalpinioideae) based on ITS sequences of nuclear ribosomal DNA. *Botanical Bulletin of Academia Sinica.* 44: 223–228.

[8] Owens, S.A, Fields, P.F. and Ewers, F.W. (1998). Degradation of the upper pulvinus in modern and fossil leaves of Cercis (Fabaceae). *American Journal of Botany,* 85: 273-284.

[9] Pan, K.Y., Lu, A.M. and Wen, J. (1990). Characters of leaf epidermis in Hamamelidaeae (*s.l*) *Acta Phytotaxonomica Sinica*, 28: 10- 26.

[10] Yousaf, Z., Shinwari, Z.K., Qureshi, R.A. and Perveen, A. (2008). Leaf epidermal anatomy of selected *Allium* species, family Alliaceae from Pakistan. *Pak. J. Bot.*, 40(1): 77-90.

[11] Gillani, S.S., Khan, M.A., Shinwari, Z.K. and Yousaf, Z. (2002). Leaf epidermal anatomy of selected *Digitaria* species, Tribe Paniceae, Family Poaceae of Pakistan. *Pak. J. Bot.*, 34: 257-273.

[12] Baranova, M. (1992). Systematic anatomy of leaf epidermis in the Magnoliaceae and some related families. *Taxon,* 21: 447-469.

[13] Jones, J.H. (1986). Evolution of the Fagaceae: the implication of foliar featuers. *Ann. Misouri Bot. Garden*, 73: 228-275.

[14] Theobald, W.L., Krahulik J.L. and Rollins, R.C. (1979). Trichome descriptionand classification. In: Metcalfe CR, Chalk L (eds) *Anatomy of the Dicotyledons*, 2nd edn. Oxford Science, Oxford, pp 40–53.

[15] Payne, W.W. (1978). A glossary of plant trichom terminology. *Brittonia* 30: 239–255.

[16] Ascensão, L., Marques, N. and Pais, M.S. (1995). Glandular trichomes on vegetative and reproductive organs of *Leonotis leonurus* (Lamiaceae). Annals of Botany, 75: 619–626.

[17] Werker, E., Putievsky, E., Ravid, U., Dudai, N. and Katzir, I. (1993). Glandular hairs and essential oil in developing leaves of *Ocimum basilicum* L. (Lamiaceae). *Annals of Botany,* 71: 43–50.

[18] Fahn, A. (1988). Secretory tissues in vascular plants. New Phytologist, 108229–257.

[19] Werker, E., Ravid, U. and Putievsky, E. (1985). Glandular hairs and their secretion in the vegetative and reproductive organs of Salvia sclarea And Salvia dominica. *Israel Journal of Botany,* 34, 239–252.

[20] Gershenzon, J., Maffei, M. and Croteau, R. (1989). Biochemical and histochemical localization of monoterpene biosynthesis in the glandular trichomes of spearmint (*Mentha spicata*) Plant Physiol., 89: 1351–1357.

[21] Stace, C.A., (1984). The taxonomic importance of the leaf surface. In *Current concepts in plant Taxonomy* (V. H. Heywood and D. M. Moore Eds.), Academic Press, London: 67 – 94.

[22] Metcalfe, C. R., (1968). Current development in systematic plant anatomy. In *Modern methods in plant taxonomy* (V.H. Heywood Ed.) Academic Press, London, New York, 45 – 57.

[23] Carlquist, S. (1961). *Comparative plant anatomy.* Holt, Rinehart & Winston, New York.

[24] Barthlott, W. (1981). Epidermal and seed surface characters of plants: systematic applicability and some evolutionary aspects. *Nordic Journal of Botany* 1: 345–355.

[25] Heslop-Harrison, Y. (1970). Scanning electron microscopy of fresh leaves of *Pinguicula. Science*, 167: 172-174.

[26] Metcalfe, C.R. and Chalk, L. (1979). *Anatomy of the Dicotyledons, Systematic Anatomy of the leaf and stem.* Vol. I. 2nd Ed. Clarendon Press, Oxford.

[27] Pettigrew, C.J. and Watson, L. (1973). On the identification of sterile Acacia's and the feasibility of establishing and automatic key-generating system. *Aust. J. Bot.,* 21: 141-150.

[28] Hetherington, A.M., and Woodward, F.I. (2003). The role of stomata in sensing and driving environmental change. *Nature* 424 901–908.

[29] Nadeau, J.A., and Sack, F.D. (2002). Stomatal development in Arabidopsis. In The Arabidopsis Book, C.R. Somerville and E.M. Meyerowitz, eds (Rockville, MD: American Society of Plant Biologists, doi/10.1199/tab.0066.

[30] Woodward, F.I., Lake, J.A. and Quick, W.P. (2002). Stomatal development and CO_2: Ecological consequences. *New Phytol.,* 153 477–484.

[31] Scheres, B., and Benfey, P.N. (1999). Asymmetric cell division in plants. Annu. Rev. Plant Physiol. Plant Mol. Biol., 50 505–537

[32] Jan, Y.N., and Jan, L.Y. (1998). Asymmetric cell division. Nature 392 775–778.

[33] Horvitz, H.R., and Herskowitz, I. (1992). Mechanisms of asymmetric cell division: Two Bs or not two Bs, that is the question. Cell 68 237–255

[34] Abid, R., Sharmeen, S.and Perveen, A. (2007). Stomatal types of monocots within Flora of Karachi, Pakistan. *Pak. J. Bot.*, 39(1): 15-21

[35] Adeniji, K.A., and Ariwaodo, J.O. (2012). Comparative foliar epidermal studies of genus *Pericopsis* (Papilionaceae) in Nigeria. *Phytologia Balacania.* 18(1): 37- 41.

[36] Bhagat, A. P. and Bhuktar, A.S. (2012). Anatomical Standardization of *Capparis Zeylanica* L. *International Research Journal of Pharmacy.* 3(7):336-342.

[37] Vatén, A. and Bergmann, D.C. (2012). Mechanisms of stomatal development: an evolutionary view. *Journal of Evolutionary developmental biology.*3:11.

[38] Khan, F., Yousaf, Z., Rani, S. and Khan, F. (2011). Taxonomic treatment of medicinally important arboreal flora of tropical and subtropical region based on leaf epidermal anatomical markers. *Journal of Medicinal Plants Research.* 5(28): 6439-6454.

[39] Shaheen, S., Ahmad, M., Khan, F., Zafar, M., Khan, M. A., Bano, A., Sultana, S., Jamil, M., Jan, G., and Leghari, M.K. (2011). Micormorphological diversity in leaf epidermal anatomy of *Brachiaria* species using elemental dispersive spectrophotometer analysis. *Journal of Medicinal Plants Research.* 5(17): 4279-4286.

[40] Shaheed, S.A. and Illoh, H.C. (2010). Taxonomic study of some species in *Cassiae* (Leguminoceae) using leaf epidermal characters Notulae Botanicae. *Horti Agrobotanici Botanical Club.* 38(1):20-27.

[41] Yasmin, G., Khan, M. A., Shaheen, N. and Hayat, M.Q. (2010). Taxonomic significance of leaf epidermal anatomy of selected Persicaria Mill. Species of family Polygonaceae from Pakistan. *African Journal of Biotechnology.* 9(25): 3759-3768.

[42] Ahmad, K., Khan, M.A., Ahmad, M., Shaheen, N. and Nazir, A. (2010). Taxonomic diversity in epidermal cells of some sub-tropical plant species. *International journal of agriculture and biology.* 12: 115–118.

[43] Shaheen, N., Khan, M.A., Yasmin G. and Hayat, M.Q. (2009). Diversity of foliar trichomes and their systematic. Relevancen in the Genus *Hibiscus* (Malvaceae). *International journal of agriculture and biology.* 11(3):279 -284.

[44] Ahmad, K., Khan, M.A., Ahmad, M., Zafar, M., Arshad, M. and Ahmad, F. (2009). Taxonomic diversity of stomata in dicot flora of a district tank (N.W.F.P.) in Pakistan. *African Journal of Biotechnology.* 8 (6): 1052-1055.

[45] Zou, P., Liao, J. and Zhang, D. (2008). Leaf epidermal Micromorphology of *Cercis* (Fabaceae: Caesalpinioideae). *Botanical Journal of the Linnean Society.* 158: 539-547.

[46] Abid, R., Sharmeen, S. and Perveen, A. (2007). Stomatal types of Monocots within flora of Karachi, Pakistan. *Pakistan journal of Botany.* 39(1): 15-21.

[47] Perveen, A., Abid, R. and Fatima, R. (2007). Stomatal types of some dicots within flora of Karachi, Pakistan. *Pakistan journal of Botany.* 39(4):1017-1023

[48] Wang, H., Ngwenyama, N., Liu, Y., Walker, J. C. and Zhangb, S. (2007). Stomatal development and Patterning are regulated by environmentally responsive Mitogen-

Activated Protein Kinases in Arabidopsis. *American Society of Plant Biologists*. 19: 63–73.

[49] Agbagwa, I.O. and Okoli, B.E. (2006). Leaf epidermal Micromorphology in the systematics of *Abrus* (Papilionaceae) in parts of Tropical West Africa. *Asian Journal of Plant Sciences*. 5(1): 41-49.

[50] Ozdemir, C. and Altn, Y. (2006). Morphological and anatomical investigation of some endemic *alkana* species. *Pakistan journal of Botany*. 38(3):527-537.

[51] Klimko, M. and Truchan, M. (2006). Morphology variability of the leaf epidermis in selected taxa of the genus *Ficus* L. (Moraceae) and its taxonomic implication. *Acta Socities botanicorum poloniae*. 75 (4): 309-324.

[52] Bergmann, D.C. and Sack, F.D. (2006). Stomatal Development. *Annual review of plant biology*.19:33.

[53] Liscovsky, I.J. and Cosa, M.T. (2005). Comparative anatomy of leaf and stem in species of Cestreae G. Don (Solanaceae) from Argentina. *Gayana Botanica*. 33 – 43.

[54] Carpenter, R. J., Hill, R. S. and Jordant, G. J. 2005. Leaf cuticular morphology links Platanaceae and Proteaceae. *International journal of plant science*. 5:846-855.

[55] Kanademir, N. (2003). The morphological anatomical and karyological properties of endemic *Saliva hypargeia* Fich and Mey. (Lamiaceae) in Turkey. *Pakistan journal of Botany*. 35(2):219-236.

[56] Snir, P. and Sachs, T. (2002). The evolution of epidermal development: Example from the Fabaceae. *Israel Journal of Plant Sciences*. 50: 129- 139.

[57] Kong, H. (2001). Comparative morphology of leaf epidermis in the Chloranthaceae. *Botanical Journal of the Linnean Society*. 136: 279-294.

[58] Idu, M., Olorunfemi, D.I. and Omonhinmin, A.C. (2000). Systematics value of stomata species of Fabaceae in some Nigerian hardwood. *Plant Biosystematic*. 134 (1): 53-60.

[59] Sheteolu, A.O. and Ayodele, A.E. (1997). Epidermal morphology of the genus *Dialium* (Fabaceae: Caesalpinoideae). *Feddes Repertorium.3-4:151-158.*

[60] Ogundipe, O.T. (1992). Leaf epidermal studies of genus *Dautura* (Linn.) Solanaceae. *Phytomorphology*. 42(34): 207-217.

[61] Kagan, M.L. and Sachs, T. (1991). Evidence for epigenetic selection of a spacing pattern. *Israel journal of Developmental Biology*.146(1):100–105

[62] Lackey, J.A. (1978). Leaflet anatomy of Phaseoleae (Leguminosae; Paplionoidae) and its relation to taxonomy. *Botanical gazette*. 139(4): 436-446.

[63] Hardin, J.W. (1990). Variation patterns and recognition of varieties of *Tilia Americana* Système. *International system of botany*. 15: 33-48.

[64] Awordine, D.O., Nwoye, D.U. and Olagoke, A.A. (2009). Taxonomic significance of foliar epidermis in in some member of Euphorbiaceae Family in *Nigeria*. Research Journal of Botany. 4: 17-28.

[65] Abdul Rahman, A.A. and Oladele, F.A. (2010). Stomatal complex types and epidermis cells in some *jatropha* species L. (Euphorbiaceae). Nigeria Global Journal of Pure and Applied Sciences. 23:2160-2163

[66] Oladele, D. A. (2000).Anatomical studies of family Euphorbiaceae. *Botanical Journal of Linnean Society*. 6:159-230.

[67] Aguiar, M. A. and Preisinger, H. (2000). Traits of Leaf Anatomy of *Corton lanjouwensis jablonski* (Euphorbiaceae) in Different Strata of the plant. German

Brazilian Workshop on Neotropical Ecosystem – Achievement and Prospects of Cooperative Research Hamburg. Research Journal of botany. 3- 8.

[68] Khan, F., Yousaf, Z., Rani, S. and Khan, F. (2011). Taxonomic treatment of medicinally important arboreal flora of tropical and subtropical region based on leaf epidermal anatomical markers. *Journal of Medicinal Plants Research.* 5(28): 6439-6454.

[69] Webster, G. L., Aguilar, M.J. and Smith, B.A.1983. Systematic distribution of foliar trichomes types in *Corton* (Euphorbiaceae). *Botanical Journal of Linnean Society.*

[70] Wang, H., Ngwenyama, N., Liu, Y., Walker, J. C. and Zhangb, S. (2007). Stomatal development and Patterning are regulated by environmentally responsive Mitogen-Activated Protein Kinases in Arabidopsis. *American Society of Plant Biologists.* 19: 63–73.

[71] Snir, P. and Sachs, T. (2002). The evolution of epidermal development: Example from the Fabaceae. *Israel Journal of Plant Sciences.* 50: 129- 139.

[72] Yukawa, T., Ando,T., Karasawa, K. and Hashimoto, K. (1992). Existence of two stomatal shape in the *Genus Dendrobium* (Orchidaceae) and its systematic significance. *American Journal of Botany.* 79: 946-952.

[73] Kagan, M.L. and Sachs, T. 1991. Evidence for epigenetic selection of a spacing pattern. *Israel journal of Developmental Biology.*146(1):100–105

[74] Senakuni, C. and Chantaranothaf, P. (2010). A morphological survey of foliar trichomes of *Corton* L. (Euphorbiaceae) in Thailand. *Botanical Journal of the Linnean Society.* 111: 299-423.

In: Leaf Epidermal Anatomical Characters …
Editors: Z. Yousaf, A. Younas and A. Aftab
ISBN: 978-1-63483-190-1

Chapter 7

TAXONOMIC EVALUATION OF FAMILY POACEAE FROM DISTRICT LAHORE BASED ON LEAF EPIDERMAL ANATOMY CHARACTERS

Sumaira Rani and N. R. Qamar*
Department of Botany,
Lahore College for Women University
Lahore, Pakistan

ABSTRACT

Morphological based taxonomic information of family Poaceae is contradictory. Therefore, there is need to investigate new taxonomic tools. Leaf epidermal anatomy is considered to be one of the effective techniques used for the generation of more authentic taxonomic information. Important anatomical characters identified for the family Poaceae were types, shapes of epidermal cells, cell membrane variation, shape and types of trichomes, oil droplets, silica bodies, stomatal cells and plasmodesmata. Cells were rectangular in shape and elongated with tapering ends. The maximum length of long cells ranging 520μm- 450μm was found in *Lolium temulentiumon* on abaxial side of epidermis. Silica bodies, oil droplets and plasmodesmata were absent from both abaxial and adaxial surface of *Setaria intermedia*. Cross shape silica bodies were observed in the genus *Digitaria* while regular to irregular were found in all other species of family Poaceae. Trichomes may be originated intracellular, from the surface of outer memberane/inner- membrane. Leaf epidermal anatomy was found to be an important tool for the identification and characterizationn of family Poaceae and it can be effectively utilized for solving existing taxonomic problems regarding this family.

* Email: nadiariaz20@gmail.com

INTRODUCTION

Grasses has great economic significance with ecological dominance [1]. Taxonomically, family poaceae is one of the largest families having 908 genera and more than 11,000 species [2, 3]. All the available taxonomic information of grasses is based on morphological characters [4]. Leaves have key importance among the morphological characters of family Poaceae. It can be distinguished on the basis of their size, thickness and coloration [5, 6]. Micro morphological characters are also important tool for systematic study of family Poaceae. For example, the absence of micro hair characterized the subfamily Poideae [7]. Micro morphological characters related to leaf epidermis also have taxonomic significance (8, 9, 10, 11, and 12]. Micro morphological characters of the floral parts of grasses are useful for dtermination of evolutionary trends and for assessment of systematic relationship of the grass species [13]. Morphological characters are directly exposed to environment, leading towards changes in phenotype of the plant species. Therefore the environmental influence contributed towards enhancement of taxonomic confusion. Other than morphometry, among other taxonomic characters leaf epidermal anatomy was found to be persistent and significant tool for identification of different grass species [14].

Leaf epidermal anatomical markers are quite important to describe different taxa in terms of phylogenetic and taxonomic consideration [15]. Grouping of various taxa from different families of plant systematics is done on the basis of leaf epidermal characters [10, 16]. Several taxonomists recognize the importance of epidermal features for delimitation of various taxa belonging to different angiospermic families [16, 17, 18, 19, and 20]. In the past anatomical studies, incorporation with morphological studies for the resolution of taxonomic problems of monocot has been used [21]. Therefore the aim of current study was to find out the solution of existing taxonomic problems of grass species, which overlap in most of their morphological characters and to elucidate systematic relationship of the critical taxa by utilization of leaf epidermal characters.

HISTORICAL BACKGROUND

Desai et al., [22] made a comparative study of two species namely *Coix aquatic* Roxb and *Coix Lacryma-jobi* L. of family Poaceae based on leaf epidermal anatomical character. The important anatomical characters identified in *Coix aquatic* Roxb and *Coix lacryma jobi* L. were macro hairs with long apex on abaxial epidermis and dumbbell shaped silica cells on abaxial and adaxial epidermis surface. The character of identification for *C. lacryma-jobi* was cross shaped and dumbbell shaped silica cells. The difference of characters on abaxial and adaxial surface leads towards the identification of different species. However this study was limited to only two species of family Poaceae, lacking information about the all other species of this family.

Yousaf et al., [23] studied leaf epidermal anatomy of selected *Allium* species. Leaf epidermal anatomy of a selected Allium species showed variation in size and shape of stomatal cells, stomatal cavity, micro and macro hairs, trichomes, silica bodies and long cells. Leaf epidermal anatomy helpful for the resolution of taxonomic confusion of *Allium consanguineum* with other species of the same genus, as this species had a great diversity for leaf epidermal anatomical features. This species has largest stomatal cell (6-14μm) and silica bodies. Presence of micro hairs was another important distinguished character for *Allium carolianum*. The length of micro hairs varies from 150-200μm. Dumbbell shape of silica bodies was observed in six different species viz., *A. dolichastylum, A. borszczewii, A. micranthum, A. consanguiem, A. stocksian* and *A. stoliczki*. Trichomes were present in *A. barszczewksi, A. borszczewii, A. micranthum, A. miserbile, A. congicallum, A. gilli and A. dalichostylum.*

Reimer and Costa-Sanchez, [24] worked out on SEM survey of leaf epidermal in *Danthonoid* grasses (Poaceae, *Danthonoideae*). This survey consists of 21 taxa and seven genera (*Austrodauthonia, cortadania, Dauthonioidea, Mermuellera, Notodanthonia, Rytidosperma and Tribolum*). Authors examined the micromorphological characters like micro/macro hairs, silica bodies and stomata. *Danthonia* is characterized by the absence of abaxial stomata and presence of bicelluar micro hairs with basal or terminal cells of equal length. Macro hairs were present in coastal region of abaxial epidermis found in four species of North America *Danthonia.* Marco hairs were reported first time in *Danthonia.* Authors studied the subfamily *Danthonioidead* for abaxial epidermis in detail and differentiated four groups of *Danthonioidead.*

Jiang et al., [25] worked out on the leaf epidermis of two species of *Flos lonicera.* Anatomical study was done by scanning electron microscope. Leaf epidermal surfaces of these species were studied to examine stomata, trichomes and epidermal cells. Leaf epidermal anatomical characters were successfully differentiated on the basis of stomatal location, epidermal cells, trichomes density and shape.

Munsif et al., [26] worked on leaf epidermal anatomy of Genera *Lantana verbena* and *vitex* of the family Verbenaceae from Pakistan. They studied abaxial and adaxial leaf epidermal anatomy of eight species of Verbenaceae. They used leaf epidermal anatomy as an aid to identify genera *Lantana, Verbana* and *Vitex* of Verbenaceae. Anatomical characters such as stomata types and size, epidermal cells shapes, size and trichomes shape were used as taxonomic tools. *Lantana camara,* L. *indica, Verbena bonariensis, Ver. Officinalis, Ver. Bipinnatifida, Ver. Hybrid, Vitex negundo and V. trifolia* were differentiated on the basis of trichomes shapes.

Rajas, [27] studied anatomy of the foliar epidermis of *Solanum nudum* species, *genus Solanum sect. germinate* (Solanaceae). Variation in epidermal cells, types, dimensions and location of stomata presence or absence and density of trichome was evaluated. Results showed difference in some anatomical characteristics between *S. nudum* species groups i.e., *S. aphyodendrom, S. campaniforme, S. nudum* and *S. tepuiense* and 34 species of *Solanum sect germinate.*

Motomura et al., [28] worked out on silica deposition on abaxial epidermis before the opening of the leaf blades of *Pleioblastus chino* (Poaceae, Bambusoideae). In Poaceae family silica deposition is the one of most important identification character. Deposition of silica, its distribution process and physiological influence in the Poaceae family was extensively investigated. Bamboo among the members of Po has fairly long life span and leaf tissues accumulate silica throughout their life span.It has been observed that process of silica deposition in relation to aging of the bamboo leaf after opening depend on the cell types. However silica depositions have never been observed during the period of development and maturation. Hence in *Pleioblastus chino* the distribution of silica on abaxial epidermis before the opening of the leaf blades was observed. Abaxial epidermis study was done by preparing slides of abaxial epidermis and observing it under scanning electron microscope equipped with an energy dispersive X-rays microanalyser.

Ozdemir and Atan, [29] investigated morphology and anatomy of some endemic *alkana* species. Two *Alkana* species were examined for morphological and anatomical differences. The aim of study was to provide information of morphological and anatomical properties of *alkana froedinii*, Rech, Fie and *A. cardifalia* L. Koch. It was observed that *A. froedinii* has some differentiating anatomical features from *A. cardifalia* e.g., numerous setiform and glandular hairs.

Liscovsky and Cosa, [30] worked on comparative anatomy of leaf and stem in species of Cestreae G. Don (Solanaceae) and thirteen native and cultivated species of *lestrum* L. and *Sessa vestioides*.The best found distinguishing characters of leaf among species were secondary vein Architecture, marginal ultimate vention, stomata in hairs, cuticle characteristics, and presence of fibers adjacent to external and internal phloem in the middle vein andscleried in the middle vein.

Kanademir [31] studied the morphological, anatomical and karyological properties of endemic Solvia hypargeia Fich and Mey (Lamiaceae). .In this study it was observed that these species have perennial root system, herbaceous stem, simple leaves, glandular and non-glandular hairs on both sides of the leaves. Anatomically the internal morphological properties of root, stem and petiole were also observed.

Gilnai et al., [14] made a study of leaf epidermal anatomy of selected *Digitaria* species of tribe Paniaceae of family Poaceae. They studied abaxial and adaxial surfaces of leaf of *digitaria* species and showed variation in size and shapes of prickles, short cells, silica bodies, microhairs with basal and distill cells, hooks, stomata and long cells. In *D. abludens* cross shaped silica bodies were found. The comparative length of distal cell and basal cell of micro hairs i.e., shorten, longer or equal to each other, were found to be useful in identifying *Digitaria* species while low or tall domed shaped, parallel and triangular subsidiary cells were also observed to be helpful in identification of *D. abludens, D. setigera* and *D. violascens* Vieria et al., [32] worked on leaf anatomy of three herbaceous Bamboo species. They studied fully developed leaves of Cryptochloa capillata (Swallen) (Poaceae Bambusoideae) by using optical microscope. The important anatomical features were midrib with complex vascular system, mesophyll consisting of tubular lobed chlorophyllous element, long cell short bi-cells micro hairs, prickles and silica bodies. Chaudary et al., [33] study leaf epidermal anatomy of four medicinal species of Islamabad, Attock and Mirpur (Azad Khasmir), Pakistan. These species showed variation due to the variation in their long cells, short cells, silica bodies, macro and micro hairs, presence or absence of stomata and shape of subsidiary cells. Keys were made by diagnosing these characters and distinguishing among *cymbopon citrates, Cynodon dactylon, Panicum summatrense* and *viteveria zizanoides*. Acedo et al., [34] studied variation of micro morphological characters of lemma and palea in genus *Bromus* (Poaceae) by scanning electron microscope. Long cells were dominant element of epidermal cell. In short cell (cork cell, crown cell, prickles and micro hairs) were observed. The objective of this work was to assess the value of micro morphological typology in systematic studies and they identify 6 genera of the *Bromus* (Poaceae) by using these micro morphological features. Corxdale [35] studied stomata patterning in monocotyledon in two sections by taking *Tradescantia* as a model system.

For this plant group the first section provided information on stomatal patterning, stomatal differentiation, monocot leaf growth, common stomatal arrangement, and variation in patterning resulting from the leaf position and environment and stomatal types. The second section provides development and mechanistic outlook to stomata patterning based on the model studies.

Table 1. List of species of Family Poaceae used for leaf epidermal anatomical study

S. NO.	Sub Families	Tribes	Specie name
1	Chloridoideae	Eragrostideae	*Eragrostis minor.* Host
2			*Eragrostis poaeoides.* P. Beauv
3			*Eleusine flagellifera* Nees.
4			*Eeusine indica* (L.) Gaertn
5			*Acrachne racemosa Heyene ex* Roem and Schult
6			*Desmostachya bipinnata* (L) Stapf
7			*Leptochloa chinensis* (L) Nees
8			*Dactyloctenium aegyptium* (L) Willed
9		Chlorideae	*Cynodon dactylon* (L) Pers
10	Panicoideae	Paniceae	*Digitaria nodosa.* Parl
11			*Digitaria setigera.* Roth ex Roem and Schult
12			*Digitaria ischaemum* (Schreb) Schreb ex Muhl.
13			*Digitaria ciliaris* (Retz.) Koel
14			*Setaria intermedia* Roem and Schult
15			*Setaria pumila* (Poir) Roem and Schult
16			*Setaria verticillata* (L.) P. Beauv.
17			*Paspalidium flavidium* (Retz) A. Camus
18			*Pennisetum orientale.* L.C. Rich
19		Andropogoneae	*Dicanthium annulatum* (Forssk) Staphf
20			*Cymbopogon jwarancusa* (Jones) Schilt
21	Poideae	Poaceae	*Lolium temulentum* L.
22		Aveneae	*Phalaris minor* Retz
23			*Koeleria phleoides* (Vill). Pers.

Ogiundipe [36] studied leaf epidermal study of genera *Datura* L. of family Solanaceae. Five West African species of *Dautura* with narcotic properties were studied. The most useful anatomical characters for identification were palisade ratio, stomatal density and index, pattern of anti dinal walls, density and types of trichomes and density and types of crystals. *D. candida, D. suaveolens, D. dinnoxia, D. metal* and *D. stramonium* were studied and differentiated on the basis of these anatomical characters.

Lackey, [37] worked on leaflet anatomy of Phaseoleae (Leguminous: sub family papilionoideae) and its relation to taxonomy. The Author studied the leaflet anatomy of the tribe phaseoleae by light microscopy of cleaned material and by scanning electron microscopy of leaflet surfaces. Anatomical results were compared with recently revised of sub trible classification of the tribe. Taxonomically important characters that were found bulbous based and vesicular glands were restricted to the sub tribe canjaninae hooked hairs were present in three generic groups. Glandular and branched hairs were particularly present.

RESULTS

1. *Acrachne Racemose* Heyene Ex Roem and Schult

a. Adaxial Side

Oil droplets were absent. Cells were rectangular in shape. Margines wavy and membrane double layered. Length of long cells was 130 µm -200 µm (Av: 165 µm) andwidth 20 µm -25 µm (Av: 22.5 µm). Length of short cell was 25 µm -100 µm (Av: 67.5 µm) andwidth of short cell 21 µm -40 µm (Av: 30.5 µm). Stomata were dumbbell shape. Number of stomata/unit area was 8-9. Length of stomatal cavity 7 µm-8.0 µm (Av: 12.5 µm). Width of stomatal cavity 6.0 µm -10.0 µm (Av: 8.0 µm). Length of stomatal cell 45 µm -50µm (Av: 47.5 µm).Width of stomatal cell 45 µm -54 µm (Av: 49.5 µm). Silica bodies of regular shape were present. Number of silica bodies/unit area 2-3. Length of silica body 40 µm -120 µm (Av: 80 µm). Width of silica body 40.1 µm -160.1 µm (Av: 100.5 µm). Trichomes were absent. Length of plasmodesmata 14 µm -16 µm (Av: 15 µm) andwidth was 9.0 µm -18.0 µm (Av: 13.5 µm).

b. Abaxial Side

Oil droplets were absent. Length of long cells was 134µm-208µm (Av: 171µm), width22.2µm-30 µm (Av: 26.1 µm). Length of short cell was 23µm-45 µm (Av: 34 µm). Cellswere rectangular, margins wavy and membrane double layered. Silica bodies were present. Number of silica bodies/ unit area 1-2. Length of silica bodies was 42 µm-124 µm (Av: 83 µm) andwidth 42.2 µm-171 µm (Av: 106.6 µm). Trichome were intracellular originated and with acute tip. Length of trichome ranged from 208 µm-361 µm (Av 284.5). Width of trichome was 24 µm-28µm (Av: 21 µm). Stomata were present. Number of stomata/unit area was 3-4.dumbbell shaped . Length of stomatal cavity was 8.5 µm–20µm (Av: 14.25 µm) andwidth 7 µm -12.2 µm (Av: 9.6 µm). Length of stomatal cell 47.0 µm-55.5 µm (Av: 51.25) andwidth 55 µm-61.4 µm (Av: 57.7 µm). In between the cells plasmodesmata connections were present. Length of plasmodesmata 16 µm-18 µm (Av: 17 µm) and width of plasmodesmata was 10.6 µm-20 µm (Av: 15.3 µm).

2. *Ergostosis Minor* Host

a. Adaxial Side

Trichome, silica bodies and connection cell type were absent. Cells were elongated with tapering ends, margined single layered and smooth. Length of long cell 31 µm -387 µm (Av: 209 µm).Width of long cell 42µm -524 µm (Av: 474 µm). Length of short cell 118 µm -229 µm (Av: 173.5 µm) andwidth 31.3 µm -51.1 µm (Av: 47.7 µm). Stomata were dumbbell shape. Number of stomata per unit area were 4-5. Length of stomatal cavity21.1 µm -27 µm (Av: 24 µm).Width of stomatal cavity 10 µm -11 µm (Av: 10.5 µm). Length of stomata 53 µm -56 µm (Av: 54.5 µm).Width of stomata 9 µm 12.0 µm (Av: 11.5 µm).

b. Abaxial Side

Oil droplets were absent. Cells were rectangular in shape. Membrane single layered with wavy margin .Length of long cells 206 μm-293.4 μm (Av: 249.5 μm). Width of long cell 18 μm -22 μm (Av: 20 μm). Length of short cell 6 μm -146 μm (Av: 76 μm). Width of short cell was 16 μm-214 μm (Av: 18.5 μm). Stomata dumbbell shaped. Number of stomata/ unit area was 50-52. Length of stomatal cavity 17 μm -20 μm (Av: 18.5 μm) andwidth was 7.1 μm - 10.1 μm (Av: 8.5 μm). Length of stomata 51 μm -56 μm (Av: 53.5 μm). Width of stomata 32 μm -35 μm (Av: 3.5 μm). Silica bodies were regular in shape .Number of silica body per unit area 1-2. Length of silica body 75.0 μm -75.6 μm (Av: 75.3 μm).

Trichomes originate from outer membrane, scale like length of trichome 73.8 μm -80.5 μm (Av: 77.15 μm). Width of trichome 40.5 μm -58.7 μm (Av: 44.8 μm). Number of trichomes per unit area were 22-24. Plasmodesmata connection cell types were present. Length of connection cells 25.1 μm -35.1 μm (Av: 30.1 μm). Width of connection cell type 21.1 μm -23.1 μm (Av: 22.1 μm).

3. *Digitaria Ischaemum* (Schreb) Schreb Ex Muhl

a. Adaxial Side

Stomata, silica bodies, Oil droplet and connection cells types were absent. Cells were rectangular, margined wavy and singled layered. Length of long cell 87 μm -95.1 μm (Av: 91.05 μm). Width of long cells 26 μm -33 μm (Av: 29.5 μm). Length of short cells 40 μm -43 μm (Av: 41.5 μm). Width of short cells 25 μm -34 μm (Av: 29.5 μm). Trichomes were intracellular and acute from tips. Number of trichomes per unit area were 11-12. Length of trichome 432 μm -466 μm (Av: 449 μm). Width of trichome 9.2 μm -11.4 μm (Av: 10.3 μm).

b. Abaxial Side

Oil droplets and connection cells types were absent. Cells were rectangular,margined, single layered and wavy. Length of long cell 193 μm -235 μm (Av: 214 μm). Width of long cell 21.0 μm -31 μm (Av: 26 μm). Length of short cell 108 μm -115 μm (Av: 111.5 μm). Width of short cells 23 μm -29 μm (Av: 26.0 μm). Trichomes were intracellular in Origin. Number of trichomes per unit area were 1-2. Length of trichome 54 μm -60.0 μm (Av. 57.05 μm). Width of trichome 24.1 μm -23.4 μm (Av: 23.5 μm). Stomata were present in dumbbell shaped. Number of stomata per unit area 8-9. Length of stomatal cavity 15 μm -20 μm (Av: 17.5 μm). Width of stomatal cavity 7.0 μm -10.1 μm (Av: 8.55 μm). Length of stomata 52 μm -54 μm (Av: 53 μm). Width of stomata 20.0 μm -25.4 μm (Av: 22.5 μm). Silica bodies were cross shaped. Number of silica bodies per unit area. Length of silica body 30 μm -33 μm (Av: 31.5 μm). Width of silica bodies 45 μm -49 μm (Av: 47 μm.).

4. *Digitaria Ciliaris* (Retz.)Koel

a. Adaxial Side

Plasmodesmata connections were absent. Cells were rectangular, margined smooth and double layered. Length of long cell 69.2 μm -82 μm (Av: 75.6 μm). Width of long cell 19 μm

-20 μm (Av: 19.5 μm). Length of short cell 25.1 μm was33.1 μm (Av: 29.1 μm). Width of short cell 29 μm -31 μm (Av: 30 μm). Trichomes were intracellular in origin, acute tip. Number of trichomes per unit area 9-10. Length of trichomes range from 209 μm -317 μm (Av: 263 μm). Width of trichomes 6.3 μm -7.8 μm (Av: 7.05 μm). Stomata were present dumbbell shaped. Number of stomata per unit area 7-8. Length of stomatal cavity 6 μm -10 μm (Av: 8 μm). Width of stomatal cavity 9 μm -10 μm (Av: 9.5 μm). Length of guard cell 16 μm -21 μm (Av: 8.5 μm), width of guard cell 16 μm -25 μm (Av: 20.5 μm). Silica bodies were present irregular in shape. Number of silica bodies per unit area 2-3. Length of silica bodies 29 μm -37.9 μm (Av: 33.85 μm). Width of silica bodies 23 μm -37.6 μm (Av: 33.3 μm). Oil droplets were present. Number of droplet per unit area 1-2. Length of oil droplet 19 μm -20 μm (Av: 19.5 μm) , width of oil droplet 18.1 μm -21.0 μm (Av: 19.05 μm).

b. Abaxial Side

Plasmodesmata connection and oil droplets were absent. Cells were rectangular, margined smooth and double layered. Length of long cells 75 μm -85 μm (Av: 77.5 μm), width 28 μm -31 μm (Av: 29.5 μm). Length of short cell 36 μm -40 μm (Av: 38.0 μm), width 31 μm -35 μm (Av: 33 μm). Trichomes were intracellular in origin and acute from tips. Number of trichome per unit area was 1-2. Length of trichome 439 μm -608 μm (Av: 523.5 μm), width of trichome 7 μm -17 μm (Av: 12 μm). Stomata were present dumbbell shaped. Number of stomata per unit area 3-4. Length of stomatal cavity 6 μm -11 μm (Av: 8.5 μm), width of stomatal cavity 10 μm -11 μm (Av: 10.5 μm). Length of guard cell 26 μm -31 μm (Av: 28 μm), width of guard cell 26 μm -28 μm (Av: 27 μm). Silica bodies were present irregular in shape. Number of silica bodies per unit area 2-3. Length of silica bodies 95 μm - 153 μm (Av: 124 μm). Width of silica bodies was 86 μm -136 μm (Av: 111 μm) (Figure 1).

Figure 1. Leaf epidermal Anatomy of Digitaria ciliaris (Abaxial).

5. *Digitaria Nodosa*.Parl

a. Adaxial Side

Silica bodies, oil droplets and plasmodesmata connection were absent. Cells were rectangular double layered and smooth margined. Length of long cell 137 μm -156 μm (Av146.5 μm) and width 27 μm -38 μm (Av32.5 μm). Length of short cell 17.1 μm -22.1 μm (Av19.6 μm). Trichomes were present and originate from intra cellular membrane and acute from tips. Number of trichome per unit area was 1-2. Length of trichome45 μm -49.5 μm (Av:47.25 μm), width of trichome 21 μm -24.2 μm (Av22.6 μm) and stomata were dumbell shaped. Number of stomata per unit area 50-55. Length of stomatal cavity 14 μm -15 μm (Av: 14.5 μm) width of stomatal cavity 10 μm -13 μm (Av11. μm 5) .Length of stomata 47 μm -5 μm 1 (Av: 48.5 μm), width of stomata 20 μm -23 μm (Av 21.5 μm) (Figure 2).

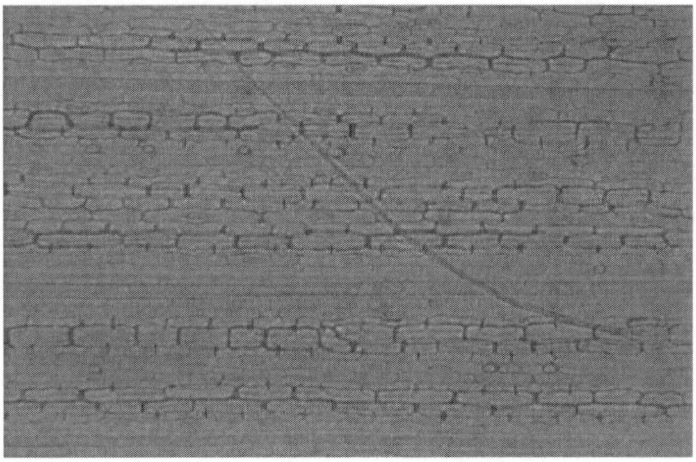

Figure 2. Leaf epidermal Anatomy of Digitaria nodosa (Adaxial side).

b. Abaxial Side

Silica bodies, oil droplets and plasmodesmata connections were absent. Cells were rectangular double layered and smooth margined. Length of long cell 180 μm -194 μm (Av: 187 μm), width 26 μm -28 μm (Av: 27.05 μm). Length of short cell 85 μm -108 μm (Av: 96.5 μm). Trichomes originate from inner membrane and acute from tips. Number of trichome per unit area was 2-3. Length of trichome 165 μm -171 μm (Av: 168 μm), width was 24 μm -26 μm (Av: 25 μm) and dumbell shaped stomata were present. Number of stomata per unit area was 7-8. Length of stomatal cavity 15 μm -17.1 μm (Av: 16.0 μm 5), width of stomatal cavity was 9.1 μm -12 μm (Av10.1 μm).

6. *Digitaria Setigera*.Roth Ex Roem and Schult

a. Adaxial Side

Trichomes, plasmodesmata cell and silica bodies were absent. Cells were rectangular, single layered and wavy margined. Length of long cell 14 μm -209 μm (Av: 111.5 μm). Width of long cell 19 μm -20 μm (Av: 19.5 μm). Length of short cell 89μm -101 μm (Av: 95

μm). Width of short cell 17 μm -36 μm (Av: 26.5 μm). Dumbell shape stomata were present. Number of stomata /unit area was 17-18. Length of stomatal cavity 22 μm -30 μm (Av: 26 μm). Width of stomatal cavity 10μm -15 μm (Av: 12.5 μm). Length of stomatal cell 65 μm - 70 μm (Av: 68.5 μm). Width of stomatal cell 48 μm -55 μm (Av: 51.5 μm). Oil droplets were present.Number of oil droplets/unit area 4-5. Length of oil droplet 19μm-21μm (Av: 20 μm) andwidth of oil droplet 18 μm -20 μm (Av: 19μm).

b. Abaxial Side

Trichome and silica bodies were absent. Cells were rectangular, margin single layered and wavy. Length of long cell 16 μm -218 μm (Av: 117 μm). Width of long cell 22 μm -24 μm (Av: .23 μm). Length of short cell was 89 μm -105 μm (Av: 97 μm). Width of short cell 17 μm -35 μm (Av: 26 μm). Stomata were present. Number of stomata/ unit 16-17. Length of stomatal cavity 24 μm -34 μm (Av: 29 μm). Width of stomatal cavity 9 μm -14 μm (Av: 11.5 μm). Length of stomatal cell 64 μm -68.1 μm (Av: 66.05 μm). Width of stomatal cell 47 μm - 56 μm (Av: 51.5 μm). Oil droplets were present. Number of oil droplets /unit area 4-5. Length of oil droplet 17 μm -20 μm (Av: 18.5 μm). Width of oil droplet 17 μm -19 μm (Av: 18 μm). Plasmodesmata were present. Length of plasmodesmata cells 28 μm -46 μm (Av: 35.5 μm). Width of plasmodesmata cell 21 μm -45 μm (Av: 33 μm).

7. *Leptochloa Chinensis* (L.)Nees

a. Adaxial Side

Plasmodesmata connection cell type and trichomes were absent. Cells were rectangular singled layered and wavy margined. Length of long cell 247 μm -276 μm (Av: 261.5 μm).Width of long cells 30 μm -39 μm (Av: 34.5 μm). Length of short cell 72 μm -113.0 μm (Av: 92.5 μm). Width of short cell 20.4 μm -25 μm (Av: 22.7 μm). Dumbbell shaped stomata were present. .Number of stomata per unit area 9-10. Length of stomatal cavity 15 μm -17.3 μm (Av: 12.95 μm). Width of stomatal cavity 12.6 μm -13.3 μm (Av: 12.95 μm). Length of stomata 60 μm -62 μm (Av: 61 μm). Width of stomata 50 μm -54.5 μm (Av: 52.25 μm). Silica bodies were regular solid form . Number of silica bodies per unit area 1-2. Length of silica bodies 29 μm -37.2 μm (Av: 33.1 μm), width of silica bodies 17 μm -30 μm (Av: 23.5 μm). Oil droplets were present. Number of oil droplet per unit area 11-12. Length of oil droplet 29 μm -37.5 μm (Av: 33.1 μm). Width of oil droplet 17 μm -30 μm (Av: 23.5 μm).

b. Abaxial Side

Plasmodesmata connections were absent. Cells were rectangular, Single layered and wavy margined. Length of long cell 229 μm -294 μm (Av: 261.5 μm). Width of long cell 30 μm -35 μm (Av: 32.5 μm). Length of short cell 117 μm -170 μm (Av: 143.5 μm). Width of short cell 28 μm -30 μm (Av: 29 μm). Prickle like trichomes originate from outer membrane. Number of trichomes per unit area was 3-4. Length of trichomes 136 μm -217 μm (Av: 176.5 μm). Width of trichomes 86 μm -103.4 μm (Av: 94.7 μm). Stomata were dumbbell shaped. Number of stomata per unit area 4-5. Length of stomatal cavity 20 μm -34.1 μm (Av: 27.1 μm). Width of stomatal cavity 14 μm -16 μm (Av: 15 μm). Length of stomata 79 μm -87.1 μm (Av: 83.1 μm). Width of stomata 60.3 μm -74.4 μm (Av: 67.35 μm). Silica bodies are present. Number of silica bodies per unit area 3-4. Length of silica bodies 50 μm -52 μm (Av: 51 μm). Width of silica bodies 45 μm -50 μm (49 μm). Oil droplets were present. Number of Oil droplet per unit area 11-12. Length of oil droplet 25 μm -33.1 μm (Av: 29.1 μm). Width of oil droplet 30 μm -33 μm (Av: 31.5 μm).

8. *Cymbopogon Jwarancusa* (Jones) Schult

a. Adaxial Side

Trichomes and plasmodesmata connections were absent. Cells were rectangular double layered and wavy margined. Length of long cell 220 μm -245 μm (Av: 232.5 μm) and width was 30 μm -36 μm (Av: 33 μm). Length of short cell 110 μm -130μm (Av: 120 μm) and width 36 μm -40 μm (Av: 38 μm).

Dumbbell shaped stomata were present . Number of stomata per unit area 8-9. Length of stomatal cavity 30 μm -36 μm (Av: 33 μm) and width of stomatal cavity 9 μm -12 μm (Av: 10.5 μm). Length of stomata 83 μm -85μm (Av: 84 μm), width of stomata 46 μm -50 μm (Av: 48 μm). Regular shaped silica bodies were present. Number of silica bodies per unit area. Length of silica bodies' 19 μm -20 μm (Av: 19.5 μm) width of silica bodies 19 μm -23 μm (Av: 21 μm). Oil droplets were present. Number of oil droplet per unit area. Length of oil droplet 13 μm -17 μm (Av: 15 μm) and width of oil droplet 16 μm -18 μm (Av: 17 μm) (Figure 3).

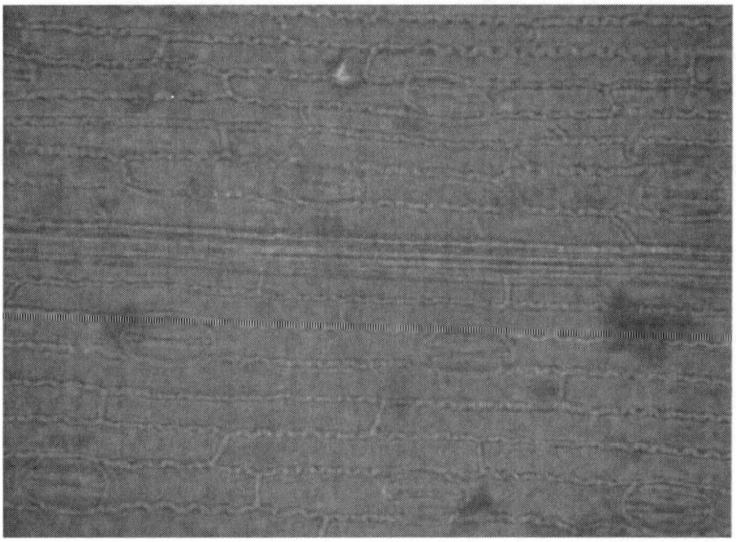

Figure 3. Leaf epidermal Anatomy of Cymbopogon jwarancusa (Jones) Schult (Adaxial side).

b. Abaxial Side

Trichomes and plasmodesmata connections were absent. Cells were rectangular double layered and wavy margined. Length of long cell 227 μm -255 μm (Av: 241 μm) and width of long cell was 32 μm -35 μm (Av: 33.5 μm). Length of short cell 112 μm -134 μm (Av: 123 μm) and width was 38 μm -43 μm (Av: 40.5 μm). Dumbbell shaped stomata were present . Number of stomata per unit area 8-9. Length of stomatal cavity 29.5 μm -35 μm (Av: 32.25

μm) and width of stomatal cavity 10 μm -13 μm (Av:11.5 μm). Length of stomata 85 μm -87 μm (A86 μm) and width was 48 μm -52 μm (Av50 μm). Regular solid form of silica bodies were present. Number of silica bodies per unit area. Length of silica bodies' 19 μm -21 μm (Av: 20 μm) and width 18 μm -22 μm (Av: 20 μm). Oil droplets were present .Number of oil droplet per unit area. Length of oil droplet 14 μm -19 μm (Av: 16.5 μm) and width was 14 μm -16 μm (Av: 15 μm).

9. Phalaris Minor Retz

a. Adaxial Side

Trichomes, oil droplets, plasmodesmata connection were absent. Cells were rectangular single layered and smooth margined. Length of long cell 341 μm -486 μm (Av: 413.5 μm) width of long cell 38 μm -52 μm (Av: 45 μm). Length of short cell 192 μm -194 μm (Av: 193 μm) width of short cell 30 μm -32 μm (Av: 31 μm). Dumbell shaped stomata were present . Number of stomata per unit area 35-40. Length of stomatal cavity 50 μm -52μm (Av: 51 μm) width of stomatal cavity 36 μm -37 μm (Av: 36.5μm). Length of stomata 132 μm -135 μm (Av: 133.5 μm) width of stomata 76 μm -80 μm (Av: 78 μm). Regular shaped silica bodies were present. Number of silica bodies per unit area 8-9. Length of silica bodies 30.0μm -41.1 μm (Av: 35.55 μm) width of silica bodies 23 μm -37 μm (Av: 30 μm).

b. Abaxial Side

Trichomes, oil droplets, plasmodesmata connection were absent. Cells were rectangular single layered and smooth margined. Length of long cell 343 μm -492 μm (Av: 417.5 μm) width of long cell 39 μm -54 μm (Av: 46.5 μm). Length of short cell 194 μm -197 μm (Av: 195.5 μm) width of short cell 32 μm -34 μm (Av: 33 μm). Dumbbell shaped stomata were present. Number of stomata per unit area 5-6. Length of stomatal cavity 51 μm -53 μm (Av: 52 μm) width of stomatal cavity 34 μm -36.1 μm (Av: 35 μm). Length of stomata 131 μm - 133 μm (Av: 82 μm) width of stomata 77 μm -81 μm (Av: 79 μm). Regular shaped silica bodies were present. Number of silica bodies per unit area 5-6. Length of silica bodies 30.1 μm -40.1 μm (Av: 35.1 μm) width of silica bodies 22 μm -38 μm (Av: 30 μm.

10. *Dactyloctenium Aegyptium* (L.)Willed

a. Adaxial Side

Plasmodesmata connection and stomata were absent. Cells were rectangular single layered and smooth margins. Length of long cell 130 μm -142 μm (86 μm). Width of long cell 17 μm -29 μm (Av:23 μm). Length of short cell 66.1 μm -67 μm (Av:66.55 μm). Width of short cell 27 μm -28 μm (Av:27.5 μm). Trichome were present and acute from the tips. Number of trichome /unit area 2-3. Length of trichome 100 μm -103 μm (Av: 101.5 μm). Width of trichome 47 μm -50 μm (Av: 48.55 μm). Regular shape Silica bodies were present . Number of silica bodies /unit area 2-3. Length of silica bodies 20 μm -44 μm (Av: 32 μm). Width of silica bodies 17.5 μm -52 μm (Av: 34.75 μm). Oil droplets were present. Number of oil droplet/unit area. Length of oil droplet30 µm -31 µm (Av: 30.5 µm). Width of oil droplet 28 µm -30 µm (29 µm).

a. Abaxial Side

Palsmodesmata conncetion were absent. Trichomes were absent. Cells were elongated with tapering end. Margined smooth and single layered. Length of long cell 243 µm -284 µm Av: (263.5 µm). Width of long cell 34 µm -44 µm (Av:56 µm).Length of short cell 5 µm 9-81 µm (70 µm). Width of short cell 26 µm -47 µm (Av:36.5 µm).

Dumbell shaped stomata were present. Number of stomata /unit area 10-12. Length of stomatal cavity 15 µm µm -21 µm µm (Av: 18 µm). Width of stomatal cavity 14 µm -14 µm (Av: 14 µm). Length of stomatal cell 57 µm -68 µm (Av: 62.5 µm). Width of stomatal cell 23.1 µm -3 µm 1(Av: 28.05 µm). Silica bodies were present irregular in shape. Number of silica bodies / unit area 2-3. Length of Silica bodies 28 µm -36 µm (Av: 32 µm). Width of silica body 32 µm -38.1 µm (Av: 35.05 µm). Oil droplet were present. Number of oil droplet/ unit area 20-25. Length of oil droplet 18 µm -24 µm (Av: 21 µm). Width of oil droplet 19 µm -24 µm (Av: 21.5 µm) (Figure 4).

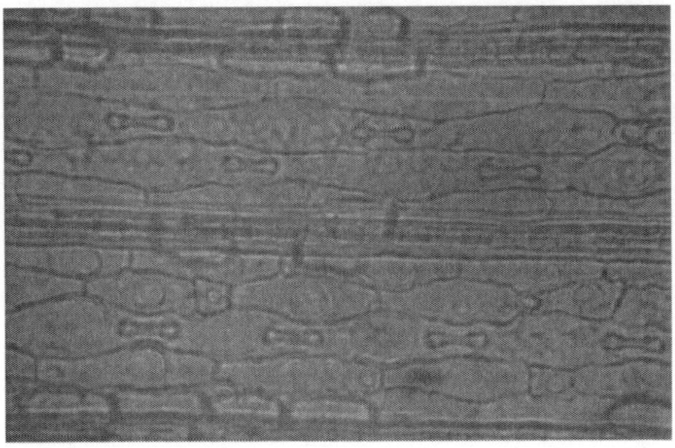

Figure 4. Leaf epidermal Anatomy of *Dactyloctenium aegyptium* (L.) Willed (Abaxial).

11. *Paspalidium Flavidium* (Retz) A.Camus

a. Adaxial Side

Oil droplet, plasmodesmata cells and trichomes were absent. Cells were rectangular single layered and smooth margins. Length of long cell83 µm -120 µm (Av: 101.5 µm). Width of long cell32 µm -42 µm (Av: 37 µm). Length of short cell 41 µm -67.4 µm (Av: 54.2 µm). Width of short cell 25 µm -40 µm (Av: 32.5 µm). Dumbell shape stomata were present . Number of stomata /unit area. Length of stmatal cavity20 µm -23.1 µm (Av: 21 µm).).Width of stomatal cavity 10.1 µm -14.1 µm (Av: 12.1 µm). Length of stomatal cell 47 µm -51 µm (Av: 49 µm). Width of stomatal cell 33.1 µm -42 µm (37.55 µm). Silica bodies were present. Regular in shape. Number of silica bodies/unit area 4-5. Length of silica bodies 31 µm -34.1 µm (Av: 32.55 µm). Width of silica bodies34.1 µm -36 µm (Av: 35.05 µm) (Figure 5).

b. Abaxial Side

Oil droplets were absent. Cells were rectangular double layered and wavy margins. Length of long cell 258 μm -314 μm (Av: 286 μm). Width of long cell 39 μm -50 μm (Av: 44.5 μm). Length of short cell 156 μm -189 μm (Av: 172.5 μm). Width of short cell37 μm - 41 μm (Av: 39 μm). Trichomes were originate from inner membrane acute from tips. Number of trichome /unit area 3-4. Length of trichome 129.1 μm -166 μm (Av: 147.5 μm). Width of trichome 80 μm -83.3 μm (Av: 80.65 μm). Stomata were present dumbell in shape. Length of stmatal cavity23 μm-127.1 μm (Av: 25.1 μm). Width of stomatal cavity 16 μm -22 μm (Av: 19.05 μm). Length of stomatal cell 70.2 μm -79.1 μm (Av: 74.33 μm). Width of stomatal cell 49 μm -53 μm (51). Silica bodies were present cross shape. Number of silica bodies /unit area is 1-2. Length of silica bodies 34 μm -41.4 μm (Av: 37.5 μm). Width of silica bodies 37 μm - 39.1 μm (Av: 38.05 μm). Plasmodesmata were present. Length of plasmodesmata cell 26 μm -31 μm (Av: 28.5 μm). Width of plasmodesmata cell 24 μm -28 μm (Av: 26 μm). (Figure 5).

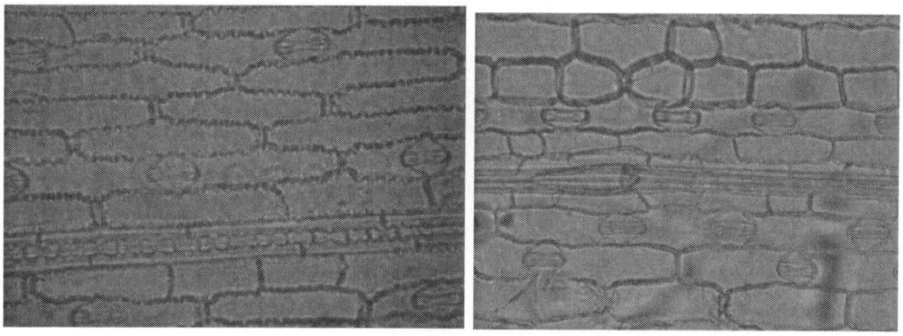

Figure 5. Leaf epidermal Anatomy of Paspalidium flavidium (Retz) A.Camus. Abaxial side (Right) Adaxial side (Left).

12. *Koeleria Phleoides* (Vill). Pers.

a. Adaxial Side

Oil droplets were absent. Cells were rectangular, single layered and wavy margined. Length of long cell 190μm - 220μm (Av: 205μm). Width of long cell 23μm – 39.5μm (Av: 31.25μm). Length of short cell 74μm - 86μm (Av: 80μm). Width of short cell 42μm- 56μm (Av: 49μm). Trichomes were present blent from the tips. Number of trichomes / unit area 1-2. Length of trichomes 74μm - 90μm (Av: 82μm). Width of trichome 23μm - 35μm (Av: 29μm). Dumbell shaped stomata were present. Number of stomata / unit area was 9-10. Length of stomatal cavity 17μ -18μm (Av: 17.5μm). Width of stomatal cavity 13μm - 16μm. (Av: 14.5μm). Length of stomatal cell 65μm-72μm (Av: 68.5). Width of stomatal cell 30μm - 35μm (Av: 32.5μm). Silica bodies were present regular in shape. Number of silica bodies/ unit area was 2-3. Length of silica bodies 20μm - 22μm (Av: 21.0μm). Width of silica bodies 17μm - 20μm (Av: 18.5μm). Plasmodesmata were present. Length of plasmodesmata cell 36μm - 41μm (Av: 37.5 μm). Width of plasmodesmata cell 29μm - 30μm (Av: 29.5μm).

b. Abaxial Side

Oil droplets were absent.Cells were rectangular single layered and wavy margins. Lenth of l;ong cell 195μm - 226μm (Av: 210.5μm). Width of long cell 25μm - 40.2μm (Av: 32.5μm). Length of short cell 72μm - 85μm (Av: 78.5μm). Width of short cell 40μm - 55μm (Av: 47.5μm). Trichomes were present blent from the tips. Number of trichome /unit area 3-4. Length of trichome 73 μm -88.2 μm (Av: 80.6 μm). Width of trichome 28.6-34.5(31.55). Stomata were present. Number of stomata / unit area is 8-9. Length of stomatal cavity 16μ - 17μm (Av: 16.5μm). Width of stomatal cavity 14μm - 15μm. (Av: 14.5μm). Length of stomatal cell 64μm-71μm (Av: 67.5). Width of stomatal cell 32μm - 37μm. Silica bodies were present regular in shape. Number of silica bodies/ unit area is 2-3. Length of silica bodies 20μm - 21μm (Av: 20.5μm). Width of silica bodies 18μm - 19μm (Av: 18.5μm). Plasmodesmata were present. Length of plasmodesmata cell 35μm - 40μm (Av: 37.5 μm). Width of plasmodesmata cell 30μm - 30μm (Av: 30μm).

13. *Eeusine Indica* (L.)Gaertn

a. Adaxial Side

Plasmodesmata connections were absent. Cells were rectangular single layered and smooth margined. Length of long cell 85 μm -94.2 μm (Av: 89.6 μm). Width of long cell 29 μm -46 μm (Av: 37.5 μm). Length of short cell 67 μm -70.6 μm (Av: 68.8 μm), width of short cell 38 μm -41.4 μm (Av: 39.7 μm). Trichome originate from outer membrane. Length of trichome 175 μm -195 μm (Av: 185 μm). Width of trichome 40 μm -65 μm (Av: 52.2 μm). Stomata were present dumbbell shape. Number of stomata per unit area 4-5. Length of stomatal cavity 19μm -21 μm (Av: 10.5 μm). Width of stomatal cavity 10 μm -11 μm (Av: 10.5 μm). Length of stomata 50 μm -51 μm (Av: 50.2 μm), width of stomata 31 μm -35 μm (Av: 33 μm). Silica bodies were present regular in shapes. Number of silica bodies per unit area was 1-2. Length of silica bodies 55.2 μm -69.1 μm (Av: 62.15 μm), width of silica bodies 55 μm -68 μm (Av: 61.5 μm). Oil droplets were present. Number of oil droplet per unit was 3-4. Length of oil droplet 16 μm -19 μm (Av: 17.5 μm). Width of oil droplet 19 μm 20.4 μm (Av: 19.7 μm).

b. Abaxial Side

Trichomes and oil droplets were absent. Cells were rectangular, margined wavy and double layered. Length of long cells 285 μm -312 μm (Av: 298.5 μm), width of long cells 35 μm -41 μm (Av: 38.0 μm). Length of short cells 131 μm -166 μm (Av: 148.5 μm).Width of short cells 35 μm -42 μm (Av: 410 μm). Stomata were present dumbbell shaped. Number of stomata per unit area 7-8 .Length of stomatal cavity 22 μm -28 μm (Av: 25 μm). Width of stomatal cavity 15 μm -18 μm (Av: 16.5 μm). Length of guard cell 66.0 μm-80 μm (Av: 72.0 μm). Width of guard cell 53 μm -66 μm (Av: 59.5 μm). Silica bodies were present irregular in shape. Number of silica bodies per unit area was 1-2. Length of silica bodies 50 μm -56 μm (Av: 53 μm).Width of silica bodies 47 μm -49 μm (Av: 48 μm). Plasmodesmata connection were present. Length of connection cells 38 μm -41 μm (Av: 39.5 μm). Width of connection cells 18 μm -31 μm (Av: 24.5 μm). (Figure 6).

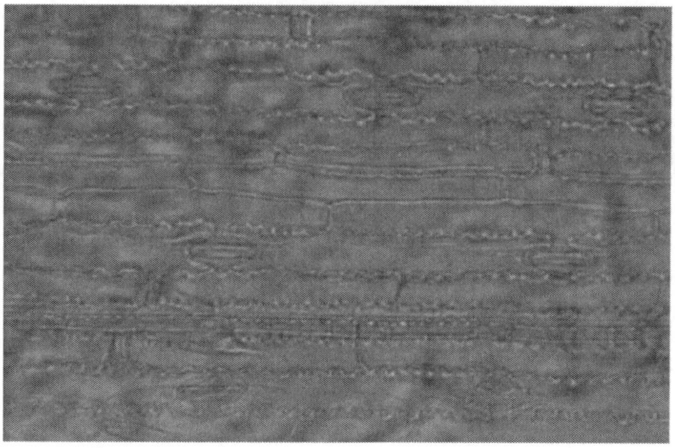

Figure 6. Leaf epidermal Anatomy of Eeusine indica (L.) Gaertn (Abaxial).

14. *Eleusine Flagellifera* Nees.

a. Adaxial Side

Stomata, trichome, plasmodesmata connection cell type were absent. Cells were rectangular to rounded in form. Length of long cell 30.5 μm -46 μm (Av: 38.25 μm). Width of long cell 18 μm -32 μm (Av20.0 μm). Length of short cell 16 μm -28 μm (Av22 μm). Width of short cell 16 μm-18 μm (Av:17 μm). Oil droplets were absent. Number of oil droplets /unit area 1-2. Length of oil droplets 20 μm-2 8 μm (Av: 24 μm). Width of oil droplets 18 μm – 33 μm (Av: 25.5 μm).

b. Abaxial Side

Stomata and plasmodesmata connection were absent, cells were rectangular and rounded in form margins single layered and smooth. Length of long cell 28.4μm-35 μm (Av:3.17 μm). Width of long cell 14 μm - 20 μm (Av:19.5 μm). Length of short cell 17 μm – 22 μm (Av:19.5 μm). Width of short cell 16 μm -20 μm (Av:18.0). Trichomes were intracellular in origin and have blunt tip.Number of trichome /unit area 3-4. Length of trichome was 121 μm -185 μm (Av: 153 μm). Width of trichome 4 μm-22 μm (Av: 13.0 μm). Oil droplets were present.Number of oil droplets /unit area 5-10. Length of oil droplet 20 μm -31 μm (Av: 25.5 μm). Width of oil droplet 19 μm - 37 μm (Av: 28.1 μm).

15. *Setaria Intermedia* Roem and Schult

a. Adaxial Side

Silica bodies, Oil droplets and connection cells types were absent. Cells were rectangular, single, and wavy margined. Length of long cells 275 μm -387.5 μm (Av: 331.25 μm). Width of long cells 37 μm -35 μm (Av: 36 μm). Length of short cell 175 μm -180 μm (Av: 177.5 μm). Width of short cells 38 μm -40 μm (Av: 39 μm). Trichomes were present intracellular in origin and acute from tips. Number of trichome per unit area was 1-2. Length of trichome 39 μm -44.6 μm (Av: 41.8 μm). Width of trichome 29 μm -33 μm (Av: 31 μm). Stomata were dumbbell shape. Number of stomata per unit area was 17-18. Length of stomatal cavity 25 μm -24 μm (Av: 24.5 μm). Width of stomatal cavity 12.4 μm -16 μm (Av: 9.2 μm). Length of stomata 64 μm -72.1 μm (Av: 68.05 μm). Width of stomata 31 μm -36 μm (Av: 33.5 μm).

b. Abaxial Side

Silica bodies, oil droplet and connection cell type were absent. Cells were rectangular single and wavy margined. Length of long cells 193 μm -153 μm (Av: 173 μm), width of long cells 26- μm -28.2.0 μm (Av: 27.1 μm). Length of short cells 71 μm -74 μm (Av: 72.5 μm). Width of short cells 20 μm -21.2 μm (Av: 20.6 μm). Trichomes were outercellular in origin .Number of trichome per unit area 5-6. Trichomeswere small scale like. Length of trichome 68.4 μm -79.3 μm (Av: 73.85 μm). Width of trichome 33 μm -40 μm (Av: 36.5 μm). Stomata were present, dumbbell shaped stomata. Number of stomata per unit area 11-12. Length of stomatal cavity 18 μm -23 μm (Av: 20.5 μm). Width of stomatal cavity 16 μm -22.1 μm (Av: 19.05 μm). Length of stomata 62 μm -67 μm (Av: 64.5 μm). Width of stomata 36 μm .40.0 μm (Av: 38 μm).

16. *Setaria Verticillata* (L.) P.Beauv.

a. Adaxial Side

Oil droplet and plasmodesmata connection were absent. Cells were rectangular wavy and double margined. Length of long cell 260 μm -320 μm (Av: 290 μm) and width of was 29 μm -30 μm (Av: 29.5 μm). Length of short cell 112 μm -150 μm (Av: 131 μm), width of short cell 28 μm -31 μm (Av: 29.5 μm). Trichomes were present acute from the tip originate from intracellular. Number of trichome per unit area was 1-2. Length of trichome 312 μm -416 μm (Av: 364 μm) and width of trichome 23 μm -25 μm (Av: 24 μm). Stomata were present dumbell shape. Number of stomata per unit area 5-6. Length of stomatal cavity 25 μm -30 μm (Av: 27.5 μm) and width of stomatal cavity 15 μm -18 μm (Av: 16.5 μm). Length of stomata 69.7 μm -76.1 μm (Av: 72.9 μm) and width of stomata 22 μm -23 μm (Av: 27.5 μm). Regular shaped silica bodies were present. Number of silica bodies per unit area 4-5. Length of silica bodies 26 μm -38 μm (Av: 31 μm) and width of silica bodies 26 μm -32 μm (Av: 29 μm).

b. Abaxial Side

Oil droplet and plasmodesmata connection were absent. Cells were rectangular wavy and double margined. Length of long cell 262.1 μm -324 μm (Av: 393.05 μm), width of long cell 29 μm -32.1 μm (Av: 30.5 μm). Length of short cell 114 μm -151 μm (Av: 132.5 μm) width of short cell 29 μm -32 μm (Av: 30.5 μm). Trichomes were present acute and intracellular. Number of trichome per unit area 2-3. Length of trichome 313 μm -418 μm (Av: 365.5 μm) and width was 24 μm -26 μm (Av: 25 μm). Stomata were present dumbell shape. Number of stomata per unit area 6-7. Length of stomatal cavity 26 μm -31 μm (Av: 27.5 μm) and width of stomatal cavity 15 μm -19 μm (Av: 17 μm). Length of stomata 69.7 μm -77.1 μm (Av: 73.4 μm) and width 23 μm -24 μm (Av: 23.5 μm). Silica bodies were present regular shaped. Number of silica bodies per unit area 2-3. Length of silica bodies 26 μm -36 μm (Av: 31 μm) and width of silica bodies was 27 μm -34 μm (Av: 30.5 μm).

17. *Setaria Pumila* (Poir) Roem and Schult

a. Adaxial Side

Silica bodies and plasmodesmata connection were absent. Cells were rectangular, margined single layered and smooth. Length of long cell 80 µm -83 µm (Av:81.5 µm).

Width of long cell 48 µm -52 µm (Av50 µm). Length of short cell 45 µm – 46 µm (Av:45.5µm). Width of short cell 36µm-40µm (Av:38µm). Trichome originated from intracellular and acute shape tip.Number of trichome /unit area 1-2. Length of trichome was 400µm–500µm (Av: 453µm). Width of trichome 6.0 µm -10.4 µm (Av:8.2 µm). Dumbell shape stomata were present. Length of stomatal cavity 40 µm – 56.1 µm (Av::48.05 µm).Width of stomatal cavity 16 µm -18 (Av:17).Length of stomata 118 µm _120 µm (Av:119 µm). Width of stomata 62 µm -65 µm (Av: 63.5 µm). Oil droplets were present. Number of oil droplets /unit area was 6-8. Length of oil droplets 30.2 µm -36.2 µm (Av:33.2 µm). Width of oil droplet 30 µm -44 µm (Av: 37 µm).

b. Abaxial Side

Silica bodies and palsmodesmata connection were absent. Cells were rectangular, margined single layered and smooth. Length of long cell 80 -87 (Av 83.5). Width of long cell 42 -51 (Av 46.5). Length of short cell 44 µm - 47 µm (Av45.5 µm). Width of short cell 37.1 µm – 39 µm (Av 38.05 µm). Trichomeoriginate from intracellular, acute form tip of trichome. Number of trichome /unit area. Length of trichome 40.2 µm -51.4 µm (Av:45.8 µm). Width of trichome 6.4 µm -11.4 µm (Av: 8.9 µm). Stomata were present dumbbell shaped. Number of stomata /unit area 40-45. Length of stomatal cavity 43.1 µm- 57.1 µm (Av: 50.1 µm). Width of stomatal cavity 15 µm-19µm (Av: 17 µm). Length of stomatal cell 119 µm -130 µm (Av: 124.5 µm).Width of stomatal cell 63µm-66µm (Av 64.5). Oil droplets were present. Number of oil droplets /unit area 6-8. Length of oil droplets 29.2 µm -38.6 µm (Av: 33.9 µm). Width of oil droplets 32 µm -40 µm (Av:36 µm).

18. *Cynodon Dactylon* (L) Pers

a. Adaxial Side

Trichomes were absent. Cells were rectangular. Single layered and wavy margined. Length of long cell 99 µm - 113 µm (Av: 106.5 µm) width of long cell 17.8µm – 21.3 µm (Av: 19.5 µm). Length of short cell 23 µm – 48µm (Av: 35.5 µm) width of short cell 20.1 µm – 28.7 µm (Av: 24.4 µm).. Dumbell shape stomata were present. Number of stomata per unit area was 16-17. Length of stomatal cavity 12µm -16 µm (Av: 14 µm) width of stomatal cavity 7.5 µm – 12µm (Av: 9.7 µm). Length of stomatal cell 39.2 µm – 43.4µm (Av: 41.3 µm) width of stomatal cell 28.8µm – 31.8µm (Av: 30.3 µm). Silica bodies were present irregular in shape. Number of silica bodies per unit area 1-2. Length of silica bodies 43.6 µm – 56.2 µm (Av: 49.9µm) width of silica bodies 32.1µm – 37.3 µm (Av: 34.9 µm). Oil droplets were present. Number of oil droplet per unit was 20-25. Length of oil droplet 13.5 µm – 22.5 µm (Av: 18 µm). Width of oil droplet 17µm – 22.7 µm (Av: 19.8 µm). Plasmodesmata connection were present. Length of connection cells 20 µm – 21 µm (Av: 20.5 µm). Width of connection cells 8 µm – 9.6 µm (Av: 8.8 µm). (Figure 7).

b. Abaxial Side

Trichomes were absent. Cells were rectangular. Single layered and wavy margined. Length of long cell 120 µm - 138 µm (Av: 129 µm), width of long cell 15µm - 20 µm (Av: 17.5 µm). Length of short cell 33.5 µm – 48µm (Av: 40.75 µm) width of short cell 24.1 µm – 29.8 µm (Av: 26.9 µm). Stomata were present. Stomata were dumbell shaped. Number of stomata per unit area was 15-16. Length of stomatal cavity 13µm -17 µm (Av: 15 µm) width of stomatal cavity 8.5 µm – 13.4µm (Av: 10.9 µm). Length of stomatal cell 42.8 µm - 51µm (Av: 46.5 µm) width of stomatal cell 24.8µm - 35µm (Av: 29.9 µm). Silica bodies were present irregular in shape. Number of silica bodies per unit area was 2-3. Length of silica bodies 48.6 µm – 57.2 µm (Av: 52.9µm) width of silica bodies 32.1µm – 70.3 µm (Av: 61.7 µm). Oil droplets were present. Number of oil droplet per unit was 25-30. Length of oil droplet 53.1 µm – 70.3 µm (Av: 61.7 µm). Width of oil droplet 44.1µm -67.1 µm (Av: 55.5 µm). Plasmodesmata connection were present. Length of connection cells 21.3 µm – 22.8 µm (Av: 22.0 µm). Width of connection cells 9.6 µm – 10.6 µm (Av: 10.1 µm).

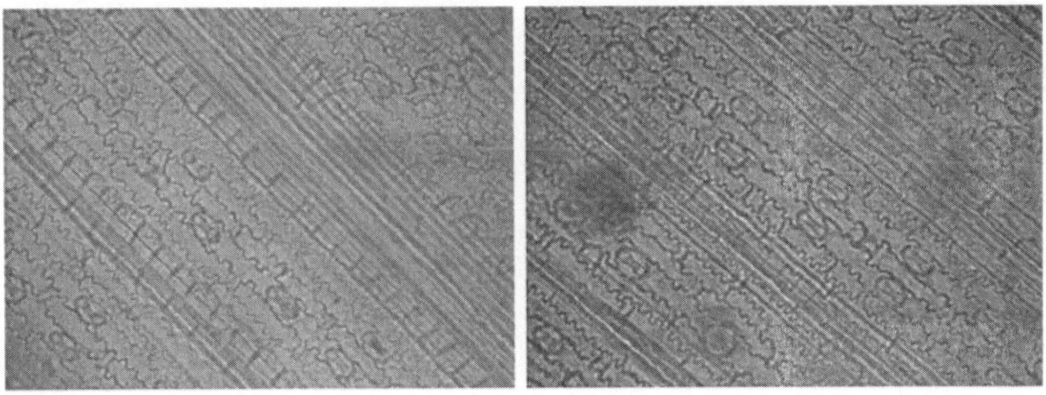

Figure 7. Leaf epidermal Anatomy of Cynodon dactyion (L) Pers. Abaxial side (Right) Adaxial side (Left).

19. *Lolium Temulentum* L.

a. Adaxial Side

Silica bodies, oil droplets and plasmodesmata were absent. Cells were rectangular. Single layered and smooth margined. Length of long cell 47 µm - 50 µm (Av: 48.5 µm) width of long cell 17µm – 19µm (Av: 18 µm). Length of short cell 15 µm – 17µm (Av: 16 µm) width of short cell 10 µm – 11 µm (Av: 10.5 µm). Trichomes were present. Number of trichome/ unit area 1-2. Length of trichome 100µm - 105µm (Av: 102.5µm). Width of trichome 15µm - 16µm (Av: 15µm).

b. Abaxial Side

Silica bodies, oil droplets and plasmodesmata were absent. Cells were rectangular, double layered and wavy margined. Length of long cell 450 μm - 520 μm (Av: 485 μm) width of long cell 48μm – 51.2 μm (Av: 49.6 μm). Length of short cell 255 μm – 285μm (Av: 270 μm) width of short cell 38 μm – 49 μm (Av: 43.5 μm). Trichomes were present and originate from outer memberane and acute from tip. Number of trichome/ unit area 1-2. Length of trichome 58μm - 77μm (Av: 67.5μm). Width of trichome 42μm - 45μm (Av: 43.5μm). Stomata were dumbell shaped. Number of stomata per unit area was 4-5. Length of stomatal cavity 43μm - 44 μm (Av: 43.5 μm) width of stomatal cavity 15 μm – 17μm (Av: 16 μm). Length of stomatal cell 101 μm – 109μm (Av: 105 μm) width of stomatal cell 43μm – 53μm (Av: 48μm).

20. *Pennisetum Orientale* L.C. Rich

a. Adaxial Side

Oil droplets were absent. Cells were rectangular, double layered and wavy margined. Length of long cell 196.5 μm – 204.6 μm (Av: 200.5 μm) width of long cell 17μm - 19 μm (Av: 18 μm). Length of short cell 112 μm – 138μm (Av: 125 μm) width of short cell 18 μm – 19 μm (Av: 18.5 μm). Trichomes were present and originate from intracellular in origion and acute from tips. Length of trichomes 32 μm – 78 μm (Av: 55 μm). Width of trichomes 21μm - 24 μm (Av: 22.5 μm). Stomata were present. Stomata were dumbell shape. Length of stomatal cavity 9μm -15 μm (Av: 12 μm) width of stomatal cavity 7 μm – 10μm (Av: 8.5 μm). Length of stomatal cell 34 μm - 37μm (Av: 35.5 μm) width of stomatal cell 22μm - 27μm (Av: 24.5 μm). Regular shaped silica bodies were present . Number of silica bodies per unit area 5-6. Length of silica bodies 19 μm – 21 μm (Av: 20μm) width of silica bodies 17μm – 18 μm (Av: 17.5 μm). Plasmodesmata connection was present. Length of connection cells 27 μm – 46 μm (Av: 36.5 μm).Width of connection cells 24μm – 43 μm (Av: 33.5 μm).

b. Abaxial Side

Silica bodies, oil droplets were absent. Cells were rectangular, double layered and wavy margined. Length of long cell 213 μm – 237.6 μm (Av: 225.5 μm) width of long cell 20μm – 22.1μm (Av: 21.5 μm). Length of short cell 90 μm – 104.8μm (Av: 97.4 μm) width of short cell 19 μm – 20 μm (Av: 19.5 μm).

Trichomes were present, intracellular in origin. Number of trichome/ unit area 3-4. Length of trichome 28μm – 38.2μm (Av: 33.1μm). Width of trichome 16μm - 24μm (Av: 20.05μm). Stomata were present. Stomata were dumbell shape. Number of stomata per unit area was 15-16. Length of stomatal cavity 12μm - 17μm (Av: 14.5 μm) width of stomatal cavity 9.0 μm – 10.0μm (Av: 9.5 μm).

Length of stomatal cell 39.2 μm – 47.3μm (Av: 43.25 μm) width of stomatal cell 24μm – 24.5μm (Av: 24.1μm). Plasmodesmata cells were present. Length of plasmodesmata cell 31μm - 42μm (Av: 36.5μm). Width of plasmodesmata cell 23μm - 25μm (Av: 24μm).

21. *Dicanthium Annulatum* (Forssk) Stapf.

a. Adaxial Side

Trichomes, Oil droplets and plasmodesmata were absent. Cells were rectangular,D double layered and wavy margined. Length of long cell 182 μm – 233 μm (Av: 207.5 μm) width of long cell 19μm - 22 μm (Av: 20.5 μm). Length of short cell 40 μm – 80μm (Av: 60 μm) width of short cell 15 μm – 20 μm (Av: 17.0 μm). Stomata were present anddumbell shaped. Number of stomata/ unit area was 4-5. Length of stomatal cavity 10μm -15 μm (Av: 7.5 μm) width of stomatal cavity 6 μm – 11μm (Av: 8.5 μm). Length of stomatal cell 40 μm - 50μm (Av: 45 μm) width of stomatal cell 20 μm - 22μm (Av: 21 μm). Silica bodies were present. Number of silica bodies per unit area 7-8. Length of silica bodies 32 μm – 50 μm (Av: 41 μm) width of silica bodies 26μm – 52 μm (Av: 39 μm). (Figure 8).

b. Abaxial Side

Trichomes and Oil droplets were absent. Cells were rectangular, double layered and wavy margined. Length of long cell 182 μm – 233 μm (Av: 207.5 μm) width of long cell 19μm - 22 μm (Av: 20.5 μm). Length of short cell 46 μm – 89μm (Av: 67.0 μm) width of short cell 16μm – 22μm (Av: 19 μm). Stomata were present. Stomata were dumbell shaped. Number of stomata/ unit area was 17-18. Length of stomatal cavity 12μm -18 μm (Av: 20 μm) width of stomatal cavity 5μm – 10μm (Av: 7.5 μm). Length of stomatal cell 46 μm - 50μm (Av: 48 μm) width of stomatal cell 20μm - 27μm (Av: 23.5 μm). Regular shape silica bodies were present . Number of silica bodies per unit area 5-6. Length of silica bodies 36 μm – 52 μm (Av: 44μm) width of silica bodies 29 μm – 52 μm (Av: 43 μm). Plasmodesmata connection was present. Length of connection cells 16 μm – 18 μm (Av: 15.3 μm). Width of connection cells 10.6μm – 20 μm (Av: 15.3 μm) (Figure 8).

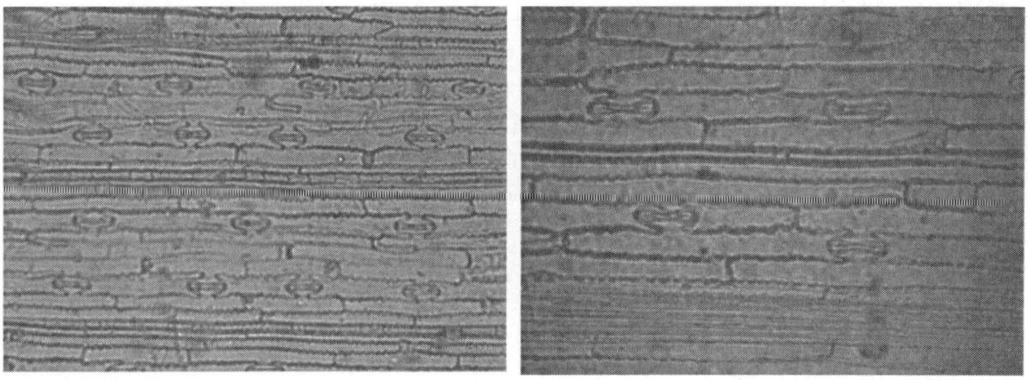

Figure 8. Leaf epidermal Anatomy of *Dicanthium annulatum* (Forssk) Stapf. Abaxial side (Right) Adaxial side (Left).

22. *Desmostachya Bipinnata* (L) Stapf

a. Adaxial Side

Oil droplets and plasmodesmata were absent. Cells were rectangular. Double layered and wavy margined. Length of long cell 180 μm - 225 μm (Av: 202.5 μm) width of long cell

19µm – 28.5 µm (Av: 23.7 µm). Length of short cell 109 µm – 115µm (Av: 112 µm) width of short cell 18 µm – 25 µm (Av: 21.5 µm). Trichomes were present and originated from intracellular. Number of trichomes per unit was 19-20. Length of trichomes 30 µm – 43 µm (Av: 36.5 µm). Width of trichomes 20 µm - 26 µm (Av: 23 µm). Stomata were present. Stomata were dumbell shaped. Length of stomatal cavity 9µm -12 µm (Av: 10.5 µm) width of stomatal cavity 7 µm – 7µm (Av: 7.0 µm). Length of stomatal cell 35 µm - 42µm (Av: 38.5 µm) width of stomatal cell 23µm - 30µm (Av: 26.5 µm). Regular shaped silica bodies were present. Number of silica bodies per unit area 5-6. Length of silica bodies 27 µm – 62 µm (Av: 44.5 µm) and width of silica bodies 23 µm – 65 µm (Av: 44 µm).

b. Abaxial Side

Oil droplets were absent. Cells were rectangular double layered and wavy margined. Length of long cell 186 µm - 232 µm (Av: 209 µm) width of long cell 19µm - 21 µm (Av: 20 µm). Length of short cell 112 µm – 119µm (Av: 115.5 µm) width of short cell 20 µm – 26.5 µm (Av: 23.1 µm). Trichomes were present and originated from intracellular and acute from tips. Number of trichomes per unit were 20-22. Length of trichomes 32 µm – 45 µm (Av: 38.5 µm).Width of trichomes 22µm - 28 µm (Av: 25 µm), Stomata were present. Stomata were dumbell shaped. Number of stomata per unit area was 20-22. Length of stomatal cavity 10µm -13 µm (Av: 11.5 µm) width of stomatal cavity 7 µm – 7.5 µm (Av: 7.15 µm). Length of stomatal cell 37 µm - 45µm (Av: 41 µm) width of stomatal cell 25µm - 32µm (Av: 28.5 µm). Irregular shaped silica bodies were present. Number of silica bodies per unit area 3-4. Length of silica bodies 29 µm – 67 µm (Av: 48 µm) width of silica bodies 27µm – 67 µm (Av: 47 µm). Plasmodesmata were present. Length of plasmodesmata 20µm - 26µm (Av: 23.4µm). Width of plasmodesmata 18µm - 25µm (Av: 21.5µm).

23. *Eragrostis Poaeoides* P.Beauv

a. Adaxial Side

Oil droplets and plasmodesmata were absent. Cells were rectangular. Double layered and smooth margined. Length of long cell 120 µm – 198 µm (Av: 184 µm) width of long cell 47µm - 50 µm (Av: 48.5 µm). Length of short cell 60 µm – 89µm (Av: 74.5 µm) width of short cell 37.1µm – 46µm (Av: 41.55 µm).

Trichomes originated from inner membrane. Number of trichomes / unit area was 3-4. Length of trichomes 152µm - 160µm (Av: 156µm). Width of trichomes 30.4µm - 35µm (Av: 32.7µm). Stomata were present. Stomata were dumbell shaped. Number of stomata/ unit area was 6-7. Length of stomatal cavity 33.1µm – 36.1 µm (Av: 34.6 µm) width of stomatal cavity 19µm – 25µm (Av: 22 µm). Length of stomatal cell 82.1 µm - 86µm (Av: 84.05 µm) width of stomatal cell 38µm - 41µm (Av: 39.5 µm). Silica bodies were present. Number of silica bodies per unit area 2-3. Length of silica bodies 24 µm – 39.4 µm (Av: 36.7µm) width of silica bodies 13 µm – 15 µm (Av: 14 µm).

b. Abaxial Side

Trichomes were absent. Cells were rectangular. Double layered and wavy margined. Length of long cell 257.1 µm – 331.3 µm (Av: 303.2 µm) width of long cell 30µm - 32 µm

(Av: 31 µm). Length of short cell 129 µm – 159µm (Av: 144 µm) width of short cell 30µm – 38µm (Av: 34 µm) Stomata were present dumbell in shape. Number of stomata/ unit area was 13-14. Length of stomatal cavity 24µm - 30 µm (Av: 27 µm) width of stomatal cavity 19µm – 21µm (Av: 20 µm). Length of stomatal cell 71 µm - 88µm (Av: 79.5 µm) width of stomatal cell 40µm - 48µm (Av: 44 µm). rRgular shape silica bodies were present. Number of silica bodies per unit area 18-20. Length of silica bodies 13 µm – 15 µm (Av: 14µm) width of silica bodies 13 µm – 18 µm (Av: 15.5 µm). Oil droplets were present. Number of oil droplets/ unit area was 4-5. Length of oil droplets 18µm - 19µm (Av: 18.5µm). Width of oil droplets 12µm - 17µm (Av: 14.5µm). Plasmodesmata connection was present. Length of connection cells 46 µm – 65 µm (Av: 55.5 µm). Width of connection cells 17µm – 30 µm (Av: 23.5 µm) (Figure 9).

Figure 9. Leaf epidermal Anatomy of *Eragrostis poaeoides*. P. Beauv (Abaxial).

Table 1 shows relationship between species of the sub family Chloridoideae. Co-relation between species of this sub family ranged 0.01-0.99. Based on anatomical characters, *Eragrostis poeoides* abaxial side was closely related to *Elucine indica*. The value of co-relation for *Eragrostis* and *Elucine* was 0.99 (Table 1). The species that showed minimum co-relation were *Eleusine flagllifera* and *Cynodon dactylon*. These two species were totally different from one another on the basis of leaf epidermal characters. This table also shows that species belonging to genus *Eragrostis* (*E. minor* and *E poaeoides*) were closely related to each other (0.83). *E. flagellifera* and *E.indica* although belonging to the same genus but their similarity index was 0.25.

Table 2 shows similarity index between the species of sub family Panicoideae. Their co-relation range from 0.02 to 0.98. *Digitaria nodosa* (abxial side) and *Setaria intermedia* showed close relationship (0.98) also species of this sub family which showed close relationship were *Digitaria ciliaris, D. setigaria, D. ischaeum* and *Cymbopogon jawarancusa.* The negative relationship ranged from -0.01 to -0.10. Minimum relationship was found between *Digitaria cillaris* and *Cymbopogon jwarancusa* (-0.01).

Table 1. Corlation between the species of sub family Chloridoideae based on leaf epidermal anatomical characters

Species Names	**1**	**2**	**3**	**4**	**5**	**6**	**7**	**8**	**9**	**10**	**11**	**12**	**13**	**14**
1	1													
2	0.66	1												
3	0.83	0.77	1											
4	0.80	0.59	0.70	1										
5	0.92	0.74	0.93	0.77	1									
6	0.26	0.06	0.04	0.62	0.18	1								
7	0.81	0.75	0.99	0.71	0.92	0.04	1							
8	0.48	0.22	0.23	0.75	0.41	0.80	0.25	1						
9	0.67	0.32	0.37	0.80	0.58	0.77	0.36	0.80	1					
10	0.87	0.78	0.95	0.72	0.91	0.06	0.94	0.27	0.41	1				
11	0.79	0.52	0.64	0.82	0.79	0.64	0.62	0.71	0.76	0.69	1			
12	0.87	0.61	0.75	0.93	0.85	0.57	0.75	0.71	0.78	0.76	0.90	1		
13	0.82	0.75	0.97	0.72	0.89	0.08	0.97	0.26	0.36	0.97	0.66	0.77	1	
14	0.70	0.60	0.76	0.47	0.71	0.01	0.70	0.08	0.29	0.83	0.54	0.57	0.80	1

Key: 1: *Eragrostis minor* abaxial; 2: *Eragrostis minor* adaxial; 3: *Eragrostis poaeoides* abaxial; 4: *Eragrostis poaeoides* adaxial; 5: *Desmostachya bipinnata*; 6: *Eleusine flagellifera*; 7: *Eleusine indica* abaxial 8: *Elucine indica* adaxial; 9: *Acachne racemose*; 10: *Dactyloctenium aegyptium* abaxial; 11: *Dactyloctenium aegyptium* adaxial; 12: *Leptochola chinensis* abaxial; 13: *Leptochola chinesis* adaxaial; 14: *Cynadon dactylon.*

Table 2. Co-relation between the species of sub family Panicoideae based on leaf epidermal anatomical characters

Species Names	1	2	3	4	5	6	7	8	9	10	11	12	13	14	15	16	17
1	1																
2	0.74	1															
3	0.42	0.22	1														
4	0.52	0.09	0.97	1													
5	0.93	0.58	0.62	0.74	1												
6	0.96	0.81	0.25	0.38	0.89	1											
7	0.66	0.73	-0.10	0.03	0.51	0.71	1										
8	0.95	0.85	0.27	0.37	0.84	0.96	0.69	1									
9	0.49	-0.01	0.94	0.97	0.73	0.36	-0.01	0.31	1								
10	0.96	0.77	0.33	0.47	0.92	0.98	0.69	0.93	0.44	1							
11	0.93	0.83	0.11	0.25	0.81	0.97	0.78	0.96	0.21	0.94	1						
12	0.85	0.46	0.79	0.87	0.96	0.77	0.39	0.75	0.85	0.82	0.67	1					
13	0.46	0.04	0.90	0.95	0.71	0.36	0.06	0.28	0.95	0.46	0.19	0.82	1				
14	0.92	0.78	0.11	0.25	0.80	0.95	0.74	0.94	0.22	0.92	0.97	0.67	0.19	1			
15	0.95	0.75	0.24	0.35	0.83	0.94	0.70	0.95	0.31	0.90	0.95	0.73	0.27	0.97	1		
16	0.83	0.89	0.05	0.15	0.66	0.86	0.72	0.93	0.06	0.83	0.91	0.56	0.04	0.92	0.89	1	
17	0.84	0.91	-0.01	0.13	0.7[illegible]	0.93	0.77	0.93	0.06	0.88	0.96	0.56	0.10	0.93	0.89	0.94	1

Key: 1: *Paspalidium flavidum* abaxial; 2: *Paspalidium flavidum* adaxial; 3: *Digitaria ciliaris* abaxial; 4: *Digitaria ciliaris* adaxial; 5: *Digitaria nodosa* Adaxial; 6: *Digitaria nodosa* Abaxial; 7: *Digitaria setigera* 8: *Digitaria ischaemum* abaxial; 9: *Digitaria ischaemum* adaxial 10: *Seteria intermedia* abaxial; 11: *Seteria intermedia* adaxial; 12: *Seteria verticillata*; 13: *Seteria pumila* adaxaial; 14: *Pennisetum orientales* abaxial; 15: *Pennisetum orientales* adaxial; 16: *Dichanthium annulatum*; 17: *Cybopogon jwarancusa.*

Table 3. Co-relation between the species of sub family poideae based on leaf epidermal anatomical characters

Specie name	Lolium temulentium	Phalaris minor	Koeleria phleoides
Lolium temulentium	1		
Phalaris minor	0.48	1	
Koeleria phleoides	0.43	0.89	1

Table 4. Co-relation among the quantitative leaf epidermal characters of family Poaceae based on leaf epidermal anatomical characters

Characters	**1**	**2**	**3**	**4**	**5**	**6**	**7**	**8**	**9**	**10**	**11**	**12**	**13**	**14**	**15**	**16**
1	1															
2	0.14	1														
3	0.49	0.46	1													
4	0.11	0.80	0.39	1												
5	-0.37	0.09	-0.31	0.16	1											
6	0.00	0.18	0.25	0.19	0.19	1										
7	0.34	0.66	0.56	0.42	-0.14	-0.12	1									
8	0.40	0.60	0.42	0.37	-0.22	-0.15	0.70	1								
9	0.45	0.62	0.57	0.41	-0.17	-0.10	0.97	0.80	1							
10	0.41	0.43	0.49	0.32	-0.13	-0.05	0.77	0.76	0.83	1						
11	0.03	-0.20	0.25	-0.02	0.35	0.21	-0.25	-0.10	-0.21	-0.01	1					
12	0.09	-0.16	-0.20	0.01	0.32	0.21	-0.22	-0.06	-0.08	0.06	0.96	1				
13	-0.24	-0.14	-0.30	-0.08	-0.03	-0.14	-0.02	-0.12	0.00	0.08	-0.02	0.99	1			
14	-0.29	-0.11	-0.32	-0.05	0.02	-0.09	-0.02	-0.13	-0.01	0.08	-0.01	-0.06	0.99	1		
15	0.29	-0.26	0.24	-0.02	-0.31	-0.09	0.01	0.05	0.08	0.18	0.01	-0.02	-0.03	-0.07	1	
16	0.25	-0.28	0.22	-0.06	-0.28	-0.01	-0.02	-0.01	0.05	0.17	0.04	0.03	-0.13	-0.15	0.95	1

Key: 1: length of long cell; 2: width of long cell; 3: length of short cell; 4: width of short cell; 5: length of trichome; 6: Width of trichome; 7: length of stomatal cavity 8: width of stomatal cavity; 9: length of stomatal cell; 10: width of stomatal cell; 11: length of silica body; 12: width of silica body; 13: Length of droplet; 14: width of droplet; 15: length of connection of cell type; 16: Width of connection of cell type.

Table 3 shows co-relationship of Sub family Poideae consisting of three species *Lolium temulentium, Phalaris minor* and *Koeleria phleoides* based on leaf epidermal anatomical characters. The relationship of these species ranges from 0.43-0.89. Highest relationship was found between *Phalaris minor* and *Lolium temulentum* with 89% relationship percentage. The similarity index of *Lolium temulentium* and *Koeleria phleoides* was found to be 0.43.

Table 4 shows co-relation among 16 quantitative characters of leaf epidermal anatomy of family Poaceae. The similarity index between characters of Poaceae member ranged from 0.00 to 0.99. Length of oil droplet was closely related to width of oil droplet with co-relation index 0.99. Negative relation of species ranging from -0.01 to -0.31.

DISCUSSION

Poaceae family has important economic and medicinal value. In past taxonomic information of the family poaceae was based on morphological markers, leads towards taxonomic confusion [40]. From anatomical studies it was observed that these species showed variations in size/shape of long cell, short cell, trichomes, prickles, hairs, silica bodies, oil droplets hooks and stomata [10]. Important anatomical feature possessed by species of the family Poaceae was presence of long/short cell with single or double layered, smooth or wavy margined cell membrane [14]. Same observations of anatomical features were made in the present investigation. Cells with rectangular and elongated shape with tapering ends were found in *Eragrostis minor* and *Dactyloctenium aegyptium*. All these characters were found in different species of family Poaceae. The size and shape of organelles were different on both abaxial and adaxial surface of leaf epidermal because of bifacial leaf, which also confirmed the findings of Webster [21] and Shouloang et al., [41]. The maximum length of long cell ranging 450μm- 520μm was found in *Lolium temulentium*on abaxial side of epidermis. Similarly maximum length of short cell ranging 255μm - 285μm was also part of same species which is a unique leaf epidermal character of this species and sufficient to differentiate it from other species of family Poaceae.

Presence of silica bodies, oil droplets and plasmodesmata were few more taxonomically important characters of Poaceae. All of these three anatomical charactesr were absent from both abaxial and adaxial surface of *Setaria intermedia*. The length of trichome varies from species to species. In *Setaria pumila* length of trichome ranges from 402μm- 514μm on abaxial side of the leaf epidermis. Trichomes not only showed variation in length but also for origin, it may be intracellular, from the surface of outer membrane/ inner membrane. There was also diversity in form of trichomes; it may be prickles like or hairy type. Trichome with acute or blent trichome were also present. Same observations were made by Gilani et al., [14] while they were working for the leaf epidermal anatomy of *Digitaria*.

Another important anatomical marker was types of silica bodies that could be used for delimitation of taxonomic taxa. The family Poaceae was characterized by different types of silica bodies including cross shaped, regular to irregular shaped. Cross shaped bodies were found in all four species of *Digitaria*. Regular shape was the character of Acrachne racemose, *Dichanthium annulatum, Dichanthium annulatum, Desmostachya bipinnata, Elucine flagellifera, Setaria pumila, Eragrostosis minor, Laptochola chinensis, Phalaris minor,*

Cybopogon jwarancusa. Irreglar shaped were found in *Dactyloctenium aegyptium, Paspalidium flavidium* and *cynodon dactylon.*

Trichomes were absent from both abaxial and adaxial surfaces of *Phalaris minor, Cymbopogon jwarancusa* and *Digitaria setigera*. Stomata were absent from both abaxial and adaxial epidermis of *Eleusine flagillifera* and *Digitaria ischaemu* and from adaxial side of *Dactylochletenium aegyptium.* The present study shows that stomata were dome, parrel or triangular in shape. Dome shaped were found in *Lolium temulentium, Leptochloa chinensis, Digitaria nodosa, Cynodon dactylon, Dichanthium annulatum, Desmostachya bipinnata, Eleucine flagellifera, Setaria pumila, Eragrostosis minor, Laptochola chinesis, Phalaris minor* and *Cymbopogon jwarancusa.* Irregular shaped were found in *Dactyloctenium aegyptium, Paspalidium flavidium* and *Cynodon dactylon.* Trichomes were absent from both abaxial and adaxial side of *Phalaris minor, Cymbopogon jwarancusa* and *Digitaria setigera.*

Stomata were absent from abaxial and adaxial epidermis of *Eleusine flagillifera, Digitaria ischaemum* and *Dactylochletenium aegyptium*. Present investigation indicated that stomata were of dome, parrel or triangular in shape. Dome shaped were found in *Lolium temulentium, Leptochloa chinesis, Digitaria nodosa, Cynodon dactylon, Sertaria verticellata, Cymbopogon jwarancusa* and triangular shaped in *Acrachne racemosa. Digitaria setigera, Pennisetum orientales* and parrell shaped in *Eleusine indica* From the results of present study it was concluded that leaf epidermal anatomical features were important taxonomic markers and could be effectively utilized for solving the persistent taxonomic problems of family Poaceae.

REFRENCES

[1] Shantz, S.(1954). The place of grassland in the earth cover of vegetation. *Ecology*. 35:143-145.

[2] GPWG (grass phylogeny working group), 2001. Phylogeny and sub families classification of the grasses (Poaceae). *Ann. Missouri Bot Garden*. 88: 373-457.

[3] Tzvelev, N.(1989). The system of grasses (Poaceae) and their evolution. *The Bot. Rev*. 55(3): 141-204.

[4] Stace, C.E. (1980). Plant taxonomy and biosystematic Edward Ardnold, London. PP: 279.

[5] Carpuran, R. (1962). Revision des Rhopalocarpacees. *A dansonia*. 2: 228-267.

[6] Laory, J.F. (1973). Recherchur Sur a Speciation et I' endemisme dans la flora malgache- III. Note sure Ie genre Dialyceras R. Capuron (Sphaerosepalacees). *Adansonial* 3: 37-53.

[7] Clayton, W.D. and Renvoize, and S.A. (1986): Genera Germanium: grasses of world Royal Bot. Gardens Kew Bull, *Addl. Ser*. 13: 389.

[8] Part, H. (1932). L' des graminees. Etude anatomique et sysematique. *Ann. Sci. Nat. Bot. Ser*. 10. 14: 117-324.

[9] Part, H. (1936): General features of the epidermis in zea mays. *Ann. Missouri Bot. Garden*. 35:341-351.

[10] Metacalf, C.R. (1960). Anatomy of monocotyledons I. Gramineae, Oxford university press, London: 1-711

[11] Ellis, R.P. (1979): A procedure for standardizing comparative leaf anatomy in the Poaceae. *The epidermis as seen in surface view Bothala* 12: 641-671.

[12] Baum, B. R., Tulloch, A.P. and Bailly, and L.G. (1989): Epicuticular waxes of the genus Horedum: A survey of their chemical composition and ultra structure. *Can. J. Bot* 67: 3219-3226.

[13] Amarasinghe, V. and Watson, L. (1990): Taxonomic significance of micro hairs in the genus Eragrostis Beauv. (Poaceae) *Taxon* 39: 59-65.

[14] Gilani, S.S., M.A., Khan, Z.K., Shinwari and Yousaf., Z. (2002). Leaf epidermal anatomy of selected Digitaria species, Tribe Paniceae, Family Poaceae of Pakistan. *Pak. J. Bot* 34 (3): 257-273.

[15] Raole, V.M. and Deasi, R.J. (2009). Epidermal studies in some member of Andropogoneae (Poaceae). *Not. Bot. Hort. Agrrobot. Cluj*. 37(1): 59-64.

[16] Stebbins, G.L. and Kush, G.S. (1961). Variation in the organization of the stomatal complex in the leaf epidermis of monocotyledons and its bearing on their phylogeny. *Annuals journal of Botany*. 48: 51-59.

[17] Inamdar J.A., Mohan, J.S.S. and Subramanium, R.B. (1986). Stomatal classification. *A Review Feddes Report*. 97 (3 and 4): 147-160.

[18] Pareveen, N.S., Murthy, K.S.R. and Pullaiah, T. (2002). Leaf epidermal characters in cotolaria spp. Paloiliono from Easten. *Ghart. Phytomorphology*. 50: 205-212.

[19] Edeoga,H.O. and Ikem C.I. (2001). Comparative Morphology of leaf epidermis in three species of *Boerhevia* L J *Econ Tax Bot* 19:197-205

[20] Rammaya and Rao, S.S. (1987). Trichome types and their systematic importance in tiliaceae. *Indian Journal of Botany*. 10: 65-73.

[21] Webster, R.D. (1983). A revision of Digitaria Haller (Paniaceae, Poaceae) in Australia. *Brnnonia*. 6: 131-216.

[22] Desai, R., Raoleand, V.M. and Arya, A. (2009). Comparative foliar epidermal studies in Coix Lacryma Jobi L. and Loix aquatic Roxb. (Poaceae). *Not. Sci. Bio*. 11 (1): 37-40.

[23] Yousaf, Z., Shinwari, Z.K., Asghar, R.and Perveen, A. (2008). Leaf epidermal anatomy of selected Allium species, Family Alliaceae from Pakistan. *Pakistan Journal of Botany*. 40(1): 77-90.

[24] Reimer, E. and Costa-Sanches, J.H. (2007). An SEM survey of the leaf epidermis in danthonioid grasses (Poaceae: Danthonlodeae). *Syst. Bot*. 32: 60-70.

[25] Jiang, Y.L., Li, Y., Deng, W. Li, Tang, L.M. and Lua, C.J. (2007). Leaf epidermal anatomy of *Lonicera japonica* and *Lonicera confuse. Journal of Forestry Research*. 18(2): 103-108.

[26] Munsif, S., Khan, M.A., Ahmad, M., Zafar, M., Shana G.M. and Shaheen. S. (2007). *Journal of Agriculture Society of Science*. 3 (2): 43-46.

[27] Rajas, C.B. (2007). Anatomy of the foliar epidermis of the *Solanum nudum* species group of *Solanum sect germinate* (Solanaceae). *Acta Hort. (ISHS)* 745: 287-294.

[28] Motomura, H., Fujii, T. And Suzuki., M. (2006). Silica deposition in abaxial epidermis before the opening of leaf blades of Pleiblastus chino (Poaceae, Bombusoideae) *Annals of Botant*. 97:513-519.

[29] Ozdemir, C. and Atan., Y. (2006). Morphological and anatomical investigation of some endemic alkanna species. *Pakistan journal of Botany*. 38 (3): 527-537.

[30] Liscovsky, I.J. and Cosa, M.T. (2005). Comparative anatomy of leaf and stem in species of Cestreae. G. Don (Solanaceae) from Argentina. *Gayana Botanica*: 33-43.

[31] Kanademir, N. (2003). The morphological, anatomical and Karyological properties of endemic *Saliva hypargeia* Fich and Mey. (Lamiaceae) in turkey. *Pak journal of Botany* 35 (2): 219-236.

[32] Vieira, R.C., Gomes, D.M.S., Sarahyba, L.S. and Arruda, R.C. (2002). Leaf anatomy of three herbaceous Bamboo species. *Brazilian Journal of Biology*. 62 (4): 1519-6984.

[33] Chaudary, M.I., Mumtaz, A.S. and Khan, M.A. (2001). Leaf epidermal anatomy of medicinal grasses of Islamabad Attock and Mirpur. *Pak. J. of Bio. Sci.* 4(12): 1466-1469.

[34] Acedo, C. and Liamas, F. (2001). Variation of micromorphological characters of lemma and paleain the genus Bbrumus (Poaceae). *Ann. Bot. Fennici.* 38:114.

[35] Corxdale, J. (1998). Stomatal patterning in monocotyledons: Tradescantia as a model system. *J. of Exp. Bot*. 49: 279-292.

[36] Ogiundipe, O.T. (1992). Leaf epidermal studies of genus Datura (Linn) Solanaceae. *Phytomorphology*. 42 (34): 207-217.

[37] Lackey, J.A. (1978). Leaf let anatomy of Phaseoleae (Leguminosae; Paplionoidae) and its relation tp taxonomy. *Bot. Gaz*. 139 (4): 434-446.

[38] Cope, T. (1982) Poaceae. In: E. Nash and S.I. Ali (Eds.), *Flora of Pakistan.* Karachi.

[39] Shouloang, C., Yuexing, J., Zhujun, W.and Xintian, L. (1996). Systematic evolutionary study of Poaceae (Graminaceae) and its relatives using leaf epidermis. *Proc. IFCD*: 417-425.

In: Leaf Epidermal Anatomical Characters …
Editors: Z. Yousaf, A. Younas and A. Aftab

ISBN: 978-1-63483-190-1

Chapter 8

TAXONOMIC EVALUATION OF TREES BASED ON LEAF EPIDERMAL ANATOMICAL CHARACTERS FROM TROPICAL AND SUBTROPICAL AREAS OF PAKISTAN

Faiza Khan and Arusa Aftab*
Department of Botany,
Lahore College for Women University,
Lahore, Pakistan

ABSTRACT

Present study was conducted for taxonomic evaluation of the related thirty tree species of dicot flora belonging to eight orders and fifteen families. The experiment was performed in molecular taxonomy lab of Lahore College for Women University, Jail road Lahore. Anatomical observations showed variation in size and shape of epidermal cells, and their arrangement. Variation in stomata types, number of subsidiary cells, different shapes of silica bodies were also observed. Leaf epidermal anatomy was found to be important tool for identification of dicot species. The average length of the organelles of leaf epidermis provided more clear difference between the species The result showed substantial, variation in the type and shape of epidermal cells from straight to polygonal with, smooth and wavy cell membrane. The comparative length of glandular and non-glandular trichomes was considered as an important taxonomic marker for dicot flora. Non-glandular trichomes were found only on adaxial surface of *Ficus glomerata*, Roxb and *Murraya koenigii* L. Within dicot flora, eight different shapes of stomata were observed (anomocytic, anomotetracytic, actinocytic, amphianisocytic, brachyparacytic, anisocytic, amphicyclocytic and staurocytic). Anomocytic stomata were the most dominant in trees flora of Lahore District. Staurocytic stomata were found in *Ficus religiosa*, L. and *Erythrina subrosa*, L. Round and irregular shape of silica bodies were also found on adaxial surface of species *Artocarpus integrifolia*, Spreng, *Azadirachta indica* L., *Syzygium cumini*, L. The present investigation shows that the shape of leaf epidermal cells could not be utilized significantly delimiting the closely related taxa.

* Corresponding author: E. mail: arusa_ravi@yahoo. com.

INTRODUCTION

Taxonomy is the backbone of all sciences. However most of the taxonomic information of plants accumulated so far is based solely upon morphometry. As most of the taxonomist were agreed that similarities and dissimilarities of plants could be measured by using morphological markers [1]. Although morphological characters are directly exposed to environment, cause change in morphology and leads toward the taxonomic confusions. Morphometery could not provide solution of complex taxonomic problems. Therefore taxonomist involved other biological techniques like leaf epidermal anatomy, cytology, electrophoresis as taxonomic tools [1].

Anatomical markers of leaf epidermis were found to be an important tool for taxonomic evaluation [2] on inter and intra generic level. The epidermis is a system of cells, variables in structure and function that together constitute the covering of the plant body in the primary state [3]. The characteristics of leaf epidermis are being increasingly important in resolving existing taxonomical confusions [1, 4] As leaf epidermis anatomical features such as stomata, trichomes and other markers have been proved useful anatomical tool of potential taxonomical importance [5, 6]. Hence anatomical features are of particular value to identify small parts of plant materials [7]

Pakistan lies in temperate zone. The climate is generally arid, characterized by hot summer and cold winter. There is a little rainfall (Sagged et al., 2009). Lahore is bounded by Sheikhupura district on North, West, by Wagah on east and Kasur on south district. The Ravi River flows on north side of Lahore. Lahore city covers a total land area of 404 km^2 and is still growing. There is large varieties of tree that are native of Lahore region[8].

Although, there is no information available on leaf epidermal anatomy of trees in Lahore district from Pakistan. The present study is the first attempt that was to find out the solution of existing taxonomic problem of tree species that are found in Lahore district flora which overlap in most of their morphological characters and to evaluate the relationship of the critical taxa by characterization of leaf epidermal characters.

OBJECTIVE

The main objective of present investigation was:

1) To explore leaf epidermal character of trees flora present in Lahore district to solve existing taxonomic problem.

HISTORICAL BACKGROUND

Ahmad et al.,[9] worked on taxonomic diversity in epidermal cell of some sub-tropical plant species. For this purpose they selected 40 species belonging to 38 genera of 22 families the shape and type of large epidermal cells among these varies from straight and polygonal. They concluded the large variation in epidermal cells configuration at different taxonomic level But they ignored another anatomical characters that could provide an intergenic and intragenic variation in many species. Although there is a need of more comprehensive work on the shape and type of leaf epidermal cells for their further elaboration on a taxonomic characters.

Sahreen et al., [10] worked on leaf epidermal anatomy of the genus *Silene caryophyllaceous* from Pakistan. They selected 16 species of the genus *Silene*. Among these species, they studied the different types of stomata, shape and size of epidermal cells, trichomes and crystal at specific level. Although they observed the most important anatomical characters. But they did not observed most variety of the anatomical characters (arrangement of epidermal cells, size of stomata and silica bodies). If they did that could be provided better result.

Yasmin et al., [11] worked on the micro morphological investigation of foliar anatomy of *Fagopyrum* Mill., and *Rumex*. L. of Polygonaceae. They selected the specimens from herbarium of Quaid-e-Azam University Islamabad. They selected eleven species belonging to *Fagopyrum* Mill and Rumex. L. of the family Polygonaceae. They studied the anatomical features such as shape of epidermal cells, stomatal pattern, their distribution on abaxial and adaxial on the leaf surface and variation in trichomes types. Among of these species, they reported five different stomatal (paracytic, pericytic, anomocytic, staurocytic, anisocytic). Although it is good piece of work but they emphasized only on shape of stomata and trichome types and ignored all these important characters (arrangement of epidermal cells, variation of trichomes and silica bodies etc. Which could be proved very important taxonomic tool for identification.

Yasmin et al., [12] worked out on the micro morphological investigation of foliar anatomy of genera *Acononon* and *Bistorta* family Polygonaceae. They collected the material from the herbarium of Quaid-e-Azam University Islamabad Pakistan. They selected the six species belonging to two genera of the family Polygonaceae. They studied anatomical features such as shape of epidermal. stomatal distribution on abaxial and adaxial surface and trichome types under the light microscopy. Among of these species they found the six types of stomata (paracytic, cyclocytic, staurocytic, anisocytic, anomocytic, pericytic) and variation among glandular and non glandular trichome. They just emphasized on the anatomical features of these species. They ignored the arrangement of epidermal cell, stomatal variety and variation in trichome which could provide very important taxonomic tool for identification of species.

According to Carmen et al., [13] identified the morphology of foliar epidermis in two groups of Solanum section *Geminata* (Solanaceae). They selected the eight species from the genus *Solanum*. They studied the quantitative and qualitative anatomical features on abaxial and adaxial epidermis in these species they observed the large variation in the observation of leaf segment. They just revealed the anatomical character. But they did not describe the vast variety of epidermal cell, stomata and trichome. they just emphasized the presence of these anatomical character and they ignored the other anatomical characters like arrangement of epidermal cell, variation in trichome types and silica bodies etc.

Ahmad et al., [14] evaluated taxonomic diversity of stomata in dicot flora from (N. W. F. P) in Pakistan. They selected a total 36 dicot species of 34 genera and 20 families They observed the stomatal types in different species. Among of these species they recognized seven (anisocytic, amphianisocytic, axillocytic, anomotetracytic, actinocytic, diacytic, staurocytic) types of stomata. Wherein amphianisocytic stomata was the dominant. But they ignore other anatomical character. They just emphasized on stomata anatomical marker.

Shaheen et al., [15] worked on diversity of foliar trichome and their systematic relevance in the genus *Hibiscus* (Malvaceae). They collected the material from the different localities of Pakistan and dried leaves from the herbarium of Quaid-e-Azam University Islamabad. They examined qualitative and quantitative morphological characters. They distributed the systematic relevance of both glandular and eglandular foliar trichome in species of genus *Hibiscus*. Glandular capitate trichome was the most dominant on leaf surface. Among of these species they just emphasized only on trichome anatomical marker. They ignored other anatomical character that could provide better result for systematical studies.

Mbagwu et al., [16] worked on leaf epidermal features of *Canarium schwenfurthii* and *Ddaeryodaedus* (Brusiaceae). They collected thre species from different L. G. A of Imo; Ebony and Anambar states of Nigeria. They studied the epidermal cells and stomatal cells. Although it is a good piece of work yet they ignored another anatomical character like arrangement of epidermal cells, different shape of stomata types, variation in trichome and silica bodies' etc. which could play a vital role in a characterizing on taxonomic tool for identification of species.

Mbagwu et al., [17] worked on leaf epidermal characteristics of four species of the genus *Citrus* (Rutaceae). They collected the material from the different garden of Oweri, Imo state, Nigeria. They studied four species of *Citrussinensis*, Climon C. *aurantifolia* and *C parandist* and observed that stomata and trichomes are absent on adaxial surface. Although it was a good piece of work yet they oversight another anatomical character like shape of epidermal cells, arrangement of epidermal cells, stomatal cavity and silica bodies. That could be better for identification of species at taxonomic level?

Perveen et al., [18] worked on stomata types of some dicots within flora of Karachi, Pakistan. For this purpose they selected the 69 dicot species distributed in 64 genera and 24 families. They examined the stomatal morphology by using light and scanning electron microscope. Among of these species, six types of stomata were recognized. During study they ignored the other anatomical characters (shape of cells, size of cell, type of trichome etc) at taxonomic level. If they observed the rest of character then their work could be provided better result.

Klimko et al., [19] worked on the morphological variability of the leaf epidermis in selected taxa of the genus *Ficus* L. (Moraceae) and its taxonomic implication. They collected the material from the botanical garden, Adam Mickiewicz University, pozan. They selected the leaf epidermal of 36 taxa (species, subspecies, varieties and cultivars) of the genus *Ficus* (subgenera *Ficus*, *synoecia*, *sycamores*, *pharmacosycea* and *urostigma*) that was observed under LM and SEM. They revealed the number of micro morphological characters at taxonomic level. They used the plant growth under cultivation condition to minimize the modifying effect of environment condition on the epidermal leaf traits. They emphasis on leaf epidermal characters and ignored other anatomical characters. If they observed these anatomical characters that could be provided better result.

Carpenter et al., [20] worked onleaf cuticular morphology links planteaceae and proteaceae. They collected the specimen from the Florida museum of natural history, university of Florida. For this purpose, they used the light and scanning electron microscope for the investigation of leaf cuticular morphology of extent species of planteaceae. They observed the anatomical characters types of stomata, epidermal cells on abaxial and adaxial surface. They studied the morphology synapomorphy linking plantanaceae and proteaceae. Although it is good piece of work yet they just emphasized on the particular characters among these species. But they ignored all other anatomical characters like shape and size of epidermal cells and arrangement of cells, variation in types of stomata cells and silica bodies etc. This could be proved very important taxonomic tool for identification of species.

Nikolic et al., [21] worked on comparative analysis of stomata and leaf trichome characteristics in *Quercusrobex* L. genotype variability of leaf trichome and stomata characteristics. Among these species, they recognized solitary glandular and uniseriate glandular hairs that were present on the adaxial leaf surface. Single glandular trichome was observed in all genotype, But they ignored all other anatomical characters and studied the impact of unfavorable factors of the environment, particularly air pollution morpho anatomical changes of leaves in different tree species.

Farooqui et al., [22] worked on use of cuticular studies for distinguishing species of *Dalbergia*. They studied cuticular characters in three species of *Dalbergia*. And observed the comparison study between two species of *Dalbergia lantifolia* and *Dalbergia sisoo*. They distinguished the papillate and non papillate cuticular characters in two species. Although it is good work yet they just emphasized on the epidermal cells. But they ignored all other anatomical characters. This could be very important taxonomic tool for identification of species.

Popa and Sipos [23] worked out epidermal formation trichomes. They selected the 15 species, belonging to 10 families of angiosperm, collected from Oradea. They observed the different types of trichomes like pluticellular. lector T shape, glandular unicellular secretor gland, scaly pluricellular, non ramified in *chrysthemummonifolium*, gazana splendes, *Gynuraauranteaea*, *Symphytumofficinale*, *Solanummelongera*, *Petuniahybridra*. Although it is a good piece of work yet they just emphasized on types of trichomes among these species. But they ignored all other anatomical characters like arrangement of epidermal cells, stomatal cavity, variation in stomatal types and silica bodies etc which could be proved better for identification of these species at taxonomic tool.

Shaheed and M. Illoh, [24] worked on taxonomic study of some species in *Cassinae* (leguminaceae) using leaf epidermal characters. They selected ten species from the genus *Cassia* L. Emend, Gaetrier Sinna miller and chanaaecusio, moench. They collected the specimen from the different localities within the South Western part of Nigeria. Among these species, the aim was to establish some useful daigonistic features. they revealed the excited featuers like guard cell area, stomtal index and frequency of various types of trichomes as well as the length of epidermal cells and types of cell wall. Although it is good piece of work yet they just emphasized on the specific featuers. But they ignored the arrangement of epidermal cells, subsidery cell and silica bodies etc which could be proved better for identification of species.

Rahaman et al., [25] worked on the stomatal complex types, stomatal density and stomatal index in some species of *Dioscorea*. They collected the fresh material from the Grain Research Unit (GRU) of the International Institute of Tropical Agriculture; Ibadan. They observed the three stomatal types like paracytic, anisocytic and tetracytic stomata. However they just emphasized on stomatal types and they ignored other important anatomical character like cell arrangement, types of cells, various types of trichomes and silica bodies etc. That could be better proved for identification of species at taxonomic level.

Hayat et al., [26] studied taxonomic implication of foliar epidermal characteristics with special reference to stomatal variation in the genus *Artemisia* (Asteraceaea) They collected the 24 taxa of the genus *Artemisia* from the herbarium of Quaid–e-Azam University,

Islamabad and herbarium of University of Peshawar. They observed four types of stomata (anomocytic, paracytic, diacytic and paratetracytic) under the light microscopy and scanning electron microscopy. They revealed the variation in epidermal cells of both abaxial and adaxial surfaces and types of stomata. Although this study indicates the anatomical characteristics at taxonomic traits yet they missed the other important character like various types of trichomes and silica bodies etc. That could be proved comprehensive work for the identification of the species of genus *Artemisia* at taxonomic level.

Table 1. List of tree species used for leaf epidermal anatomical study

S. No	Botanical Names	Common/local name	Family Name
1	*Acacia arabica,* Stewart	Waltles, thorn tree	Fabaceae
2	*Alstonia scholaris*, R. Br	Bitter bark, dita bark	Apocynaceae
3	*Albezia* lebbeck. L(bent).	sherin	Fabaceae
4	*Artocarpos integrifolia* spreng	khatal	Moraceae
5	*Averthoa caranmbola* L	kamrak	Oxalidiaceae
6	*Azadirachta indica* L.	neem	Meliaceae
7	*Callistemon lanceolatus*, DC	Bottle brush	Myrtaceae
8	*Cassia fistula* Linn.	amaltas	Fabaceae
9	*Dalbergia sisso*, Roxb	Shisham, tahli	Leguminoaceae
10	*Erythrina subrosa*, L	Coral tree	Leguminoceae
11	*Eucalyptus camaldulensis* Dehnh.	Safeeda. eucalyptus	Myrtaceae
12	*Ficus racemosa* Roxb	Gular, fig tree	Moraceae
13	*Ficus religiosa*, L	Papal	Moraceae
14	*Ficus infectoria* Roxb. sana	Sweetheart tree	Moraceae
15	*Ficus glomerata*, Roxb	Gular, fig tree	Moraceae
16	*Kigelia pinnate (*Jacq.*)*	Gul_e_fanoos	Bignoniaceae
17	*Lagerstroemia indica*, L	Crepe myrtle	Lythraceae
18	*Magnolia grandiflora* L.	Magnolia	Magnoliaceae
19	*Mangifera indica*, L	Mango	Anacardiaceae
20	*Melia azedarachta*, L	Bakain, drek	Meliaceae
21	*Murraya koenigi*, L	Curry –leaf	Rutaceae
22	*Murraya paniculata*, L	mock orange jasmine	Rutaceae
23	*Pongamia glabra,* Vent, Roxb	Suckh-chain	Fabaceae
24	*Thevetia peruviana* (pers). k schum	Kanair	Arecaeae
25	*Putranjiva roxburjhi* wall. tent. FI	Putranjivan, luck tree	Euphrobiaceae
26	*Saraca asoca*, L	Ashoq, asoka	Fabaceaea
27	*Sterculia alata* Roxb	sterculia	Sterculiaceae
28	*Syzygium cumini*, L	Jaman	Myrtaceae
29	*Sygzgium aromarticum* L.	Laung, clove	Myrtaceae
30	*Trewia nudiflora* L		Euphorbiaceae

RESULTS

1. Murraya Paniculata, L

a. Adaxial side

Epidermal cell are arranged rectangular and wavy having double membrane. length of epidermal cell is 110.6µm-56.4µm(Av:83.5µm). Width of epidermal cell 85.4 µm-32.1µm(Av:58.75µm)length of short epidermal cell 64.8µm-29.0µm(Av:46.9µm). Width of short cell 24.46µm-45.6µm(Av:20µm). Non-glandular cells originated cell. Length of trichome from the intracellular cells are 358.21µm-55.8µm(Av:257µm). Width of trichome cells are 55.1µm-20.0µm(Av:35.6µm) Trichome per unit area are 1-2. Anomotetracytic stomata are present. Number of subsidiary cell is four. Number of stomata/unit area is 8-10. Length of stomatal cavity 50.0µm-23.3µm (Av:36.65µm). Width of stomatal cavity 18.8µm-7.6µm (Av:13.2µm). Length of subsidiary cell 124.4µm-40.4µm (Av:82.4µm). Width of subsidiary cell is 62.4µm-19.0µm (Av:40.7µm). Round shape silica bodies are present. Length of silica body is 130.8. µm-260.6(Av:195.7µm). Width of silica body is 41.7µm-37.3µm (Av:39.5. µm).

b. Abaxial side

Epidermal cells are arranged as rectangular and wavy having double membrane. Length of long epidermal cells range from 116.4µm-66.1µm(Av:91.25µm). Width of long epidermal cells 85.4µm-34.1µm(Av:59.75µm). Length of short epidermal cells 60.8µm-25.1µm (Av:42.59µm). Width of short cell 22.4 µm-14.2 µm(Av:18.3µm). Trichomes originated from the intracellular cell having 2-3 cell non –glandular trichome cell. Length of trichome cells are 362.8µm-159.8µm(Av:261.3µm). Width of trichome cells 51.1µm-20.0µm(Av:35.55µm). Number of trichome/unit area is 2-3. Anomotetracytic Stomata are present.). Number of subsidiary cell is four. Number of stomata/unit area is 8-10. Length of stomatal cavity ranges from 50.6µm-23.1µm (Av:35.55µm). Width of stomatal cavity 18.0µm-7.0µm (Av:25.0µm). Length of subsidiary cell 126.4µm-44.0µm (Av:85.05µm). Width of subsidiary cell 62.8µm-19.6µm (Av:41.2µm). Round shape silica bodies are found. Length of silica body is 260.0. µm-138 (Av:199.0µm). Width of silica body 41.7µm-37.3µm (Av:39.5. µm).

2. Murraya Koenigi, L

a. Adaxial side

Trichome is presesent. Polygonal, smooth with single layered cells present. Length of long cell 260.0µm–124.4µm (Av:192.5µm). Width of long cell 85.4µm–34.1µm (Av:60.55µm). Length of short cell 90.0µm–65.3µm (Av:77.65µm). Width of short cell 54.4µm–41.2µm (Av:47.8µm). Anisocytic stomata are present. Number of stomata/unit area are 4-8. Length of stomatal cavity 50.7µm-33.3µm (Av:42.0µm). Width of stomatal cavity 15.5µm-9.0µm (Av:12.25µm). Number of subsidiary cell are three. Length of subsidiary cell 60.2µm-35.5µm (Av:47.85µm). Width of subsidiary cell 45.5µm–21.0µm (Av:33.25µm) Non-glandular cells originated cell from the intracellular cells. Trichome per unit area is 1-2.

Length of trichome are 158.2μm-55.8μm (Av:107.3μm). Width of trichome cells are 50.1μm-20.0μm (Av:35.05μm).

b. Abaxial side

Trichome and silica bodies are absent. Polygonal, smooth with single layered cells are present. Length of long cell 266.6μm–122.4μm (Av:194.5μm). Width of long cell 89.7μm–31.4μm (Av:60.55μm). Length of short cell 90.8μm–64.0μm (Av:77.4μm). Width of short cell 55.9μm–45.5μm (Av:50.7μm). Anisocytic stomata are present. Number of stomata/unit area are 4-8. Length of stomatal cavity 57.0μm-33.2μm (Av:45.1μm). Width of stomatal cavity 15.2μm-9.4μm (Av:12.3μm). Number of subsidiary cell are three. Length of subsidiary cell 62.2μm-35.8μm (Av:49.1μm). Width of subsidiary cell 45.2μm–21.5μm (Av:33.75μm).

3. *Acacia Arabica*, Stewart

a. Adxial side

Silica bodies are absent. Epidermal cells are smooth and rectangular having single membrane. Length of long cell is 189.9μm-106.8μm (Av:148.5μm). Width of long cell is 58.5μm-45.5μm (Av:52.0μm). Length of short epidermal cells are 82.3μm-42.2μm (Av:62.25μm). Width of short cell are 40.0μm-28.8μm (Av:34.4μm). Non-glandular cells are found. Length of trichomes cells are 266.1μm-168.2μm (Av:217.5μm). Width of trichome cell are 32.4μm-16.8μm (Av:24.6μm). Anisocytic stomata are present. Length of stomatal cavity is 58.8μm-18.0μm (Av:38.4μm). Width of stomatal cavity is 18.8μm-7.0μm (Av:12.9μm). Number. of subsidiary cells are three. Length of subsidiary cell is 50.9μm-29.0μm (Av:39.95μm). Width of subsidiary cells are 20.1μm-15.0μm (Av:17.55μm).

b. Abaxial side

Silica bodies are absent. Epidermal cells are arranged smooth and rectangular having single membrane. Length of long epidermal cells 196.7μm-113.5μm (Av:155.1μm). Width of long epidermal cells ranges from 57.5μm-42.5μm(Av:50.0μm). Length of short epidermal cells ranges from 83.9μm-43.5μm (Av:64.2μm). Width of short cell 40.0μm-26.4μm (Av:33.2μm). Trichome originated from cell wall having single cell non-glandular trichome cell. Length of trichome cell 260.5μm-167.5μm (Av:138.2μm). Width of trichomes cell 32.0μm-17.0μm (Av:4.5μm). Trichomes cells/unit area are 3-4. Stomata are anisocytic shape. Number of stomata/unit area are 5-10. Length of stomatal cavity is 58.1μm-18.7μm (Av:38.4μm). Width of stomatal cavity is 18.0μm-7.0μm (Av:12.5μm). Number of subsidiary cells are three. Length of subsidiary cell is 51.9μm-29.4μm (Av:40.65μm). Width of subsidiary cells are 22.1μm-15.3μm (Av:18.85μm)

4. *Alstonia Scholaris*, R. Br

a. Adaxial side

Silica bodies and stomata are absent. Smooth, pentagonal with layered cell are present. Length of long cell 90.9μm-66.0μm (Av:78.45μm). Width of long cell 53.3μm-33.0μm

(Av:43.15μm). Length of short cell 43.0μm-26.8μm (Av:34.9μm). Width of short cell 39.9μm-20.2μm (Av:30.4μm). Number of trichomes/unit area are 10-12. Length of trichome 140.4μm_80.8μm (Av:110.6μm). Width of trichome 103.3μm-55.0μm (Av:79.15μm). (Figure 1).

b. Abaxial side

Silica bodies are absent. Smooth, pentagonal with double layered epidermal cells are present. Length of long cell 99.0μm-66.1μm (Av:82.55μm). Width of long cell 503.3μm-33.5μm(Av:41.9μm). Length of short cell 43.1μm–27.6μm (Av:35.35μm). Width of short cell 40.0μm_20.0μm(Av:30.0μm). Trichome originated from the intracellular cell. Glandular shape of trichomes are present. Number of trichomes/unit area are 10-12. Length of trichome 147.8μm_85.8μm (Av:116.8μm). Width of trichome 106.3μm-55.9μm (Av:81.1μm). Anisocytic stomata are present. Number of stomata/unit area are 1-2. Length of stomatal cavity 66.4μm–31.7μm(Av:48.9μm). Width of stomatal cavity 12.6μm-8.2μm (Av:10.4μm). Number of subsidiary cell are three. Length of subsidiary cell 168.8μm-99.3μm (Av: 134.05μm). Width of subsidiary cell 93.9μm-38.2μm (Av:66.05μm) (Figure 1)

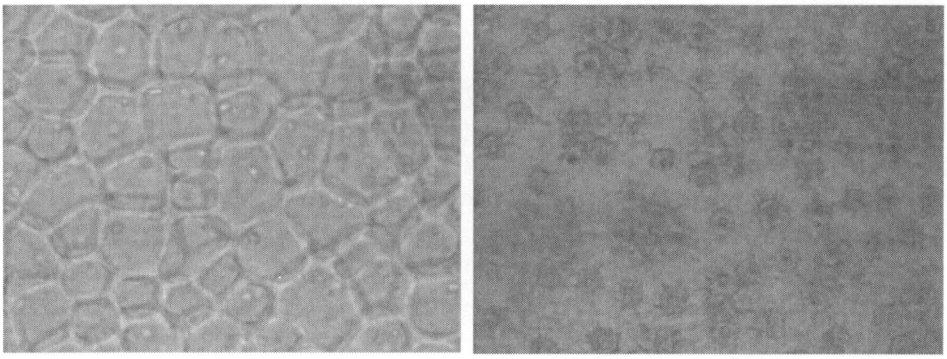

Figure 1. Leaf Epidermal Anatomy of. *Alstonia scholaris*, Abaxial side (Right) Adaxial side (Left).

5. *Albezia Lebbeck*

a. Adaxial side

Silica bodies are absent. Surface, epidermal cells are wavy and irregular cells having double membranes. Length of long cells are 212.8μm-125.9μm(Av:169.35μm). width of long cells are 109.8 μm-85.6 μm(Av:97.7 μm). Length of short cells are 83.3 μm-47.7μm (Av:65.5μm). Width of short cells are 44.1 μm-28.7μm(Av:36.4 μm). Non-glandular trichomes originated from the cell wall. Length of trichomes cell are 220.9 μm- 85.6μm (Av:153.25μm). Width of trihcomes cells 22.4 μm-14.7μm(Av:18.55μm). Brachyparacytic stomata are present. Length of stomatal cavity 30.5μm - 23.0μm (Av:26.75μm). Width of stomatal cavity 10.1 μm-6.9μm (Av:8.5μm). Number. of subsidiary cells are five. Length of subsidiary cells are 143.0μm-90.8μm (Av:116.9μm). Width of subsidiary cells are 71.0μm-55.5(Av:63.25μm).

b. Abaxial side

Silica bodies are absent Leaf epidermal cells are arranged wavy and irregular having double membrane. Length of long cell are 210.8 μm-138.7μm(average: 174.75 μm). Width of long cells are 119.2μm- 85.0 μm (average: 102.1μm). Length of short cells are 93.0μm-44.9μm (average 68.95 μm). Width of short cells are 40.0 μm- 29.0 μm(average 34.5 μm)trichomes are originated from intracellular cell and having a single cell non –glandular trichomes. trichomes per unit area are 6 -8. length of trichomes are 226.0 μm- 91.8 μm(average 158.9 μm). width of trichomes are 22.1 μm -11.2 μm(average 16.65 μm). brachyparacytic stomata are found. length of stomatal cavity are 32.5 μm- 23.1 μm(average 27.8 μm).width of stomatal cavity are 10.0 μm -7.0 μm(average 8.5 μm). No. of subsidiary cell are 5. length of subsidiary cells are 143.7 μm -91.8 μm(average 117.75 μm). width of subsidiary cells are 71.5 μm- 55.2 μm(average 63.35 μm).

6. *Artocarpos Integrifolia* Spreng

a. Adaxial side

Stomata and silica bodies are absent. Hexagonal and smooth with single layered cells are present. Length of long cell 105.0μm–69.9μm (Av:87.45μm). Width of long cell 65.5μm–43.3μm (Av:54.4μm). Length of short cell 79.0μm–58.0μm (Av:68.55μm). Width of short cell 53.3μm–41.0μm (Av; 47.15). Trichome originated from the intracellular cell. Glandualr trichome cells is present. Number of trichome/unit area are 2-3. Length of trichome are 68.8μm–40.2μm (Av:54.5μm). Width of trichome 48.8μm-30.0μm (Av:39.4μm).

b. Abaxial side

Hexagonal and smooth with single layered cells are present. Length of long cell 105.9μm–70.1μm (Av:88.0μm). Width of long cell 65.3μm–46.3μm (Av:55.8μm). Length of short cell 79.2μm–58.7μm (Av:68.95μm). Width of short cell 56.2μm–41.2μm (Av:47.2μm). Trichome originated from the intracellular cell. Glandular trichome cells are present. Number of trichome/unit area are 1-2. Length of trichome are 68.4μm–48.2μm (Av:58.3μm). Width of trichome 48.3μm-30.5μm (Av:39.4μm). Anisocytic stomata are present. Length of stomatal cavity 57.9μm–32.5μm (Av; 45.2μm). Width of stomatal cavity 26.4μm–11.7μm (Av; 19.05μm). Number of subsidiary cells is five. Length of subsidiary cell 91.9μm–54.6μm (Av:76.25μm). Width of subsidiary cell 45.1μm–19.6μm (Av:32.36μm). Round shape silica bodies are present. Length of silica bodies 287.0μm–85.0μm (Av:186μm). Width of silica bodies 233.1μm–54.3μm (Av:143.75μm)

7. *Averthoa Caranmbola* L

a. Adaxial side

Trichome and silica bodies are absent. Hexagonal and smooth with double layered cells are present Length of long cell 178.5μm-83.1μm (Av:130.8μm). Width of long cell 63.3μm - 42.6μm (Av:52.96μm). Length of short cell 51.7μm-37.2μ (Av:44.45μm). Width of short cell 38.5μm -22.4μm (Av:30.45μm). Brachyparacytic stomata are present. Number of stomata/

unit area is 5-7. Length of stomatal cavity 35.5μm-20.5μm (Av:28μm). Width of stomatal cavity 23.3μm-6.9μm (Av:15.1μm). Number of subsidiary cells are five. Length of subsidiary cells 69.1μm-26.3μm (Av:47.7μm). Length of subsidiary cells 79.1μm-28.3μm (Av:53.7μm). Width of subsidiary cells 40.2μm-20.0μm (Av:29.3μm).

b. Abaxial side

Silica bodies are absent. Hexagonal and smooth double layered are present. Length of long cell 178.5μm-82.0μm(Av130.25μm). Width of long cell 77.8μm-37.2μm (Av:57.5μm). Length of short cell 64.6μm-44.1μm (Av:54.35μm). Width of short cell 38.5μm-11.0μm (Av:24.75μm). Trichome originated from the cell wall.2-3 cell non-glandular blunt shape trichome are present, number of trichome per unit area are 2-3. Length of trichome 84.1μm-42.6μm (Av:6.35μm). Width of trichome 64.3μm-39.1μm (Av:51.7μm). Brachyparacytic stomata are present. Number of stomata/unit area are 5-7. Length of stomatal cavity 38.5μm-25.5μm(Av:32μm). Width of stomatal cavity 23.0μm -7.0μm (Av:15μm). Number of subsidiary cells are five. Length of subsidiary cells 79.1μm-28.3μm (Av:53.7μm). Width of subsidiary cells 40.2μm-21.0μm (Av:30.6μm).

8. *Azadirachta Indica* L.

a. Adaxial side

Silica bodies are absent. Smooth and pentagonal cells are present with double layered. Length of long cell 110.7μm-75.0μm(Av:92.85μm). Width of long cell 50.9μm-30.1μm (Av:40.5μm). Length of short cell 62.8μm-32.9μm (Av:47.85μm). Width of short cell 33.0μm-28.3μm (Av:30.65μm). Trichome are originated from the cell wall.2-3. cells non-glandular blunt shape trichomes are present. Length of trichome 390.1μm-210μm (Av:300.25μm). Width of trichome 40.3μm-24.0μm (Av;32.15μm). Anomocytic stomata are present. Number of stomata/unit area are 4-3. Length of stomatal cavity 43.3μm-20.1 (Av:31.7μm). Width of stomatal cavity 18.0μm-10.1μm (Av:14.05μm). Number of subsidiary cell are seven. Length of subsidiary cell 89.0μm-40.7μm (Av:64.85μm). Width of subsidiary cell 56.5μm-30.2μm (Av:43.35μm).(Figure 2)

b. Abaxial side

Trichomes and silica bodies are also present. Pentagonal and smooth double layered cells are present. Length of long cell 113.7μm-76.1μm (Av:94.9μm). Width of long cell 55.9μm-29.1μm(Av:42.5μm). Length of short cell 63.8μm-31.9μm (Av:47.85μm). Width of short cell 32.0μm-26.3μm (Av:29.15μm). Trichome originated from the cell wall.2-3 cells non-glandular blunt shape trichome are present. Number of trichome/unit area are 2-3. Length of trichome 394.1μm-209.4μm (Av:301.7μm). Width of trichome 43.3μm-21.0μm (Av: 32.15μm). Anomocytic stomata are present. Number of stomata/unit area are 4-3. Length of stomatal cavity 45.3 μm-21.1μm(Av:33.2μm). Width of stomatal cavity 18.4μm-10.4μm (Av:14.4μm). Number of subsidiary cell are seven Length of subsidiary cells silica bodies are absent. 89.4μm-44.7μm (Av:67.05μm). Width of subsidiary cell 56.7μm-36.2μm (Av: 46.65μm). Silica bodies are present. Number of silica bodies /unit area are 2-3. Round shape silica shape are present. Length of silica bodies 150.9-44.2μm (Av:97.55μm). Width of silica bodies 110.0μm-50.0μm (Av:80μm) (Figure 2)

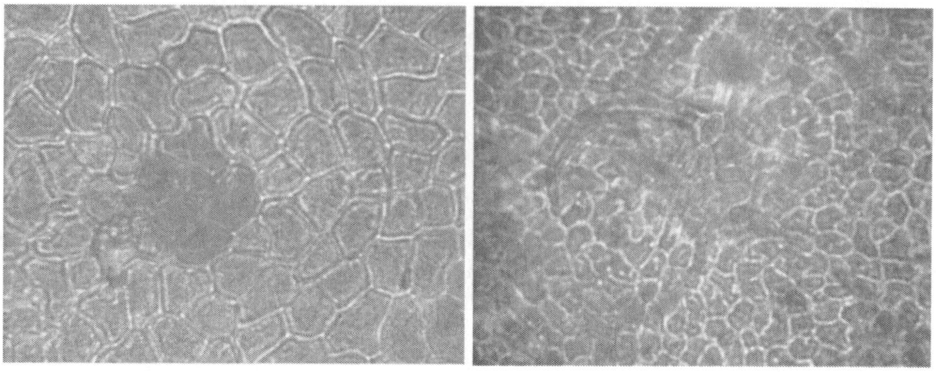

Figure 2. Leaf Epidermal Anatomy of *Azadirachta indica*. Abaxial side (Right) Adaxial side (Left).

9. *Callistemon Lanceolatus*, DC

a. Adaxial side

Silica bodies are absent. Cells is smooth and polygonal having double membranes. Length of long cell are 110.6μm-64.9μm (Av:87.75μm). Width of long cells are 72.5μm-45.5μm (Av:59.0μm). Length of short cells is 57.4 μm-33.3μm (Av:45.37μm). Width of short cells are 36.4μm-22.1μm (Av:29.25μm). Trichomes originated from the intracellular cells that are glandular trichomes. Number of trichomes/unit area is 5-6. Length of trichomes are 39.9μm-30.3μm (Av:35.1μm) Width of trichomes are 361 μm-24.4μm (Av:30.25μm). Anomocytic stomata are present. Number of stomata/unit area are 6-12. Length of stomatal cavity are 40.7μm-18.8μm (Av:29.75μm). Width of stomatal cavity are 10.0μm-6.2μm (Av:8.1μm). Length of subsidiary cells are 96.6μm-45.5μm (Av:71.05μm). Width of subsidiary cells are 51.1μm-42.6μm (Av:46.85μm) (Figure 3)

b. Abaxial side

Silica bodies are absent. Epidermal cells are smooth and polygonal having double membranes. Length of long cells 118.6μm-73.4μm (Av:96μm). Width of long cells 75.2 μm-50.5μm (Av:62.25μm). Length of short cells 57.2 μm-31.1μm (Av:44.15μm). Width of short cells is 39.7 μm-20.5μm (Av:30.1μm). Trichomes originated from the intracellular cells that are glandular trichomes. Number of trichomes/unit area is 5-6. Length of trichomes are 43.9μm-38.3μm(Av:41.1μm). Width of trichomes are 38.1μm-21.4μm(Av:29.75μm). Anomocytic stomata are present. Number of stomata/unit area are 6-12. Length of stomatal cavity are 39.7 μm-18.0μm(Av:28.85μm). Width of stomatal cavity are 10.8μm-6.7μm(Av:8.75μm). Length of subsidiary cells are 96.3μm-45.1μm(Av:70.7μm). Width of subsidiary cells are 55.1μm-43.9μm(Av:49.5 μm) (Figure 3).

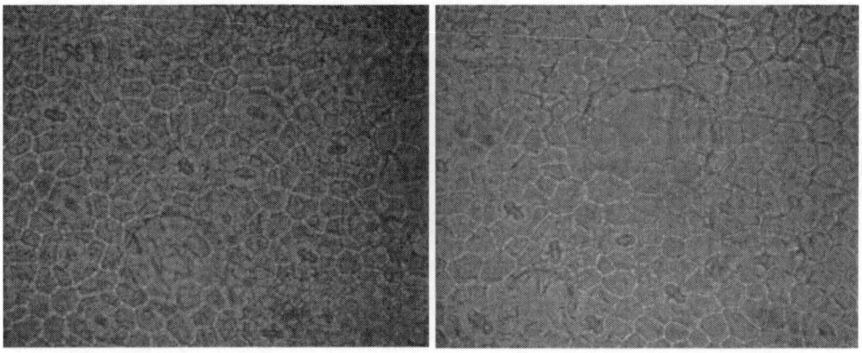

Figure 3. Leaf Epidermal Anatomy of *Callistemon lanceolatus*. Abaxial side (Right) Adaxial side (Left).

10. *Cassia Fistula* Linn

a. Adaxial side

Silica bodies and trichomes are absent. Smooth and hexagonal with single layered cells are present Length of long cell 218.0µm-120.3µm (Av:169.15µm). Width of long cell 70.9µm-54.4µm(Av:62.65µm). Length of short cell 80.4µm–58.8µm (Av:69.6µm). Width of short cell 53.5µm–18.0µm (Av:35.75µm) Anisocytic stomata are popresent. Number of stomata/unit area are 2-3. Length of stomatal cavity 59.0µm–30.9µm(Av:44.95µm). Width of stomatal cavity 20.0µm-10.3µm (Av:15.15µm). Number of subsidiary cells are five. Length of subsidiary cell 220.0µm–85.5µm (Av:152.75µm). Width of subsidiary cell 53.3µm-26.0µm (Av:39.65µm).

b. Abaxial side

Silica bodies are absent. Smooth, hexagonal with single layered cells are present. Length of long cell 218.3µm-123.0µm (Av:170.5µm). Width of long cell 73.9µm-56.4µm (Av:65.15µm). Length of short cell 83.4µm–58.0µm (Av:70.7µm). Width of short cell 57.0µm–17.0µm (Av:37.0µm) Trichome originated from the cell wall. Single cell non-glandular trichomes are present. Number of trichome/unit area are 1-2. Length of trichome 463.2µm-22.4µm (Av:242.6µm). Width of trichome 61.1µm-29.0µm (Av:45.05µm). Anisocytic stomata are popresent. Number of stomata/unit area are 3-4. Length of stomatal cavity 59.2µm–31.9µm (Av:45.55µm). Width of stomatal cavity 20.2µm-11.3µm (Av:15.75µm). Number of subsidiary cells are five. Length of subsidiary cell 220.6µm–80.5µm (Av:150.55µm). Width of subsidiary cell 53.0µm-26.2µm (Av:39.6µm).

11. *Dalbergia Sisso*, Roxb

a. Adaxial side

Polygonal, smooth and single layered cells are present. Length of long cell 150.9µm-46.6µm (Av:98.75µm). Width of long cell 56.6µm-20.5µm (Av:38.55µm). Length of short cell 79.0µm-31.0µm (Av:55µm). Width of short cell 23.0µm-13.3µm (Av:18.15µm).

Glandular trichome is present. Length of trichome 220.7μm-60.2μm (Av:140.45μm). Width of trichome 32.2μm-9.1μm (Av:20.5μm). Number of trichome/unit area 1-2. Anisocytic stomata are present. Number of stomata. unit area are 2-3. Length of stomatal cavity 41.0μm-18.0μm (Av:29.5). Width of stomatal cavity 12.0μm-6.9μm (Av:9.45μm). Number of subsidiary cell are three. Length of subsidiary cell 54.5μm-25.0μm (Av:39.75μm). Width of subsidiary cell 22.0μm-13.6μm (Av:17.8μm). Silica bodies are present. Round shape silica bodies are present. Length of silica bodies 50.6μm-28.8μm (Av:39.7μm). Width of silica bodies 73.3μm-36.0μm (Av:54.65μm). (Figure 4)

b. Abaxial side

Trichomes and silica bodies are present. Polygonal, smooth and single layered cells are present. Length of long cell 159.0μm-46.0μm (Av:102.5μm). Width of long cell 56.0μm-25.0μm (Av:40.5μm). Length of short cell 79.4μm-30.1μm (Av:54.75μm). Width of short cell 23.3μm-13.6μm (Av:18.45μm). Trichomes originated from the intracellular cell. Glandular trichome is present. Length of trichome 221.7μm-66.2μm (Av:146.95μm). Width of trichome 34.2μm-9.0μm (Av:21.6μm). Number of trichome/unit area 1-2. Anisocytic stomata are present. Number of stomata. unit area are 2-3. Length of stomatal cavity 41.2μm-18.8μm(Av:30μm). Width of stomatal cavity 12.2μm-7.9μm(Av:9.6μm). Number of subsidiary cell are three. Length of subsidiary cell 54.5μm-25.2μm (Av:39.85μm). Width of subsidiary cell 22.4μm-13.06μm (Av:17.73μm). Silica bodies are present. Round shape silica bodies are present. Length of silica bodies 57.6μm-28.4μm (Av:46μm). Width of silica bodies 73.4μm-36.4μm (Av:55.1μm) (Figure 4)

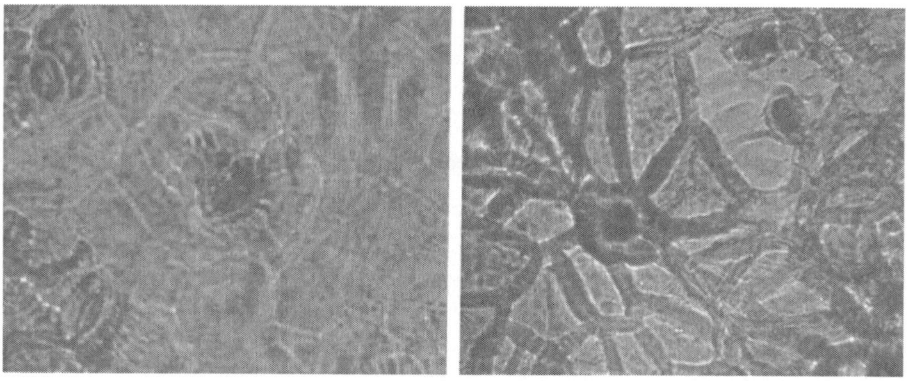

Figure 4. Leaf Epidmeral Anatomy of *Dalbergia sisso*. Abaxial side (Right) Adaxial side (Left).

12. *Erythrina Subrosa*, L

a. Adaxial side

Polygonal and smooth with single layered cells are present. Length of long cell 260.0μm–94.4μm (Av:177.2μm). Width of long cell 102.2μm–60.9μm (Av:81.55μm). Length of short cell 69.9μm-40.9μm (Av:55.4μm). Width of short cell 30.9μm-20.9μm (Av:25.9μm). Trichome originated from the intracellular cell. Glandular trichome are present. Length of trichome 105.0μm-54.4μm (Av:79.7μm). Width of trichome 85.5μm-42.0μm (Av:63.75μm)

Round shape of silica bodies are present. Length of silica bodies 108.0μm-54.0μm (Av:81μm). Width of silica bodies 80.1μm-40.6μm (Av:60.35μm).(Figure 5)

b. Abaxial side

Polygonal and smooth with single layered cells are present. Length of long cell 267.0μm–98.4μm (Av:182.7μm). Width of long cell 100.2μm–69.9μm (Av:85.05μm). Length of short cell 67.9μm-40.7μm (Av:54.3μm). Width of short cell 31.0μm -20.0μm (Av:25.5μm). Trichome originated from the intracellular cell. Glandular trichome are present. Length of trichome 105.1μm-54.8μm (Av:79.95μm). Width of trichome 89.5μm-42.5μm (Av:66μm). Staurocytic stomata are present. Length of stomatal cavity 30.1μm–18.0μm (Av:24.05μm). Width of stomatal cavity 11.2μm-8.2μm (Av:9.7μm). Number of subsidiary cell are four. Length of subsidiary cell Length of subsidiary cell 54.7μm-39.4μm (Av:47.05μm) Width of subsidiary cell 33.4μm–21.0μm (Av:27.2). Round shape of silica bodies are present. Length of silica bodies 108.1μm-54.7μm (Av:81.4μm). Width of silica bodies 88.1μm-42.6μm (Av:65.35μm).(Figure 5).

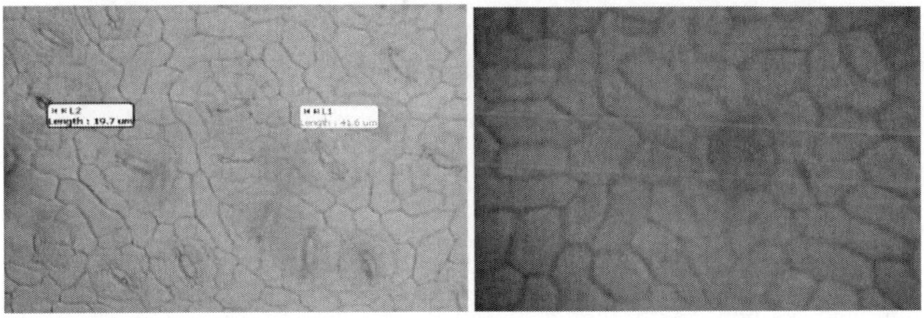

Figure 5. Leaf Epidermal Anatomy of *Erythrina subrosa*. Abaxial side (Right) Adaxial side (Left).

13. *Eucalyptus Camaldulensis* Dehnh.

a. Adaxial side

Silica bodies are absent. Rectangular, smooth and double layered cells are present. Length of long cell 109.5μm-76.1μm (Av:92.8μm). Width of long cell 53.2μm-34.1μm (Av:43.65μm). Length of short cell 72.5μm-30.1μm (Av:51.30μm). Width of short cell 44.0μm-19.1μm (Av:31.55μm). Trichomes originated from the intracellular cell. Number of trichomes/unit area are 1-2. Length of trichome 83.3μm–55.0μm (Av:69.15μm). Width of trichome 48.8μm–50.9μm (Av:49.5μm). Brachyparacytic stomata are present. Number of stomata/unit area are 2-4. Length of stomatal cavity 60.9μm–35.5μm (Av:48.2μm) Width of stomatal cavity18.0μm-13.1μm (Av:15.5μm). Number of subsidiary cell are five. Length of subsidiary cell 149.0μm–50.7μm (Av:99.85μm). Width of subsidiary cell 76.0μm–43.3μm (Av:59.65μm). (Figure 6)

b. Abaxial side

Rectangular, smooth and double layered cells are present. Length of long cell 110.8μm-77.5μm (Av:94.15μm). Width of long cell 52.4μm-36.5μm (Av:44.4μm). Length of short cell 70.5μm-31.3μm (Av:50.65μm). Width of short cell 45.0μm-18.0μm (Av:31.5μm). Trichomes originated from the intracellular cell. Number of trichomes/unit area are1-2. Length of trichome 83.5μm–55.2μm (Av:69.35μm). Width of trichome 48.5μm–53.9μm (Av:51.2μm). Brachyparacytic stomata are present. Number of stomata/unit area are 2-4. Length of stomatal cavity 67.9μm–35.5μm (Av:51.7μm) Width of stomatal cavity18.4μm-13.0μm (Av:15.7μm). Number of subsidiary cell are five. Length of subsidiary cell 149.5μm–57.7μm (Av:103.6μm). Width of subsidiary cell 76.2μm–40.3μm (Av:59.25μm). Irregular shape of silica bodies are present. Length of silica bodies 118.9μm–92.7μm (Av:105.8μm). Width of silica bodies 76.4μm-51.6μm (Av:64.0μm). (Figure 6)

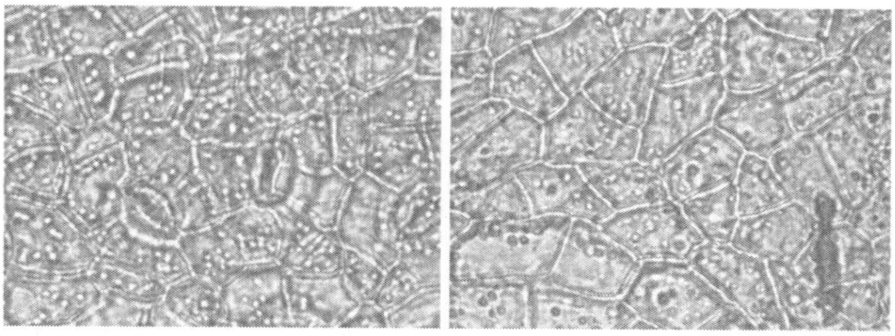

Figure 6. Leaf Epidermal Anatomy of Eucalyptus camaldulensis. Abaxial side (Right) Adaxial side (Left).

14. *Ficus Religiosa*, L

a. Adaxial side

Hexagonal and smooth with single layered cells are present. Length of long cell 190.3μm-96.6μm (Av:143.45μm). Width of long cell 99.0μm-52.2μm (Av:75.6μm). Length of short cell 70.8μm-44.4μm ((AV:75.6μm). Width of short cell 39.0μm-20.1μm (Av:29.55μm). Trichome originated from the cell wall.2-3 cell non-glandular and blunt shape trichome are present. Length of trichome 134.0μm-50.9μm (Av:92.4μm). Width of trichome 73.3μm-18.0μm(Av:45.6m). Staurocytic stomata are present. Number of stomata/unit area are 10-20. Length of stomatal cavity 42.2μm-25.0μm(Av:33.6μm). Width of stomatal cavity 15.0μm-9.0μm(Av:12. μm). Number of subsidiary cell are four. Length of subsidiary cell 140.0μm-36.3μm(Av:88.15m). Width of subsidiary cell 62.2μm-28.8μm(Av:45.5μm). Silica bodies are present. Irregular silica bodies are present. Length of silica bodies 190.1μm-162.2μm (Av:176.15μm). Width of silica bodies 150.9μm-14.4μm (Av:82.65μm) (Figure 7)

b. Abaxial side

Smooth and hexagonal with single layered cells are present. Length of long cell 198.6μm-96.2μm (Av:147.4μm). Width of long cell 99.8μm-59.1μm (Av:76.1μm). Length of short cell 79.8μm-44.8μm (Av:62.3μm). Width of short cell 32.0μm023.0μm (Av:7.55μm). Trichome originated from cell wall.2-3 cell non-glandular and blunt shape trichome are present. Length of trichome 134.5μm-59.9μm (Av:97.2μm). Width of trichome 73.4μm-18.4μm (Av:45.9μm). Staurocytic stomata are present. Number of stomata/unit area are 4-6. Length of stomatal cavity 42.3μm-25.1μm (Av:33.7μm). Width of stomatal cavity 15.0μm-9.1μm (Av:12.05μm). Number of subsidiary cell are four. Length of subsidiary cell 140.9μm-39.3μm (Av:96.1μm). Width of subsidiary cell 62.0μm-28.3μm(Av:49.1μm). Silica bodies are present. Irregular shape of silica bodies is present. Length of silica bodies 197.1µm-162.0µm (Av:179.55µm). Width of silica bodies 153.9µm-14.2µm (Av:84.05µm) (Figure 7)

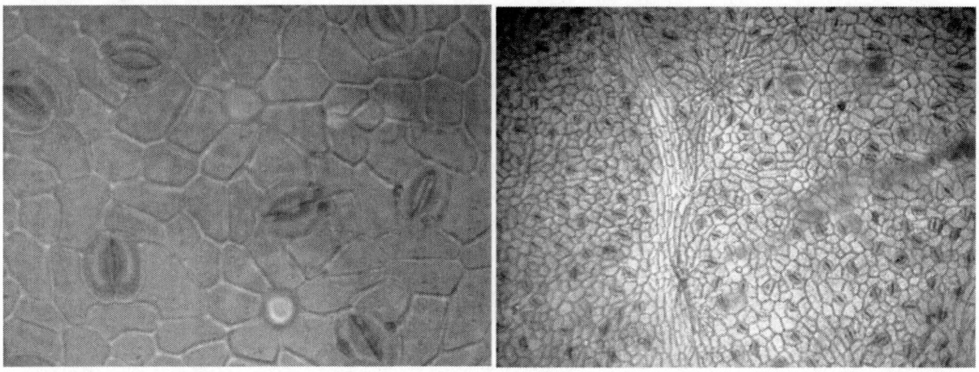

Figure 7. Leaf Epidermal Anatomy of *Ficus religiosa.* Abaxial side (Right) Adaxial side (Left).

15. *Ficus Infectoria* Roxb. Sana

a. Adaxial side

Silica bodies and stomata are absent. Polygonal, smooth with layered cells are present. Length of long cell 160.4µm-80.40µm (Av:120.2µm). Width of long cell 130.60µm–43.3µm (Av:86.65µm). Length of short cell 80.0µm–50.4µm (Av:65.2µm). Width of short cell 53.3µm-34.7µm (Av:44.0µm).

b. Abaxial side

Silica bodies are absent. Polygonal, smooth with layered cells are present. Length of long cell 167.4µm-80.4µm (Av:123.9µm). Width of long cell 130.6µm–45.3µm (Av:87.95µm). Length of short cell 86.0µm–51.4µm (Av:68.7µm). Width of short cell 53.8µm-34.7µm (Av:44.25µm). Trichome originated from the intracellular cell. Glandular trichome cell is present. Length of trichome 148.9µm–71.0µm (Av:109.95µm). Width of trichome 84.1µm–49.8µm (Av:66.95µm). Anomocytic stomata are present. Number of stomata/unit area are 3-5 Length of stomatal cavity 54.5µm-36.5µm (Av:45.5µm). Width of stomatal cavity 22.5µm-15.3µm (Av:18.9µm) Number of subsidiary cell are seven. Length of subsidiary cell 110.2µm-85.4µm (Av:97.8µm). Width of subsidiary cell 64.5µm–24.8µm (Av:44.65µm).

16. *Ficus Glomerata*, Roxb

a. Adaxial side

Stomata and silica bodies are present. Polygonal, smooth with double layered cells are present. Length of long cell 103.0µm–62.2µm (Av:82.6µm). Width of long cell 74.4µm–28.8µm (Av:51.6µm). Length of short cell 39.0µm–29.0µm (Av:34µm). Width of short cell 30.9µm–18.8µm (Av:24.85µm) Trichome originated from the intracellular cell. Glandular trichome cell is present. Length of trichome 140.9µm–71.9µm (Av:106.4µm). Width of trichome 84.0µm–48.8µm (Av:66.45µm). (Figure 8)

b. Abaxial side

Trichome and silica bodies are absent. Polygonal, smooth with double layered cells are present. Length of long cell 103.5μm–62.5μm (Av:83μm). Width of long cell 74.0μm–28.7μm (Av:51.35μm). Length of short cell 39.1μm–29.3μm (Av:34.2μm). Width of short cell 30.3μm–18.6μm (Av:24.45μm). Anomotetracytic stomata are present. Number of stomata/unit area are 4–8 Length of stomata41.2μm–29.4μm (Av:35.3μm). Width of stomatal cavity 22.0μm–13.5μm (Av:17.75μm). Number of subsidiary cell are four. Length of subsidiary cell 39.8μm–26.8μm (Av:33.3μm). Width of subsidiary cell 18.6μm–12.0μm (Av:15.3μm) (Plate 31) (Figure 8)

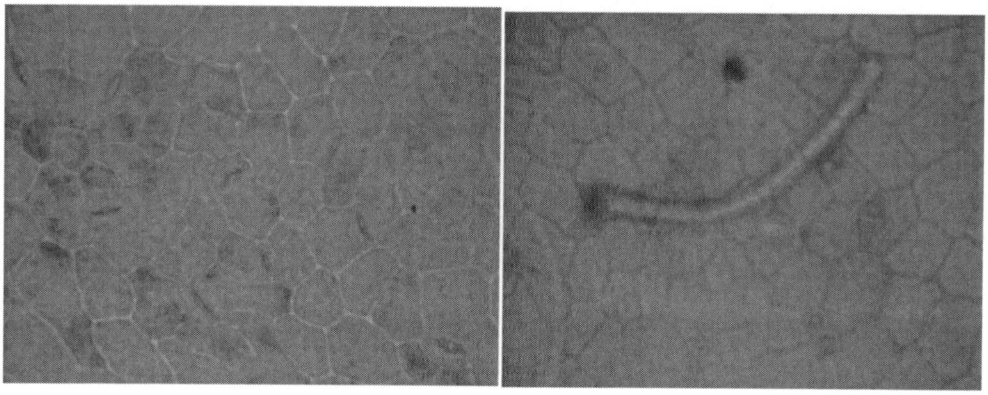

Figure 8. Leaf Epidermal Anatomy of Fiscus glomerata. Abaxial side (Right) Adaxial side (Left).

17. *Ficus Racemosa* Roxb

a. Adaxial side

Silica bodies and stomata are absent Polygonal and smooth with double layered cells are present. Length of long cell 113.0μm-61.0μm (Av:87.0μm). Width of long cell76.0μm–58.8μm (Av:67.4μm). Length of short cell 83.3μm-49.0μm (Av:66.15μm). Width of short cell 44.4μm-26.0μm (Av:35.2μm).. Trichome originated on the cell wall. Glandular trichomes is present. Number of trichome/unit area is 5-6. Length of trichome 70.9μm–32.2μm (Av:55.15m). Width of trichome 40.9μm-30.5μm (Av:35.7).(Figure 9)

b. Abaxial side

Polygonal and smooth with double layered cells are present. Length of long cell 113.5μm-61.1μm (Av:87.3). Width of long cell 76.1μm–58.0μm (Av:67.05μm). Length of short cell 83.9μm-49.5μm (Av:66.70μm). Width of short cell 44.0μm-26.4μm (Av:35.2μm). Trichome originated on the cell wall. Glandular trichomes are present. Number of trichome/unit area is 5-6. Length of trichome 70.0μm–32.0μm (Av:36.0μm). Width of trichome 42.0μm-34.5μm (Av:38.5μm). Anomocytic stomata are present. Length of stomatal cavity 39.3μm–20.6μm (Av:29.95μm). Width of stomatal cavity 14.6μm-11.2μm (Av: 12.9μm). Length of subsidiary cell 59.1μm-44.1μm (Av:51.8μm). Width of subsidiary cell 22.8μm–15.8μm (Av:19.3μm). Round shape silica bodies are present. Length of silica bodies 67.9μm-27.0μm (Av:47.45μm). Width of silica bodies 43.0μm-13.0μm (Av:28.0μm). (Figure 9)

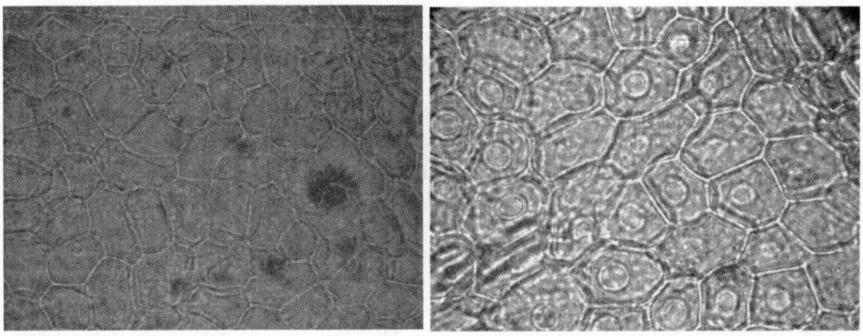

Figure 9. Leaf Epidermal Anatomy of *Ficus racemosa.* Abaxial side (Right) Adaxial side (Left).

18. *Kigelia Pinnate* (Jacq.)

a. Adaxial side

Silica bodies are absent. Smooth rectangular double cells are present. Length of long cells are 106.2μm-60.5μm (Av:86.35μm) Width of long cell 35.5μm-27.5μm (Av:31.5μm). Length of short 50.9μm-33.6μm (Av:42.25μm). Width of short cell 25.8 μm-15.8μm (Av:20.8μm). Trichomes originated from intracellular cell having non-glandular obtuse shape. Number of trichome/unit area is 2-3. Length of trichomes cell 360.0μm-219.0μm (Av:144.7μm). Width of trichomes cells 20.9μm-13.9μm (Av:17.45μm). Anomotetracytic stomata are found. Number of stomata nit area is 2-3. Length of stomatal cavity is 58.0μm-18.1μm (Av:43.1μm). Width of stomatal cavity 18.8μm-6.9μm (Av:12.85μm m). Length of subsidiary cell 50.9μm-29.0μm (Av:35.9μm). Width of subsidiary cell are 22.4μm-15.5μm (Av:18.95μm).

b. Abaxial side

Silica bodies are absent. Epidermal cells are smooth and rectangular having double layered cells are present. Length of long cell 101.2μm-61.6μm (Av:81.4μm). Width of long cells 35.5μm-25.8μm (Av:30.65μm). Length of short cell 51.9μm-32.9μm (Av:44.9μm). Width of short cell is 27.8μm-17.7μm (Av:22.75μm). Trichomes are originated from intracellular cell having non-glandular obtuse shape. Number of trichome/unit area is 5-6. Length of trichomes cell 362.0μm-219.5μm (Av:140.7μm). Width of trichomes cells 21.1μm-13.09μm (Av:17.05μm). Anomotetracytic stomata are found. Number. of stomata/unit area are 5-6. Length of stomata cavity is 58.1μm-18.7μm (Av:38.4μm). Width of stomata cavity 18.0μm-7.0μm (Av:12.5μm). Length of subsidiary cell 51.9μm-29.4μm (average: 40.6μm). Width of subsidiary cell are 22.1μm-15.8μm (average: 18.95μm)

19. *Lagerstroemia Indica*, L

a. Adaxial side

Stomata are absent. Smooth and pentagonal double layered cells are present. Length of long cell 120.1μm-73.6μm (Av:96.85μm). Width of long cell 80.0μm-31.5μm (Av:55.75μm). Length of short cell 68.8μm-32.3μm (Av:50.5μm). Width of short cell 33.3μm-20.5μm

(Av:26.9μm). Trichomes originated from the intracellular cell.2-3 non-glandular blunt shape trichome are present. Number of trichome/unit area are 2-3. Length of trichome 110.4μm-73.3μm (Av:91.85μm). Width of trichome 81.9μm-40.8μm (Av:61.5μm). Round shape silica bodies are present. Number of silica bodies/unit area are 2-3. Length of silica bodies 220.4μm-45.5μm (Av:132.45μm). Width of silica bodies 165.5μm-34.0μm (Av:99.75μm). (Figure 10).

b. Abaxial side

Trichome and silica bodies are present. Pentagonal, smooth and double layered are present. Length of long cell 120.1μm-72.7μm (Av:96.4μm). With of long cell 80.5μm-35.0μm (Av:57.75μm). Length of short cell 69.8μm-39.8μm (Av:54.8μm). Width of short cell 38.9μm-25.2μm (Av:32.05μm). Trichomes are originated from the intracellular cell 2-3 non-glandular blunt shape trichome are present. Number of trichome/unit area are 2-3. Length of trichome 111.4μm-74.3μm (Av:92.85μm). Width of trichome 81.0μm-41.8μm (Av: 61.4μm). Brachyparacytic stomata are present. Number of stomata/unit area are 1-2. Length of stomata cell 46.0μm-33.3μm(Av:39.65μm). Width of stomatal cavity 14.0μm-9.8μm(Av:11.5μm). Number of subsidiary cell are five. Length of subsidiary cell 72.6μm-46.9μm(Av:59.75μm). Width of subsidiary cell 64.1μm-39.0μm(Av:51.5μm). Round shape silica bodies are present. Number of silica bodies/unit area are 2-3. Length of silica bodies 222.4μm-46.5μm(Av:134.45μm). Width of silica bodies 165.5μm-34.0μm(Av:99.75μm). (Figure 10)

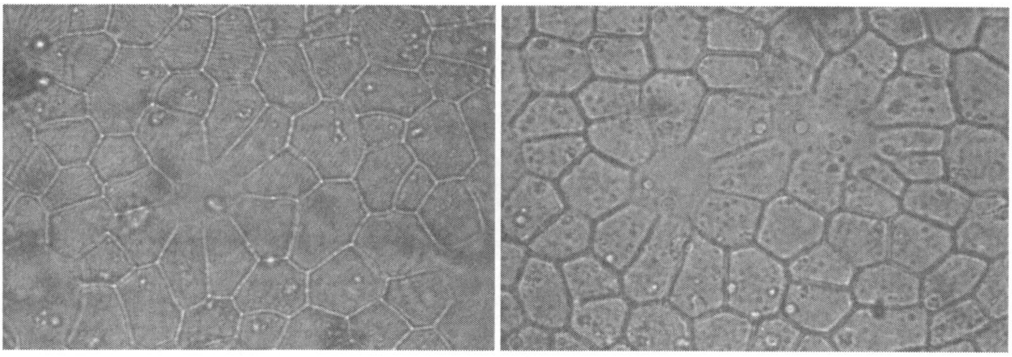

Figure 10. Leaf Epidermal Anatomy of *Lagerstroemia indica.* Abaxial side (Right) Adaxial side (Left).

20. *Magnolia Grandiflora* L.

a. Adaxial side

Epidermal cells are smooth and polygonal having single membranes. Length of long cells 84.5μm-62.4μm (Av:73.45μm). Width of long cells 45.5μm-24.1μm (Av:34.8μm). Length of short cells 52.2μm-30.9μm (Av:41.55μm). Width of short cells 36.4 μm-12.8μm (Av:24.6μm). Glandular type of trichomes originated from the intracellular cell. Length of trichomes 59.9μm-22.6μm (Av:41.25μm). Width of trichomes are 32.6μm-14.7μm (Av: 23.65μm). Anomotetracytic stomata are found. Length of stomatal cavity 30.9μm-21.2μm (Av:26.5μm). Width of stomatal cavity 10.0μm-6.9μm (Av:84.5μm). Stomata/unit area are 3-

4. Length of subsidiary cells 99.0μm-62.9μm (Av:82.5μm). Width of subsidiary cells are 67.9μm-34.7μm (Av:51.3μm).

b. Abaxial side

Silica bodies are absent. Epidermal cells are wavy polygonal having single membrane. Length of long cell 6.5μm-61.8μm (Av:74.15μm). Width of long cells 47.0μm-25.0μm (Av:36μm). Length of short cells 54.3μm-32.3μm (Av:43.3μm). Width of short cells 37.0μm 11.09μm (Av:24.04μm). Trichomes originated from intracellular cell having glandular trichomes. Length of trichomes 61.2μm-24.3μm (Av:42.75μm). Width of trichomes cells 32. 9μm-15.7μm (Av; 24.3μm). Trichomes/unit area are1-2. Anomotetracytic stomata are found. Length of stomatal cavity 31.0μm-20.0μm (Av:25.5μm). Width of stomatal cavity 10.0μm-6.0μm (Av:8μm). Stomata/unit area are 3-4. Length of subsidiary cells 99.5μm-62.19μm (Av:80.8μm). Width of subsidiary cells are 67.0μm-37.0μm (Av:52.5μm).

21. *Mangifera Indica*, L

a. Adaxial side

Trichomes and silica bodies are absent. Polygonal, smooth with single layered cells are present. Length of long cell 109.0μm-78.8μm (Av:93.05μm). Width of long cell 69.5μm-53.3μm (Av:61.4μm). Length of short cell 54.4μm-32.2μm (Av:43.35μm). Width of short cell 46.3μm–26.4μm (Av:36.35μm).

b. Abaxial side

Trichomes and silica bodies are absent. Polygonal, smooth with single layered cells are present. Length of long cell 109.2μm-78.9μm (Av:94.05μm). Width of long cell 69.9μm-55.3μm (Av:62.6μm). Length of short cell 54.8μm-30.1μm (Av:42.45μm). Width of short cell 45.0μm–29.0μm (Av:37.0μm). Anomotetracytic stomata are present. Length of stomatal cavity 57.2μm–30.9μm (Av:44.05μm). Width of stomatal cavity 22.4μm–12.2μm (Av: 17.3μm). Number of subsidiary cells are four. Length of subsidiary cell 210.6μm-56.2μm (Av:133.4μm). Width of subsidiary cell 66.9μm–22.5μm (Av:44.75μm).

22. *Melia Azedarachta*, L

a. Adaxial side

Silica bodies are absent Rectangular, smooth with double layered cells are present. Length of long cell 224.0μm–110.5μm (Av:167.25μm). Width of long cell 104.4μm–80.1μm (Av:92.25 μm). Length of short cell 90.5μm–69.6μm (Av:80.05μm). Width of short cell 65.3μm–30.4μm (Av:47.85). Anomotetracytic stomata are present. Length of stomatal cavity 36.6μm–16.4μm (Av:26.5μm). Width of stomatal cavity 17.0m–8.2μm (Av:12.66μm). Number of subsidiary cells are four. Length of subsidiary cell 55.0μm-20.6μm (Av: 37.85μm). Width of subsidiary cell 24.4μm-18.0μm (Av:20.2μm). Irregular silica bodies are present. Length of silica bodies 190.6μm_50.1μm (Av:120.35μm). Width of silica bodies 120.8μm–39.9μm (Av:85.35m).

b. Abaxial side

Rectangular, smooth with double layered cells are present. Length of long cell 224.1μm–115.4μm (Av:169.75μm). Width of long cell 104.4μm-81.1μm (Av:92.65 μm). Length of short cell 97.5μm–69.5μm (Av:83.5μm). Width of short cell 63.3μm–31.1μm (Av:47.2μm). Trichome are originated from the intracellular cell. Number of trichome/unit area are 1-2. Length of trichome cell 484.3μm–218.4μm (Av:315.35μm). Width of trichome cell 76.0μm-37.0μm (56.0μm). Anomotetracytic stomata are present. Length of stomatal cavity 39.6μm–16.4μm (Av:28.0μm). Width of stomatal cavity 17.1μm–8.0μm (Av:18.55μm). Number of subsidiary cells are four. Length of subsidiary cell 55.5μm-29.6μm(Av:42.55μm). Width of subsidiary cell 22.4μm-18.5μm (Av:20.45μm). Irregular silica bodies are present. Length of silica bodies 190.6μm_55.1μm (Av:122.85μm). Width of silica bodies 129.8μm–41.0μm (Av:85.4μm)

23. *Pongamia Glabra*, Vent, Roxb:

a. Adaxial side

Silica bodies are absent. Irregular and wavy with double layered cells are present. Length of long cell 178.1μm-72.3μm (Av:125.2μm). Width of long cell 58.5μm-28.3μm (Av: 43.4μm). Length of short cell 56.6μm–39.0μm (Av:47.8μm). Width of short cell 33.3μm – 17.8μm (Av:51.4μm). Trichomes originated from the cell wall.2-3 cells, non-glandular trichomes are present. Number of trichome/unit area are 1-2. Length of trichome 36.4μm-20.2μm (Av:28.3μm). Width of trichome 39.0μm-15.9μm (Av:27.4μm). Amphianisocytic stomata are present. Number of stomata/unit area are 2-3. Length of stomata 50.9μm-29.9μm (Av:40.45μm). Width of stomatal cavity 15.3μm-9.6μm (Av:12.55μm). Number of subsidiary cell are five. Length of subsidiary cell 75.7μm-30.8μm (Av:53.25μm). Width of subsidiary cell 33.6μm-24.0μm (Av:28.8μm).

b. Abaxial side

Trichomes and silica bodies are present. Epidermal cell are wavy and irregular in shape with double layered. Length of long cell 188.1μm-82.6μm (Av:135.35μm). Width of long cell 58.5μm–28.3μm (Av:43.4μm). Length of short cell 66.9μm-45.5μm (Av:56.2μm). Width of short cell 38.6μm-17.3μm (Av:27.95). Trichomes originated from the cell wall.2-3 cells, non-glandular trichomes are present. Number of trichome/unit area are 1-2. Length of trichome 36.7μm-23.2μm (Av:29.95μm). Width of trichome 39.4μm-15.0μm (Av:27.2μm). Amphian-isocytic stomata are present. Number of stomata /unit area are 2-3. Length of stomata 50.0μm-26.9μm (Av:38.45μm). Width of stomatal cavity 15.3μm-9.3μm (Av: 12.3μm). Number of subsidiary cell are five. Length of subsidiary cell 75.6μm-31.8μm (Av:53.7μm). Width of subsidiary cell 33.0μm-24.0μm (Av:28.5μm). Irregular silica bodies are present. Length of silica bodies 1074μm-35.0μm (Av:71.05μm). Width of silica bodies 58.9μm-25.9μm (Av:42.4μm). (Figure 11)

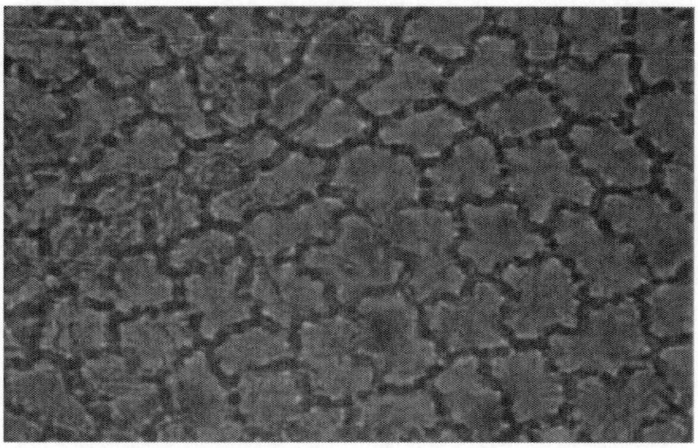

Figure 11. Leaf Epidermal Anatomy of Pongamia glabra, Abaxial side (Right) Adaxial side (Left).

24. *Putranjiva Roxburjhi* Wall. Tent. FI:

a. Adaxial side

Polygonal, smooth and double layered cells are present Length of long cell 247.0μm-62.2μm (Av:154.6μm). Width of long cell 80.0μm-51.9μm (Av:65.96μm). Length of short cell 49.0μm-29.0μm (Av:39.0μm). Width of short cell 29.9μm-12.0μm (Av:20.95μm). Trichome originated from the cell wall. Single non-glandular trichome is present. Number of trichomes/unit area are 2-3. Length of trtichome cell 190.4μm-182.2μm (Av:186.35μm). Width of trichome 17.0μm-15.5μm (Av:16.25μm). Brachyparacytic stomata are present. Number of stomata/unit area are 1-2. Length of stomatal cavity 34.4μm-17.6μm (AV: 26.05μm). Width of stomatal cavity 10.9μm-9.0μm (Av:9.95μm). Number of subsidiary cell are five. Length of subsidiary cell 169.9μm–83.3μm (Av:126.6μm). Width of subsidiary cell 85.7μm-21.0μm (Av:53.35μm). Silica bodies are present. Irregular shape of silica bodies is present. Length of silica bodies 120.7μm-76.6μm (Av:98.65μm). Width of silica bodies 88.8μm-53.5μm (Av:71.15μm). (Plate 48)

b. Abaxial side

Polygonal, smooth and double layered cells are present. Length of long cell 247.5μm-60.2μm (Av:153.85μm). Width of long cell 81.0μm-50.9μm (Av:65.95μm). Length of short cell 49.3μm-29.1μm (Av:39.2μm). Width of short cell 29.0μm-12.0μm (Av:20.5μm). Trichome originated from the cell wall. Single non-glandular trichome are present. Number of trichomes/unit area are 2-3. Length of trtichome cell 193.4μm-182.1μm (Av:187.75μm). Width of trichome 17.2μm-15.3μm (Av:16.25μm). Brachyparacytic stomata are present. Number of stomata/ unit area are 1-2. Length of stomatal cavity 31.4μm-17.1μm (AV:24.25μm). Width of stomatal cavity 10.8μm-9.1μm (Av:9.95μm). Number of subsidiary cell are five. Length of subsidiary cell 179.1μm–81.3μm (Av:130.2μm). Width of subsidiary cell 80.7μm-21.1μm (Av:50.9μm). Silica bodies are present. Irregular shape of silica bodies is present. Length of silica bodies 123.7μm-76.1μm (Av:99.9μm). Width of silica bodies 88.0μm-52.5μm (Av:70.25μm). (Figure 12)

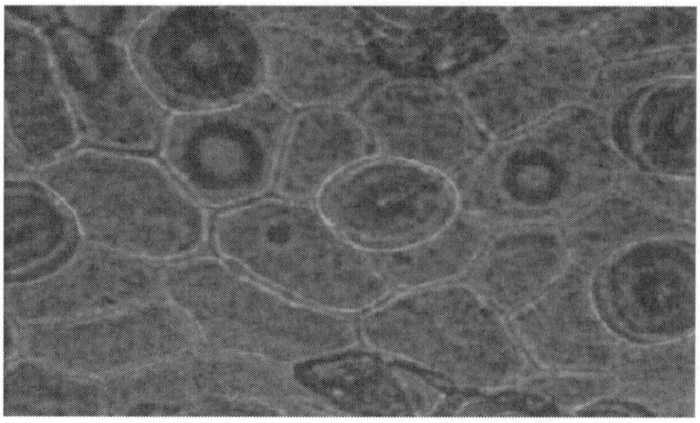

Figure 12. Leaf epidermal anatomy of Putranjiva roxburjhi (Abaxial).

25. *Saraca Asoca*, L

a. Adaxial side

Cells are wavy and rectangular having double membranes. Length of long cell 160.0μm-126.4μm (Av:143.2μm). Width of long epidermal cells 91.2μm-62.4μm (Av:76.8μm). Length of short cells 74.5μm-48.1μm (Av:61.3μm). Irregular shape of silica bodies is found. Length of silica body 220.6μm-59.0μm (Av:139.8μm). Width of silica body 140.4μm-35.5μm (Av:87.95μm) (Figure 13)

b. Abaxial side

Leaf epidermal cells are wavy and irregular having double membrane. Length of long epidermal cell 159.0μm-124.6μm (Av:141.8μm). Width of long epidermal cell are 72.5μm-43.8μm (Av:58.15μm). Trichomes cells are originated from intracellular cell that having glandular trichomes. of trichomes cells are 55.0μm-45.9μm (Av:50.45μm). Width of trichomes cells 33.1μm-14.3μm (Av:23.7μm). Stomata are brachyparacytic shape. Length of stomatal cavity 38.5μm-31.4μm (Av:34.95μm). Width of stomatal cavity 17.3μm-13.3μm (Av:15.3μm). Number of subsidiary cells is five. Length of subsidiary cell 93.05μm-5.0μm (Av:74μm). Width of subsidiary cells 45.0μm-21.9μm (Av:33.45μm). Stomata per unit area are 2-3. Irregular shape of silica bodies are found. Length of silica body is 226.0μm-59.2μm (Av:142.6μm). Width of silica body is 142.4μm-35.0μm (Av:88.7μm) (Figure 13)

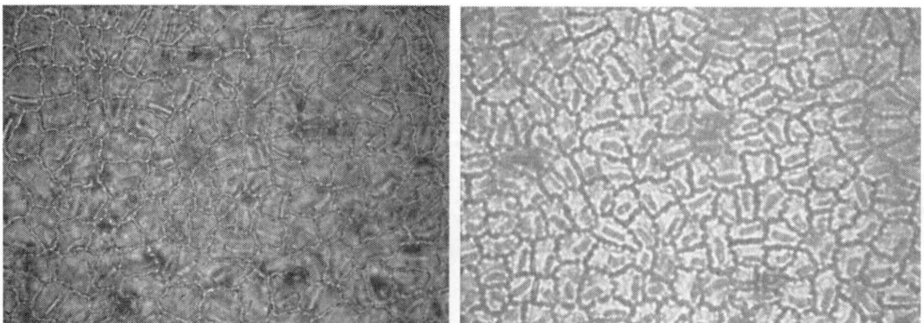

Figure 13. Leaf Epidermal Anatomy of *Saraca asoca*. Abaxial side (Right) Adaxial side (Left).

26. *Sterculia Alata* Roxb

a. Adaxial side

Trichomes and stomata are absent. Epidermal cells are polygonal and smooth having double. Length of long cell 160.0μm-122.4μm (Av:141.2μm). Width of long cell 77.9μm-43.3μm (Av:60.66μm). Length of short cell 107.7μm-66.4μm (Av:87.05μm). Width of short cell 6.0μm-30.9μm (Av:18.45μm). Round shape of silica bodies are present. Length of silica bodies 114.0μm-28.8μm (Av:71.4μm). Width of silica bodies 85.5μm–59.9μm (Av:72.7μm). (Figure 14)

b. Abaxial side

Epidermal cells are polygonal and smooth having double membrane. Length of long cell 162.0μm-120.4μm (Av:141.2μm). Width of long cell 77.0μm-44.3μm (Av:60.66μm). Length of short cell 107.6μm-62.4μm (Av:85.0μm). Width of short cell 6.0μm-31.7μm (Av:46.2μm). Trichome originated from the intracellular cell. Glandular trichome are present. Number of trichomes/unit area are 4-6. Length of trichome cell 348.5μm-83.5μm (Av:21.6μm). Width of trichome cell 49.5μm-23.4μm (Av:36.45μm). Anomotetracytic stomata are present. Number of stomata/unit area are 10-12. Length of stomatal cavity 30.7μm-19.4μm (Av:25.05μm). Width of stomatal cavity 7.3μm-6.7μm (Av:7μm). Number of subsidiary cells are four. Length of subsidiary cell 101.4μm-44.4μm (Av:72.9μm). Width of subsidiary cell53.0μm-22.6μm (Av:37.8μm). Round shape of silica bodies are present. Length of silica bodies 114.8μm-28.0μm (Av:71.4μm). Width of silica bodies 85.7μm–51.9μm (Av:68.8μm). (Figure 14)

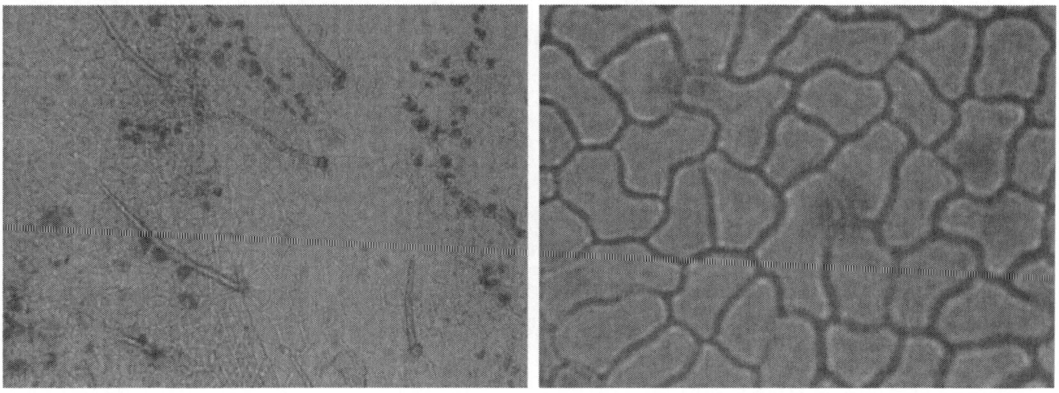

Figure 14. Leaf Epidermal Anatomy of *Sterculia alata.* Abaxial side (Right) Adaxial side (Left).

27. *Syzygium Cumini*, L

a. Adaxial side

Trichomes are absent. Irregular, smooth with single layered cells are present. Length of long cell 79.2μm–39.0μm (Av:59.1μm). Width of long cell 48.0μm–25.5μm (Av:36.75μm). Length of short cell 40.1μm-30.0μm (Av:35.05μm). Width of short cell 30.9μm–19.0μm (Av:24.95μm) Anomotetracytic stomata are present. Number of stomata/unit area are 5–6.

Length of stomatal cavity 41.4µm-24.6µm (Av:33.0µm). Width of stomatal cavity 29.9µm–15.5µm (Av:22.7µm). Number of subsidiary cell are four. Length of subsidiary cell 39.9µm–29.0µm (Av:34.45µm). Width of subsidiary cell 18.8µm-12.6µm (Av:15.7µm). (Figure 15)

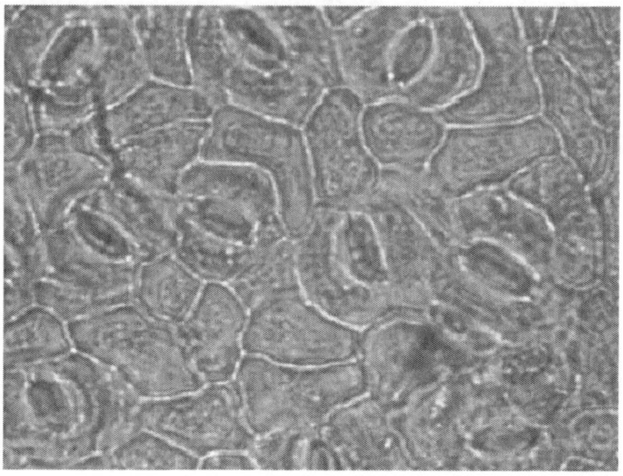

Figure 15. Leaf Epidermal Anatomy of *Syzygium cumini.* L (Adaxial).

b. Abaxial side

Trichome are absent. Irregular, smooth with single layered cells are present. Length of long cell 79.8µm–39.1µm (Av:69.45µm). Width of long cell 48.4µm–29.5µm (Av:38.95µm). Length of short cell 40.0µm-29.4µm (Av:34.7µm). Width of short cell 30.4µm–19.1µm (Av:24.75µm). Anomotetracytic stomata are present. Number of stomata/unit area are 30–40. Length of stomatal cavity 41.0µm-22.4µm (Av:31.7µm). Width of stomatal cavity 29.0µm–15.0µm (Av:22.0µm). Number of subsidiary cell are four. Length of subsidiary cell 38.9µm–29.5µm (Av:34.2µm). Width of subsidiary cell 18.0µm-12.0µm (Av:15.0µm). Round shape silica bodies are present. Number of silica bodies/unit area are 2-3. Length of silica bodies are 73.0µm–25.1µm (Av:49.05µm). Width of silica bodies 38.0µm-20.9µm (Av:29.45µm).

28. *Sygzgium Aromarticum*, L

a. Adaxial side

Irregular wavy with double layered cells are present, Length of long cell 206.0µm-99.0µm (Av:152.5µm). Width of long cell 88.0µm–33.3µm (Av:60.65µm) Length of short cell 75.5µm-30.8µm (Av:5315µm) Width of short cell 50.0µm–23.3µm (Av:36.65µm) Trichome originated from the cell wall. Number of trichome /unit area are 2-3.2-3 cells non-glandular trichome are present. Length of trichome 123.3µm-109.1µm (Av:116.2µm) Width of trichome 75.2µm–65.3µm (Av:70.25m). Irregular shape of silica bodies is present. Length of silica bodies 98.0µm-51.8µm (Av:74.9µm) Width of silica bodies 86.6µm-30.9µm (Av:58.75.).

b. Abaxial side

Irregular wavy with double layered cells are present, Length of long cell 206.3µm–90.9µm (Av:148.6µm). Width of long cell 88.9µm–33.6µm (Av:61.25µm) Length of short cell 75.3µm-36.8µm (Av:56.05µm) Width of short cell 50.9µm–23.9µm (Av:37.4µm). Trichome originated from the cell wall. Number of trichome/unit area are 2-3.2-3 cells non-glandular trichome are present. Length of trichome 122.3µm-109.9µm (Av:111.6µm) Width of trichome 77.2µm–65.2µm (Av:71.2µm). Actinocytic stomata are present, Length of stomata cavity 39.01µm–22.5µm (Av:30.75µm) Width of stoimatal cavity 18.05µm–12.3µm (Av:15.1µm) Number of subsidiary cell are 4-5. Length of subsidiary cell 91.5µm-55.6µm (Av:73.55µm). Width of subsidiary cell 45.5µm-20.9µm (Av:33.2µm). Irregular shape of silica bodies are present. Length of silica bodies 98.2µm-50.2µm (Av:74.2µm) Width of silica bodies 86.0µm-30.0µm (Av:58µm).

29. *Thevetia Peruviana* (Pers). K Schum

a. Adaxial side

Stomata and trichome are absent. Irregular, wavy and double membrane cells are present. Length of long cell 210.6µm–123.3µm (Av:166.95µm). Width of long cell 94.4µm–39.01µm (Av:66.7µm). Length of short cell 88.0µm_53.3µm (Av:70.65µm). Width of short cell 54.0µm-27.0µm (Av:40.5µm) Irregular shape of silica bodies is present. Length of silica bodies 260.2µm-152.2µm (Av:206.2µm). Width of silica bodies 125.5µm–55.5µm (Av: 90.5µm).(Figure 16)

b. Abaxial side

Irregular, wavy and double membrane cells are present. Length of long cell 216.0µm–123.0µm (Av:169.5µm). Width of long cell 94.3µm–39.7µm (Av:67.0µm). Length of short cell 88.1µm-53.7µm (Av:70.9µm). Width of short cell 54.4µm-27.2µm (Av:40.8µm). Trichome originated from the intracellular cell.. Number of trichome/unit area are 2-3. Glandular trichome are presnt. Length of trichome 68.4µm-48.2µm (Av:58.2µm). Width of trichome 64.4µm-50.5µm (Av:57.4µm). Anomotetracytic stomata are present, Number of stomata /unit area are5-6. Length of stomatal cavity 51.5µm–26.2µm (Av:38.85µm). Width of stomatal cavity 25.5µm–14.6µm (Av:20.5µm). Number of subsidiary cell are four. Length of subsidiary cell 197.6µm–94.0µm (Av:145.8µm). Width of subsidiary cell 67.0µm–30.6µm (Av:48.8µm). Silica bodies are present. Irregular shape of silica bodies is present. Length of silica bodies 261.2µm-152.0µm (Av:206.6µm). Width of silica bodies 125.0µm–55.0µm (Av:90.0µm) (Figure 16)

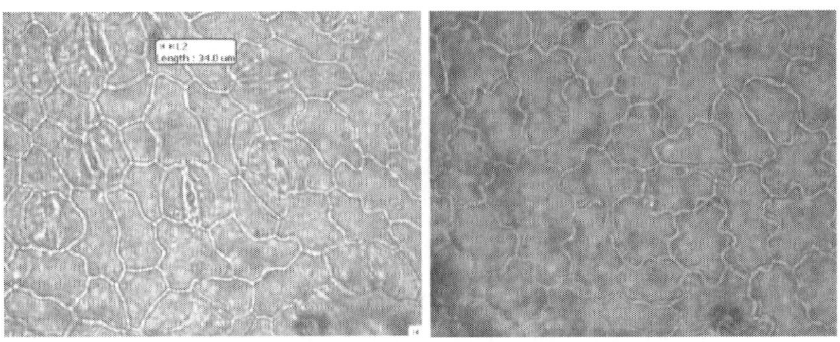

Figure 16. Leaf Epidermal Anatomy of *Thevetia peruviana* . Abaxial side (Right) Adaxial side (Left).

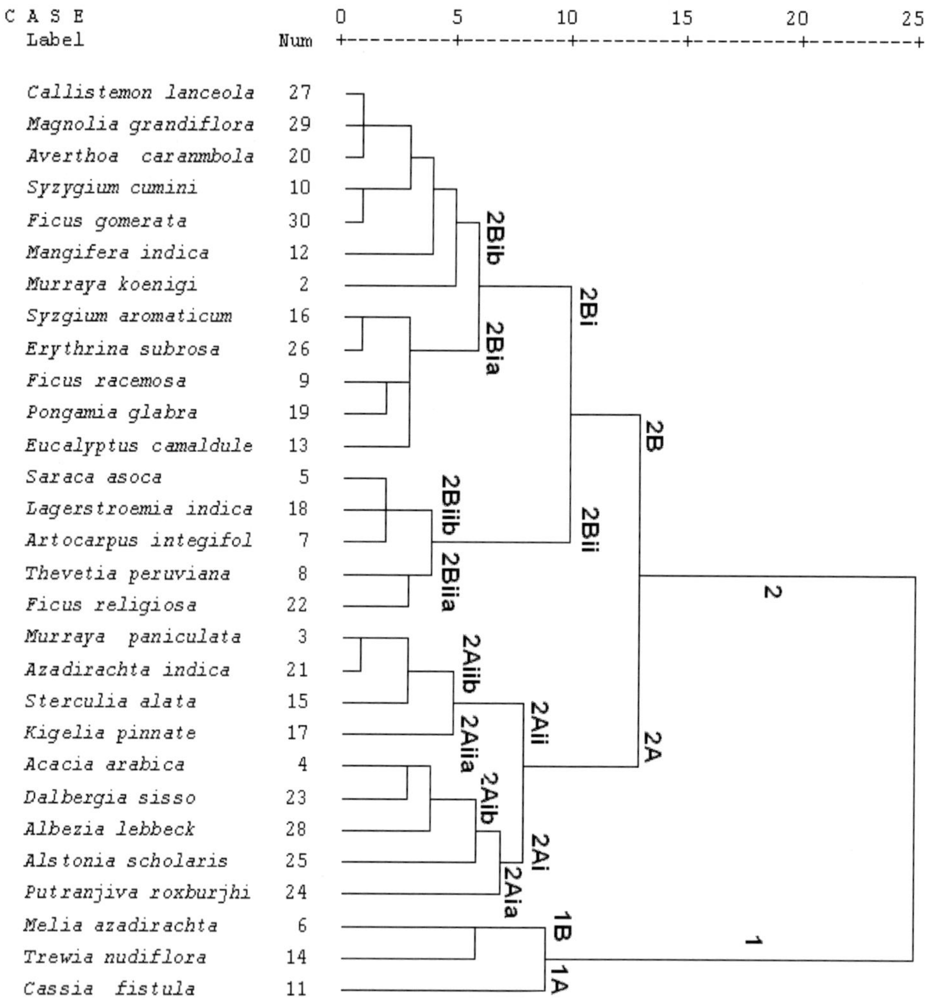

Figure 17. Cluster analysis for the tree flora of District Lahore based on leaf epidermal anatomical characters.

30. *Trewia Nudiflora* L

a. Adaxial side

Irregular, wavy and double layered cells are present. Length of long cell 191.0µm-106.6µm (Av:148.8µm). Width of long cell 94.4µm-52.2µm (Av:73.3µm). Length of short cell 96.6µm-64.4µm (Av:80.5µm). Width of short cell 64.4µm-30.9µm (Av:47.65µm) Trichcomeoriginated from the cell wall. Single non-glandular trichome are present. Length of trichome 610.6µm-123.3µm (Av:366.95µm). Width of trichome 47.7µm-10.9µm (Av: 29.3µm). Amphicyclocytic stomata is present. Length of stomatal cavity 55.2µm-21.1µm (Av:38.15µm). Width of stomata cavity 15.5µm-11.8µm (Av:13.65µm). Number of subsidiary cell are three. Length of subsidiary cell 80.9µm-52.2µm (Av66.5m). Width of subsidiary cell 40.5µm–25.7µm (Av:33.1µm) Irregular shape of silica bodies are present Length of silica bodies 76.8µm–49.0µm (Av:62.9µm) width of silica bodies 50.9µm–12.6µm (Av:31.9µm.

ab Abaxial side

Irregular, wavy and double layered cells are present. Length of long cell 191.1μm-101.6μm (Av:146.35μm). Width of long cell 94.9μm-50.2μm (Av:72.55μm) Length of short cell 96.9μm-64.5μm (Av:80.7μm). Width of short cell 64.4μm-31.0μm (Av:47.7μm) Trichcome originated from the cell wall. Single non-glandular trichome are present. Length of trichome 610.1μm-123.0μm (Av:366.55μm). Width of trichome 47.0μm-10.4μm (Av: 28.7μm). Amphicyclocytic stomata is present. Length of stomatal cavity 55.2μm-21.1μm (Av:38.15μm). Width of stomata cavity 15.2μm-11.3μm (Av:13.25μm). Number of subsidiary cell are three. Length of subsidiary cell 85.9μm-52.1μm (Av6.9μm). Width of subsidiary cell 44.5μm-25.6μm (Av:35.05μm) Irregular shape of silica bodies are present Length of silica bodies 76.4μm-49.4μm (Av:62.9μm) width of silica bodies 50.0μm-12.8μm (Av:31.4μm) (Plate 59)

Co-relation of tree species from order Fabales, Sapindales and Myrtales based on leaf epidermal anatomical character was calculated by using Microsoft Excel software and represented in Table 2. Range co-relation varies from 0.02 -0.97. *Mangiferaindica* and *Murrayapaniculata* of family Anacardiaceae and Rutaceae respectively showed maximum co-relation index 0.97. Similarly least co-relation was found between *Azadirachta indica* and *Ficusin fectoria* (0.02). *Azdirachta indica* and *Mangifera indica* were two more closely related species with similarity index of 0.96. Two important species of *Murraya* found in Lahore district are *Murraya koenigii* and *Murraya paniculata* are quite different from each other on the basis of leaf epidermal anatomical characters. The Co-relation index of these two species is 0.13. Mrytaceae economically and medicinally is important family. Two important species of this family *Syzygium cumini* and *Syzygium aromaticum* were included in the present investigation and it was found that co-relation between these two species based on leaf epidermal anatomical character is 0.60. *Cassiafistula* and *Acaciaarabica*. Although belonging to same family but they have quite different anatomical characters therefore index of correlation is only 0.24.

Table 3 shows co-relation of tree species from order oxalidales, Malvales, Magnoales, Genitales, Lamiales and Rosales based on leaf epidermal anatomical character. *Sterculiaalata* and T*rewianudifolia* showed maximum co-relation on leaf epidermal anatomical character at 0.94. *Artocarpusintegifolia* and *Kigeliapinnate* showed least co-relationship on anatomical character at 0.06. *Thevetiaperuvianal Alstonlascholaris* and *Averthoacrambola*, *Magnoliagradifolia* showed co-relation of 0.82 and 0.83 respectively. *Ficusreligiosa* and *ficusracemosa* belonging to family Moracee and co-relation index for them is 0.77.

A dendrogram was constructed on the basis of all the anatomical characters observed on both leaf abaxial and adaxial sides of tree species from Lahore district (Figure 17). At 100% linkage distance thirty species were divided into two large subclusters 1 and 2. Cluster 1 is further divided into 1A containing *Cassiafistula* and 2B with *Trewia nudifolia* and *Melia azadirachta* at 38% linkage distance. Cluster 2 divided into 2 groups 2 A and 2B at 56% genetical distance. However 36% distance divided 2A two sub groups, 2Ai and 2Aii at 36%. 2Ai also divided into two groups 2Aia having *Putranjiva roxboughii* and 2Aib with *Alstonia scholaris*, *Albezia lebeck*, *Dalbergia sisso* and *Accacia arabica* at 40%. 2Aii group further divided into 2Aiia and 2Aiib at 30% distance. 2Aiia separate into *kegelia pinnata* and 2Aiib divided the species into *Sterculia alata* and *Murraya paniculata*.2B group is further divided

Table 2. Co-relation of Tree species from order Fabales, Sapindales and Myrtales based on Leaf epidermal anatomical characters

Species Names	1	2	3	4	5	6	7	8	9	10	11	12	13	14	15	16
1	1															
2	0.13	1														
3	0.65	0.16	1													
4	0.06	0.97	0.03	1												
5	0.22	0.94	0.02	0.96	1											
6	0.41	0.61	0.59	0.54	0.52	1										
7	0.80	0.39	0.82	0.28	0.34	0.59	1									
8	0.69	0.18	0.54	0.14	0.24	0.67	0.51	1								
9	0.63	0.69	0.41	0.67	0.76	0.78	0.69	0.60	1							
10	0.65	0.73	0.61	0.65	0.70	0.56	0.86	0.37	0.76	1						
11	0.43	0.88	0.22	0.88	0.93	0.60	0.50	0.37	0.82	0.80	1					
12	0.73	0.53	0.36	0.51	0.66	0.67	0.65	0.69	0.93	0.68	0.76	1				
13	0.81	0.40	0.55	0.36	0.48	0.76	0.67	0.87	0.85	0.60	0.62	0.91	1			
14	0.50	0.82	0.26	0.80	0.86	0.42	0.61	0.19	0.75	0.88	0.88	0.64	0.49	1		
15	0.31	0.90	0.35	0.86	0.84	0.49	0.59	0.07	0.69	0.87	0.85	0.51	0.37	0.93	1	
16	0.47	0.44	0.36	0.44	0.51	0.78	0.41	0.85	0.75	0.41	0.52	0.77	0.85	0.30	0.24	1

Key: 1. *Murrayakoenigi*; 2. *Murrayapaniculata*; 3: *Ficusinfectoria*; 4: *Mangifera* indica; 5; *Azadirachta indica*; 6; *Meliaazadirachta* 7. *Eucalyptuscamaldulensis*; 8. *Callistemonlanceolatus*; 9; *Syzygiumcumini*; 10. *Syzgiumaromaticum*; 11. *Albezialebbeck*; 12. *Dalbergiasisso*; 13. *Erythrinasubrosa*; 14. *Pongamiaglabra*; 15. *Acaciaarabica* 16. *Cassiafistula* and *Murrayapaniculata* shows maximum.

Table 3. Co-relation of Tree species from order Oxalidales, Malvales, Magnoales, genitianales, Limiales and Rosales based on Leaf epidermal anatomical characters

Species Names	1	2	3	4	5	6	7	8	9	10	11	12	13	14
1	1.000													
2	0.90	1.00												
3	0.70	0.66	1.00											
4	0.06	0.24	0.51	1.00										
5	0.30	0.44	0.65	0.73	1.00									
6	0.35	0.50	0.79	0.79	0.77	1.00								
7	0.14	0.15	0.39	0.10	0.51	0.38	1.00							
8	0.30	0.43	0.67	0.93	0.78	0.84	0.18	1.00						
9	0.71	0.59	0.65	0.06	0.35	0.41	0.37	0.30	1.00					
10	0.16	0.27	0.72	0.81	0.71	0.91	0.41	0.82	0.36	1.00				
11	0.53	0.47	0.62	0.10	0.47	0.46	0.68	0.29	0.87	0.46	1.00			
12	0.82	0.94	0.76	0.46	0.65	0.66	0.26	0.63	0.61	0.50	0.56	1.00		
13	0.53	0.53	0.57	0.12	0.59	0.54	0.79	0.35	0.72	0.47	0.83	0.62	1.00	
14	0.36	0.45	0.75	0.80	0.65	0.45	0.55	0.70	0.40	0.75	0.50	0.70	0.60	1.00

Key: 1: *Kigeliapinnate*; 2: *Trewianudiflora*; 3: *Putranjivaroxburjhi;* 4: *Artocarpusintegifolia;* 5: *Ficusracemosa;* 6: *Ficusreligiosa;* 7: *Ficusglomerata;* 8: *Lagerstroemiaindica;* 9: *Alstoniascholaris;* 10: *Thevetiaperuviana;* 11: *Magnoliagrandiflora;* 12: *Sterculiaalata;* 13: *Averthoacaranmbola;* 14: *Ficus infectoria.*

into two groups 2Bi and 2Bii at 40%. 2Bi group is further divided into 2 groups, 2Bia and 2Bib at 26%. 2Bii group is further divided into 2 group, 2Biia and 2Biib at 18%. 2Biia separate the species *Ficus religiosa* and *Thevetia peruviana*.2Biib separate into *Artocrpus integifolia* and *Saraca asoca*. Nine species like, *Callistemon lanceolatus,* DC., *Magnolia grandiflora* L., *Averthoa caranmbola* L, *Ficusglomerata*, Roxb, *Sygzgium aromarticum* L., Syzygium *cumini*, L, *Erythrina subrosa*, L., *Murraya paniculata*, L., *Azadirachta indica* L. were separated from each other just at the genetical distance of 4%. This difference is negligible and these species consider similar on the basis of leaf epidermal anatomical characters.

The role of anatomical data in traditional taxonomy has been long recognized since the variation within the species, genera or a family is usually reflected in anatomical features as well [27]. Leaf epidermal anatomical features such as shape of stomata, types of trichome and other characters are useful taxonomical tools. The present study dealt with the diversity in the types and shape of leaf epidermal features in tree species of the district Lahore (Punjab) Pakistan. Although studies on gross morphology and wood anatomy of the plant have proved valuable information for the identification of the plants yet this identification criteria would be incomplete without foliar epidermis morphology [28].

Trichomes are highly variable appendages of the epidermis including glandular and non-glandular, blunt, hairs and scales types. They are found on all parts of plant in leaf, stomata and root [29]. Trichomes vary widely in structure intra and interspecifically. They are sometimes remarkably uniform and may be used for taxonomic purposes. Same observations were made by Metcalfe and Chalk [30] for dicotyledons. In the tree species trichomes were observed on both abaxial and adaxial surface. Glandular trichomes are originated from the intracellular region. This type of trichomes were found on abaxial and adaxial surface of *Alstonia scholaris, Artocarpos integrifolia*, *Callistemon lanceolatus, Dalbergia sisso, Erythrina subrosa, Eucalyptus camaldulensis, Magnolia grandiflora, Melia azedarachta* However in *Saracaasoca* glandular trichome were present only on abaxial surface. *Ficusinfectoria* found different from all other species as no trichome was found on adaxial leaf surface. Shaheen et al., [15] investigated the diversity and distribution of glandular and non-glandular trichomes within the genus *Sida* L. (Malvaceae). They recorded sellate, peltate, long stalked, short stalked and capitate trichomes. But they did not study other anatomical markers like cells types; arrangement of epidermal cells various types of stomatal types and silica bodies etc. Glandular trichomes were present only on adaxial surface in *Ficus glomerata* and *Murraya koenigii* and were absent from abaxial surface.

Single cell, non-glandular trichome are originated from the cell wall is another important anatomical feature which can be utilized for the delimitation of various taxa of dicotyledons. Single cells, non-glandular trichome were present on abaxial and adaxial surface of *Albezia lebbeck*, *Kigelia pinnate*, *Lagerstroemia indica*, *Murraya paniculata*, *Trewia nudiflora* and *Putranjiva roxburjh.* Single cells. In *Cassia fistula* and *Sterculia alata* trichome present on abaxial surface andwere absent from abaxial surface. Mbagwu et al., [17] utilized trichomes for characterization of four species of the genus *Citrus* (Rutaceae). Non-glandular trichomes with 2-3 cells originated from the cell wall found to be the unique character of tree species. These trichomes were of blunt in shape. These were recorded from abaxial and adaxial surface of *Pongamia glabra, Azadirachta indica, Ficus religiosa* and *Sygzgium aromarticum*. These non-glandular trichomes were present on the abaxial surface of *Averthoa caranmbola* and *Thevetia Peruvian.* Trichome were entirely absent from *Syzygium cumini* and *Mangifera indica.*

Another important anatomical marker is presence of silica bodies. Mostly two, round and irregular shapes of silica bodies were found in the present investigation. Round shape silica bodies were present on abaxial and adaxial surface of *Lagerstroemia indica*, *Murraya paniculata*, *Dalbergia sisso*, *Erythrina subrosa* and *Sterculia alata.* Whereasround shape of silica bodies were present only on abaxial surface of *Ficus racemosa*, *Artocarpos integrifolia*, *Azadirachta indica* and *Syzygium cumini* and were not found on adaxial surface. Irregular shape of silica bodies were present on abaxial and adaxial surface *Sygzgium aromarticum*, *Trewia nudiflora*, *Thevetia peruviana*, *Putranjiva roxburjhi*, *Melia azedarachta* and *Ficus religiosa.* These Irregularsilica bodies were only recorded on abaxial surface of *Eucalyptus camaldulensis*, *Saraca asoca* and *Pongamia glabra.* Silica bodies were absent from *Acacia arabica, Alstonia scholaris, Albezialebbeck, Callistemon lanceolatus, Cassia fistula, Ficus infectoria, Ficus glomerata, Kigelia pinnate, Magnolia grandiflora, Mangifera indica* and *Murraya koenigi.*

Different types of stomata were first of all studied by Stresburger [31], followed by Vesque [32], recognized four broad categories of stomata based on the presence and arrangement of accessory cells as well as their mode of development.. Stomatal patterning is related to ordered placement of stomata on the leaf surface. The pattering process involves the selection of undifferentiated cells to become stomata [32]. It is not concerned with the physical events of their differentiation. The opening pores are present on the surface of leaf epidermis bounded by two specialized epidermal cells called the guard cells, by which change in shape bring about the opening and closure of the aperture. The features appearances play a role in the opening and closing of the stomatal pore. On the basis of arrangement the guard epidermal cell neighboring the guard cell more than 25 main types of stomata in dicot have been recognized [30].

In present study nine different types of stomata were recognized like anomocytic, Amphianisocyic, Brachyparacytic, Anomotetracytic, Anisocytic, Amphicyclocytic, Ammphianisocytic, actinocytic and staurocytic. Ahmad et al., 2009 recognized seven (anisocytic, amphianisocytic, axillocytic, anomotetracytic, actinocytic, diacytic, staurocytic) types of stomata. They just emphasized on stomata anatomical marker but they ignore other anatomical characters. Anomocytic stomata were present on both abaxial and adaxial surface in *Callistemon lanceolatus* and *Azadirachta indica.* This type of stomata is only present on abaxial surface of *Ficus infectoria* and *Ficus racemosa*. Perveen et al., 2007 were recognized six type of stomata (anomocytic, paracytic, diacytic, parallelocytic, cyclocytic and anisocytic). In which anomocytic stomata were most dominant found in 54 taxa of 69 dicot species. Amphianisocytic stomata were present on both abaxial and adaxial surface of *Pongamia glabra.* Brachyparacytic stomata were present on both abaxial and adaxial surface in *Averthoa caranmbola*, *Albezialebbeck, Eucalyptus camaldulensis* and *Putranjiva roxburjhi.* On abaxial surface; brachyparacytic stomata are present in *Saraca asoca* and *Lagerstroemia indica.* Anomotetracytic stomata were present on both abaxial and adaxial surface in *Murraya paniculata*, *Syzygium cumini, Melia azadirachta., Magnolia grandiflora, Kigelia pinnate,* and *Syzygium cumini.* These stomata were present only on abaxial surface in *Sterculia alata, Mangifera indica*, and *Ficus glomerata.*

On abaxial surface, Actinocytic stomata were present in *Sygzgium aromarticum.* Staurocytic stomata are present on both abaxial and adaxial surface of Ficus *religiosa.* On abaxial surface, these types of stomata were found in *Erythrina subrosa.* Anisocytic stomata were present on both abaxial and adaxial surface in *Murraya koenigi*, *Dalbergia sisso*, *Cassia fistula* and *Acacia arabica*. On abaxial surface, this type of stomata was present in *Alstonia scholaris* and *Artocarpos integrifolia*. Amphicyclocytic stomata were present on both abaxial and adaxial surface of *Trewia nudiflora.*

Epidermal cells are quite variable in the configuration of the characteristic markers, which have been proved to be systematic value in cuticular characters, epidermis, stomata, subsidiary cells and trichome [33]. So the taxonomic significance of epidermal morphology is well documented in botanical literature. Considering the anatomical characters, there are five different types of epidermal cells are identified. Klimko et al., [19] revealed the number of micro morphological characters at taxonomic level in selected taxa of the genus *Ficus* L. (Moraceae). Irregular type of cells was found in *Pongamia glabra, Saraca asoca, Syzygium cumini, Sygzgium aromarticum, Trewia nudiflora, Thevetia peruviana* and *Albezia* lebbeck. Pentagonal epidermal cells are found in *Alstonia scholaris, Azadirachta indica* and *Lagerstroemia indica.*

Polygonal shape of epidermal cells were found in *Murraya koenigii, Magnolia grandiflora, Mangifera indica, Putranjiva roxburjh, Ficus racemosa, Ficus glomerata, Ficus infectoria, Erythrina subrosa, Dalbergia sisso, Callistemon lanceolatus and Sterculia alata.* Farooqui et al., [34] studied cuticular characters in three species of *Dalbergia..* They distinguished the papillate and non papillate cuticular characters in two species. They just emphasized on the epidermal cells and ignored other anatomical characters. Rectangular shape of epidermal cells is found in *Murraya paniculata, Melia azedarachta, Kigelia pinnate, Eucalyptus camaldulensis and Acacia arabica.* Hexagonal shape of epidermal cells was found *Averthoa caranmbola, Artocarpos integrifolia, Cassia fistula and Ficus religiosa.*

On the basis of the cell arrangement, thirty species were separated at different category. Smooth and single layered cells separate *Acacia arabica Artocarpos integrifolia, Cassia fistula, Dalbergia sisso, Erythrina subrosa, Ficus infectoria, Ficus religiosa, Mangifera indica* and *Syzygium cumini.* Smooth and double layered cells are present in *Alstonia scholaris, Averthoa caranmbola, Azadirachta indica, Callistemon lanceolatus, Eucalyptus camaldulensis, Ficus glomerata, Ficus glomerata, Sterculia alata, Melia azedarachta, Lagerstroemia indica, Kigelia pinnate, Putranjiva roxburjhi.* Wavy and single layered cells are present in *Magnolia grandiflora.* Wavy and double layered cells are present in *Murraya paniculata, Pongamia glabra, Thevetia peruviana, Thevetia peruviana, Sygzgium aromarticum, Trewia nudiflora* and *Albezia* lebbeck. Leaf epidermal characters like trichome types, difference of origin, variation of stomatal types, and arrangement of cells and presence of silica bodies were considered important features for the characterization of tree species of dicotyledons.

REFERENCES

[1] Stace, C. A. (1991). *New Flora of the British Isles*, 1st ed. Cambridge University Press, Cambridge.

[2] Gilani, S. S., Khan, M. A., Shinwari, Z. K. & Yousaf, Z. (2002). Leaf epidermal anatomy of selected Digiteria species, tribe paniceae, family poaceae of Pakistan. *Pakistan Journal of Botany*. 257-273.

[3] Esau, K. (1960). Anatomy of Seed Plants, John Wiley & Sons, Inc., NY

[4] Metcalfe, C. R. & Chalk, L. (1950). Anatomy of the dicotyledons. I Clarendon press, oxford.

[5] Jones J. H. (1986). Evolution of the Fagaceae: the implication of foliar features. *Ann. Misouri, Bot. Garden*, *73*, 228-275.

[6] Baranova, M. (1992). Systematic anatomy of leaf epidermis in the Magnoliaceae and some related families. *Taxon*, *21*, 447-469.

[7] Stace, C. (1980). *Plant taxonomy and Dicotyledonous*. vol. *I*. Clarendon press, Oxford. 23-25.

[8] Jamal, A., Ko, K., Kim, H. S., Cho, Y. K., Joung, H. & Ko, K. (2009) Role of genetic factors and environmental conditions in recombinant protein production for plant molecular biofarming. *Biotech Advan*, *27*, 914-923

[9] Ahmad, K., Ahmad, M. A., Shaheen, N. & Nazir, A. (2010). Taxonomic diversity in epidermal cells of some sub-tropical plant species. *Int. J. Agri. Bio.*, 12 (1), 115-118.

[10] Sahreen, S., Khan, M. A., Khan, M. R. & Khan, R. A. (2010). Leaf epidermal anatomy of the genus *Silene* (Caryophyllaceae) from Pakistan. *Biol Divers Conserv*, *3*(1), 93-102.

[11] Yasmin, G., Khan, M. A., Shaheen, N. & Hayat, M. Q. (2010). The micro morphological investigation of foliar anatomy of fagopyrum Mill, and Rumex. c of Polygonaceae. *Pakiatan Journal of Botany.*, *42*(1), 47-57.

[12] Yasmin, G., Khan, M. A., Shaheen, N. & Hayat, M. Q. (2009). The micro morphological investigation of foliar anatomy of Genus Aconogonon and Bistorta of family polygonaceae. *Int. J. of Agri & Bio.*, *11*(3), 285-289.

[13] Carmen, E., Rojas, B. D. & Maria, F. S. (2009). Morphology of foliar epidermis in two groups of Solanum section Geminata (Solanaceae). *Caldasia.*, *31*(1), 31-40.

[14] Ahmad, K., Khan, M. A., Ahmad, M., Zafar, M., Arshad, M. & Ahmad, F. (2009). Taxonomic diversity of stomata in dicot flora of district tank (N. W. F. P) in Pakistan. *African Journal of Botany.*, *8*(6), 1052-1055.

[15] Shaheen, N, M., Khan, A., Yasmin, G. & Hayat, M. Q. (2009). Diversity of foliar trichomes and their systematic relevance in the Genus Hibiscus (Malvaceae). *Int. J. Agri and Bio*, *11*(3), 279-284.

[16] Mbagwu, F. N, Osuji, C., Ezeibekwe, I. O. & Okafor, U. V. (2008). Leaf epidermal features of *Canarium schwenfurthii* and *Dacryodes edulis* (Burseraceae). *International Science Research Journal*, *1*(2), 113 – 115.

[17] Mbagwu, F. N., Nwachukwu, C. U. & Okoro, O. O. (2007). Comparative Leaf epidermal Studies on *Solanum macrocarpum* and *Solanum nigrum*. *Nature and Science*, *5*(3). 1-4.

[18] Perveen, A. R., Abid, & Fatima, R. (2007). Stomatal types of some dicots within flora of Karachi. Pakistan. *Pakistan Journal of Botany*. *39*(4), 1017-1023.

[19] Klimko, M. & Truchan, M. (2006). Morphology variability of the leaf epidermis in selected taxa of the Genus *Ficus*. L. (Moraceae) and its taxonomic implication. *Acta. societatis botanicorum poloniae*, *75* (4), 309-324.

[20] Carpenter, R. J., Hill, R. S. & Jordant, G. J. (2005). Leaf cuticular morphology links platanaceae and proteacaeae. *Int. J. Plant Sci.*, *5*, 846-855.

[21] Nikolic, N. P., Meskulov, L. S., Kestica, B. D. & Orlovic, S. S. (2003). A comparative analysis of stomata leaf trichome characteristics in *Quercusrobur* L. *Genotype. Proc. Nat. Sci, Martica srpska Novi Sad*, no: 105, 51-59.

[22] Farooqui, P., Venkalasubramanian, N. & Nallasamy, V. K. (1989). Use of cuticular studies in distinguishing species of *Dalbergia*, Proc. *Indian Acad. Sci. (Plant Sci)*, *99*, 7-14.

[23] Popa. R. M. & Sipos. (2009). Epidermal formation trichomes. *Biharean Biologist. 13*(1) 27-32.

[24] Shaheed, S. A. & Illoh, H. C.(2010). Taxonomic study of some species in cassinae (Leguminoaceaea) using leaf epidermal characters. *Not. Bot. Hort. Agro. Bot. Clu*, *38*(1), 20-27.

[25] Rahaman, A. A. A., Egbedo, F. O. & Oladele. F. A. (2009). Stomatal complex types, stomatal density, and the stomatal index in some species of *Dioscorea. Arch. Biol. Sci., Belgrade*, *61* (4), 847-851.

[26] Hayat, M. Q., Ashraf, M., Jabeen, S., Shaheen, N., Yasmin, G. & Khan, M. A. (2010). Taxonomic implication of foliar epidermal characteristics with special reference to stomatal variation in the genus Artemisia (Asteraceae). *Int. J. Agri. Bio.*, *12* (2), 221-226.

[27] Metcalfe, C. R. & Chalk, L (1960). Anatomy of the dicotyledons. Oxford at the Clarendon Press, 2, 1259-1271.

[28] Kadiri, A. B., Olowokudejo, J. D. & Ogundipe, O. T. (2006). Some Aspects of Foliar Epidermal Morphology of *Cylicodiscus gabunensis* (Taub.) Harms (Mimosaceae). *J. Sci. Res.*, *10*, 33–38.

[29] Cowen J. M. (1950). The Rhododendron leaf. A study of the epidermal appendages. *Q. Rev. Biol.*, *27*(2), 214-215.

[30] Metcalf, C. R. & Chalk. (1960). Analysis of the Dicotyledons. Clarendon Press, Oxford. 1067-1074.

[31] Stresburger, E. (1866). Ein Beitrag zur Entwicklungsgeschichte der spaltoffnungen, *Jahrb. Wiss. Bot.*, 5, 297-342.

[32] Vesque, M. J. (1889). De I' employ des caracteres anatomiques dans la classification des vegetaux.

[33] Ellis, R. P. (1976). A Procedure for standardizing Comparative leaf Anatomy in Poaceae 1. The leaf blade as viewed in transverse section, *Bothalia*, *12*, 65-109.

[34] Farooqui, P., Venkalasubramanian, N. & Nallasamy, V. K. (1989). Use of cuticular studies in distinguishing species of *Dalbergia. Proc. Indian Acad. Sci. (Plant Sci.)*, *99*, 7-14.

EDITORS' CONTACT INFORMATION

Zubaida Yousaf, PhD
Assistant Professor, Department of Botany
Lahore College for Women University
Email: Z4zubaida@yahoo.com

Afifa Younas, MS
Research Scholar, Department of Botany,
Lahore College for Women University
Email: botanistafifa@yahoo.com

Arusa Aftab, MPhil
Lecturer, Department of Botany,
Lahore College for Women University
Email: Arusa_ravi@yahoo.com

INDEX

L

M

N

O

P

S

T

U

V

W

X